Contents

W9-AHY-299

Glossary   130

Regents Examinations, Answers, and Student Self-Appraisal Guides   167

BARRON'S

Regents Exams and Answers

Living Environment
2020

GABRIELLE I. EDWARDS
Former Science Consultant
Board of Education
City of New York

Former Assistant Principal Supervision
Science Department
Franklin D. Roosevelt High School
Brooklyn, New York

MARION CIMMINO
Former Teacher, Biology and Laboratory Techniques
Franklin D. Roosevelt High School
Brooklyn, New York

FRANK J. FODER
Teacher, Advanced Placement Biology
Franklin D. Roosevelt High School
Brooklyn, New York

G. SCOTT HUNTER
Former Teacher, Biology and Advanced Biology
Former Consultant, State Education Department Bureaus of Science and Testing
Former School Business Administrator, Schodack, NY
Former Superintendent of Schools
Mexico and Chatham, New York

© Copyright 2020, 2018, 2017, 2016, 2015, 2014, 2013, 2012, 2011, 2010, 2009, 2008, 2007, 2006, 2005, 2004, 2003, 2002, 2001, 2000, 1999, 1998, 1997, 1996 by Kaplan, Inc., d/b/a Barron's Educational Series

All rights reserved.
No part of this book may be reproduced in any form or by any means without the written permission of the copyright owner.

Published by Kaplan, Inc., d/b/a Barron's Educational Series
750 Third Avenue
New York, NY 10017
www.barronseduc.com

ISBN: 978-1-5062-5391-6

Printed in Canada

10 9 8 7 6 5 4 3 2 1

Kaplan, Inc., d/b/a Barron's Educational Series print books are available at special quantity discounts to use for sales promotions, employee premiums, or educational purposes. For more information or to purchase books, please call the Simon & Schuster special sales department at 866-506-1949.

How to Use This Book

ORGANIZATION OF THE BOOK

STUDY QUESTIONS

Section 1 of the book consists of 173 questions taken from past Regents examinations in biology. Most of the questions require that you select a correct response from four choices given. A few questions provide you with a list of words or phrases from which to select the one that best matches a given description. Others are constructed-response, graphical analysis, or reading comprehension questions. You should become familiar with the formats of the questions that appear on the Living Environment Regents examination.

Each question in Section 1 of this book has a well-developed answer. Each answer provides the number of the correct response, the reason why the response is correct, and explanations of why the other choices are incorrect.

A useful feature of Section 1 is the student Self-Appraisal Guide. This device allows you to determine where your learning strengths and weaknesses lie in the major topics of each unit. For specific topics within the units, the numbers of related questions are given. As you attempt to answer each question in Section 1, you may wish to circle on the appraisal form the numbers of the questions that you are unable to answer. The circled items then help you to identify at a glance subject matter areas in which you need additional study.

NEW YORK STATE LEARNING STANDARDS

Commencement Standards

There are several commencement standards required of students in New York State public schools regarding their performance in math, science, and technology. The Core Curriculum for The Living Environment addresses two of these standards:

Standard 1: **Students will use mathematical analysis, scientific inquiry, and engineering design, as appropriate, to pose questions, seek answers, and develop solutions.**

Standard 4: **Students will understand and apply scientific concepts, principles, and theories pertaining to the physical setting and living environment and recognize the historical development of ideas in science.**

The Core Curriculum for The Living Environment was built from these two commencement standards. It is important to recognize that the Core Curriculum is not a syllabus. It does not prescribe what will be taught and learned in any particular classroom. Instead, it defines the skills and understandings that you must master in order to achieve the commencement standards for life science.

Instead of memorizing a large number of details at the commencement level, then, you are expected to develop the skills needed to deal with science on the investigatory level, generating new knowledge from experimentation and sharpening your abilities in data analysis. You are also expected to read and understand scientific literature, taking from it the facts and concepts necessary for a real understanding of issues in science. You are required to pull many facts together from different sources to develop your own opinions about the moral and ethical problems facing modern society concerning technological advances. These are thinking skills that do not respond to simple memorization of facts and scientific vocabulary.

Key Ideas, Performance Indicators, and Major Understandings

Each commencement standard is subdivided into a number of Key Ideas. Key Ideas are broad, unifying statements about what you need to know. Within Standard 1, three Key Ideas are concerned with laboratory investigation and data analysis. Together, these unifying principles develop your ability to deal with data and understand how professional science is carried out in biology. Within Standard 4, seven Key Ideas present a set of concepts that are central to the science of biology. These Key Ideas develop your understanding of the essential characteristics of living things that allow them to be successful in diverse habitats.

Within each Key Idea, Performance Indicators are presented that indicate which skills you should be able to demonstrate through your mastery of the Key Idea. These Performance Indicators give guidance to both you and your teacher about what is expected of you as a student of biology.

Performance Indicators are further subdivided into Major Understandings. Major Understandings give specific concepts that you must master in order to achieve each Performance Indicator. It is from these Major Understandings that the Regents assessment material will be drawn.

Laboratory Component

A meaningful laboratory experience is essential to the success of this or any other science course. You are expected to develop a good sense of how scientific inquiry is carried out by the professional scientist and how these same techniques can assist in the full understanding of concepts in science. The Regents requirement of 1,200 minutes of successful laboratory experience, coupled with satisfactory written reports of your findings, should be considered a minimum.

Students are required to complete four laboratory experiences required by the New York State Education Department and tested on the Regents Examination. See Barron's *Let's Review: Biology—The Living Environment* for a complete treatment of this requirement.

REGENTS EXAMINATIONS

Section 2 of the book consists of actual Biology Regents examinations and answers. These Regents examinations are based on the New York State Core Curriculum for the Living Environment.

ASSESSMENTS: FORMAT AND SCORING

The format of the Regents assessment for The Living Environment is as follows, based on actual Regents examinations.

Part A: Variable number (usually 30) of multiple-choice questions that test the student's knowledge of specific factual information. **All** questions must be answered on Part A.

Part B: Variable number (usually 25) of questions, representing a mixture of multiple-choice and constructed-response items. Questions may be based on the student's direct knowledge of biology, interpretation of experimental data, analysis of readings in science, and ability to deal with representations of biological phenomena. **All** questions must be answered on Part B1–B2.

Part C: Variable number (usually 17) of multiple-choice and constructed-response questions. Questions may be based on the student's direct knowledge of biology, interpretation of experimental data, analysis of readings in science, and ability to deal with representations of biological phenomena. **All** questions must be answered on Part C.

Part D—Laboratory component: Variable number (usually 13) of multiple-choice and constructed-response items. This component of the examination aims to assess student knowledge of and skills on any of four required laboratory experiences supplied to schools by the New York State Education Department. The content of these questions will reflect specific laboratory experiences. You are strongly encouraged to include review of these laboratory experiences as part of your year-end Regents preparation activity.

The following chart summarizes the current laboratory requirement for New York State public schools:

Laboratory Requirements

Laboratory Title	Description
The Beaks of Finches	Explores the adaptive advantages of beaks with different physical characteristics
Relationships and Biodiversity	Explores the relationship between DNA structure and the biochemistry of inheritance
Making Connections	Explores the effects of physical activity on human metabolic activities
Diffusion Through a Membrane	Explores the nature of cross-membrane transport in living cells
Adaptations for Reproductive Success in Flowering Plants*	Not yet available
DNA Technology*	Not yet available
Environmental Conditions and Seed Germination*	Not yet available

*Not yet available as of this writing.

Studying questions from past Regents examinations is an invaluable aid in developing a mind set that will enable you to approach questions with understanding. Although exact questions are not repeated, question types are repeated. If you practice questions that require interpretation, problem solving, and graph construction, you will do well on the entire exam. During the school year, the 30 required laboratory lessons teach you certain manipulative skills. Questions involving identification, measurement, and other laboratory procedures are based on the laboratory exercises. Review of past materials gives you insight as to the types of questions that you may be asked to answer. Study the questions in the Regents exams in this book diligently.

HOW TO STUDY

GENERAL SUGGESTIONS FOR STUDY

You've spent all year learning many different facts and concepts about biology —far more than you could ever hope to remember the "first time around." Your teacher has drilled you on these facts and concepts; you've done homework, taken quizzes and tests, performed laboratory experiments, and reviewed the material at intervals throughout the year.

Now it's time to put everything together. The Regents exam is only a few weeks away. If you and your teacher have planned properly, you will have finished all new information about 3 weeks before the exam. Now you have to make efficient use of the days and weeks ahead to review all that you've learned in order to score high on the Regents.

The task ahead probably seems impossible, but it doesn't have to be! You've actually retained much more of the year's material than you realize! The review process should be one that helps you to recall the many facts and concepts you've stored away in your memory. Your Barron's resources, including Barron's *Let's Review: The Living Environment*, will help you to review this material efficiently.

You also have to get yourself into the right frame of mind. It won't help to be nervous and stressed during the review process. The best way to avoid being stressed during any exam is to be well rested, prepared, and confident. We're here to help you prepare and to build your confidence. So let's get started on the road to a successful exam experience!

To begin, carefully read and follow the steps outlined below.

1. Start your review early; don't wait until the last minute. Allow at least 2 weeks to prepare for the Regents exam. Set aside an hour or two a day over the next few weeks for your review. Less than an hour a day is insufficient time for you to concentrate on the material meaningfully; more than 2 hours daily will yield diminishing returns on your investment of time.
2. Find a quiet, comfortable place to study. You should seat yourself at a well-lighted work surface, free of clutter, in a room without distractions of any kind. You may enjoy watching TV or listening to music curled up in a soft chair, but these and other diversions should be avoided when doing intense studying.
3. Make sure you have the tools you need, including this book, a pen and pencil, and some scratch paper for taking notes and doing calculations. Keep your class notebook at hand for looking up information between test-taking sessions. It will also help to have a good review text, such as Barron's *Let's Review: The Living Environment*, available for reviewing important concepts quickly and efficiently.

4. Concentrate on the material in the "Study Questions and Answers" section of this book. Read carefully and thoughtfully. Think about the questions that you review, and try to make sense out of them. Choose the answers carefully. (See the following section, "Using This Book for Study," for additional tips on question-answering techniques.)

5. Use available resources, including a dictionary and the glossary in this book, to look up the meanings of unfamiliar words in the practice questions. Remember that the same terms can appear on the Regents exam you will take, so take the opportunity to learn them now.

6. Remember: study requires time and effort. Your investment in study now will pay off when you take the Regents exam.

USING THIS BOOK FOR STUDY

This book is an invaluable tool if used properly. Read carefully and try to answer *all* the questions in Section 1 and on the practice exams. The more you study and practice, the more you will increase your knowledge of biology and the likelihood that you will earn a high grade on the Regents exam. To maximize your chances, use this book in the following way:

1. Answer all of the questions in the section entitled "Study Questions and Answers." Check your responses by using "Answers to Topic Questions," including "Wrong Choices Explained," following each question set. Record the number of correct responses on each topic in the "Self-Appraisal Guide" at the end of the section to identify your areas of strength and weakness. Use a good review text, such as Barron's *Let's Review Regents: Living Environment,* to study each area on which you did poorly. Finally, go back to the questions you missed on the first round and be sure that you fully understand what each question asks and why the correct answer is what it is.

2. When you have completed the questions in "Study Questions and Answers," go on to the examination section. Select the first complete examination and take it under test conditions.

3. Interpret the term *test conditions* as follows:

   • Be well rested; get a good night's sleep before attempting *any* exam.
   • Find a quiet, comfortable room in which to work.
   • Allow no distractions of any kind.
   • Select a well-lighted work surface free of clutter.
   • Have your copy of this book with you.
   • Bring to the room a pen, a pencil, some scratch paper, and a watch or alarm clock set for the 3-hour exam limit.

4. Take a deep breath, close your eyes for a moment, and RELAX! Tell yourself, "I know this stuff!" You have lots of time to take the Regents exam; use it to your advantage by reducing your stress level. Forget about your plans for later. For the present, your number 1 priority is to do your best, whether you're taking a practice exam in this book, or the real thing.

5. Read all test directions carefully. Note how many questions you must answer to complete each part of the exam. If test questions relate to a reading passage, diagram, chart, or graph, be sure you fully understand the given information before you attempt to answer the questions that relate to it.

6. When answering multiple-choice questions on the Regents exam, TAKE YOUR TIME! Pay careful attention to the "stem" of the question; read it over several times. These questions are painstakingly written by the test preparers, and every word is chosen to convey a specific meaning. If you read the question carelessly, you may answer a question that was never asked! Then read each of the four multiple-choice answers carefully, using a *pencil* to mark in the test booklet the answer you think is correct.

7. Remember that three of the multiple-choice answers are *incorrect*; these incorrect choices are called "distracters" because they seem like plausible answers to poorly prepared or careless students. To avoid being fooled by these distracters, you must think clearly, using everything you have learned about biology since the beginning of the year. This elimination process is just as important to your success on the Living Environment Regents exam as knowing the correct answer! If more than one answer seems to be correct, reread the question to find the words that will help you to distinguish between the correct answer and the distracters. When you have made your best judgment about the correct answer, circle the number in *pencil* in your test booklet.

8. Constructed-response questions appear in a number of different forms. You may be asked to select a term from a list, write the term on the answer sheet, and define the term. You may be asked to describe some biological phenomenon or state a biological fact using a complete sentence. You may be asked to read a value from a diagram of a measuring instrument and write that value in a blank on the answer paper. When answering this type of question, care should be exercised to follow directions precisely. If a complete sentence is called for, it must contain a subject and a verb, must be punctuated, and must be written understandably in addition to answering the scientific part of the question accurately. Values must be written clearly and accurately and include a unit of measure, if appropriate. Failure to follow the directions for a question may result in a loss of credit for that question.

9. A special type of constructed-response question is the essay or paragraph question. Typically, essay or paragraph questions provide an opportunity to earn multiple credits for answering the question correctly. As in the constructed-response questions described above, it is important that you follow the directions given if you hope to earn the maximum number of credits for the question. Typically, the question outlines exactly what must be included in your essay to gain full credit. Follow these directions step by step, double-checking to be certain that all question components are addressed in your answer. In addition, your essay or paragraph should follow the rules of good grammar and good communication so that it is readable and understandable. And, of course, it should contain correct information that answers all the parts of the question asked.

10. Graphs and charts are a special type of question that requires you to organize and represent data in graphical format. Typically for such questions, you are expected to place unorganized data in ascending order in a data chart or table. You may also be asked to plot organized data on a graph grid, connect the plotted points, and label the graph axes appropriately. Finally, questions regarding data trends and extrapolated projections may be asked, requiring you to analyze the data in the graph and draw inferences from it. As with all examination questions, always follow all directions for the question. Credit can be granted only for correctly following directions and accurately interpreting the data.

11. When you have completed the exam, relax for a moment. Check your time; have you used the entire 3 hours? Probably not. Resist the urge to quit. Go back to the beginning of the exam, and, in the time remaining, *retake the exam in its entirety*. Try to ignore the penciled notations you made the first time. If you come up with a different answer the second time through, read over the question with extreme care before deciding which response is correct. Once you have decided on the correct answer, mark your choice in ink in the answer booklet.

12. Score the exam using the Answer Key at the end of the exam. Review the "Answers Explained" section for each question to aid your understanding of the exam and the material. Remember that it's just as important to understand why the incorrect responses are incorrect as it is to understand why the correct responses are correct!

13. Finally, focus your between-exam study on your areas of weakness in order to improve your performance on the next practice exam. Complete all the practice exams in this book using the techniques outlined above.

14. Be sure to sign the declaration on your answer sheet. Unless this declaration is signed, your paper cannot be scored.

Test-Taking Tips—
A Summary

The following pages contain seven tips to help you achieve a good grade on the Living Environment Regents exam.

TIP 1

Be confident and prepared.

SUGGESTIONS

- Review previous tests.
- Use a clock or watch, and take previous exams at home under examination conditions (i.e., don't have the radio or television on).
- Get a review book. (The preferred book is Barron's *Let's Review Regents: Living Environment.*)
- Talk over the answers to questions on these tests with someone else, such as another student in your class or someone at home.
- Finish all your homework assignments.
- Look over classroom exams that your teacher gave during the term.
- Take class notes carefully.
- Practice good study habits.
- Know that there are answers for every question.
- Be aware that the people who made up the Regents exam want you to pass.
- Remember that thousands of students over the last few years have taken and passed a Biology Regents. You can pass, too!
- Complete your study and review at least one day before taking the examination. Last-minute "cramming" may hurt, rather than enhance, your performance on the exam.
- Visit *www.barronseduc.com* or *www.barronsregents.com* for the latest information on the Regents exams.
- Use Barron's website to communicate directly with subject specialists.

- Be well rested when you enter the exam room. A good night's sleep is essential preparation for any examination.
- On the night before the exam day: lay out all the things you will need, such as clothing, pens, and admission cards.
- Bring with you two pens, two pencils, an eraser, and, if your school requires it, an identification card. Decide before you enter the room that you will remain for the entire 3-hour examination period, and either bring a wristwatch or sit where you can see a clock.
- Once you are in the exam room, arrange things, get comfortable, and attend to personal needs (the bathroom).
- Before beginning the exam, take a deep breath, close your eyes for a moment, and RELAX! Repeat this technique any time you feel yourself "tensing up" during the exam.
- Keep your eyes on your own paper; do not let them wander over to anyone else's paper.
- Be polite in making any reasonable demands of the exam room proctor, such as changing your seat or having window shades raised or lowered.

TIP 2

Read test instructions carefully.

SUGGESTIONS

- Be familiar with the format of the examination.
- Know how the test will be graded.
- Read all directions carefully. Be sure you fully understand supplemental information (reading passages, charts, diagrams, graphs) before you attempt to answer the questions that relate to it.
- Underline important words and phrases.
- Ask for assistance from the exam room proctor if you do not understand the directions.

TIP 3
Read each question carefully and read each choice before recording your answer.

SUGGESTIONS

- When answering the questions, TAKE YOUR TIME! Be sure to read the "stem" of the question and each of the four multiple-choice answers very carefully.
- If you are momentarily "stumped" by a question, put a check mark next to it and go on; come back to the question later if you have time.
- Remember that three of the multiple-choice answers (known as "distracters") are incorrect. If more than one answer seems to be correct, reread the question to find the words that will help you to distinguish between the correct answer and the distractors.
- When you have made your best judgment about the correct answer, circle the appropriate number in pencil on your answer sheet.

TIP 4
Budget your test time (3 hours).

SUGGESTIONS

- Bring a watch or clock to the test.
- The Regents examination is designed to be completed in 1½ to 2 hours.
- If you are absolutely uncertain of the answer to a question, mark your question booklet and move on to the next question.
- If you persist in trying to answer every difficult question *immediately*, you may find yourself rushing or unable to finish the remainder of the examination.
- When you have completed the exam, relax for a moment. Then go back to the beginning, and, in the time remaining, *retake the exam in its entirety*. Pay particular attention to questions you skipped the first time. Once you have decided on a correct response for multiple-choice questions, mark an "X" in ink through the penciled circle on the answer sheet.
- Plan to stay in the room for the entire three hours. If you finish early, read over your work—there may be some things that you omitted or that you may wish to add. You also may wish to refine your grammar, spelling, and penmanship.

TIP 5

Use your reasoning skills.

SUGGESTIONS

- Answer *all* questions.
- Relate (connect) the question to anything that you studied, wrote in your notebook, or heard your teacher say in class.
- Relate (connect) the question to any film, demonstration, or experiment you saw in class, any project you did, or to anything you may have learned from newspapers, magazines, or television.
- Look over the entire test to see whether one part of it can help you answer another part.

TIP 6

Don't be afraid to guess.

SUGGESTIONS

- In general, go with your first answer choice.
- Eliminate obvious incorrect choices.
- If still unsure of an answer, make an educated guess.
- There is no penalty for guessing; therefore, answer ALL questions. An omitted answer gets no credit.

TIP 7

Sign the Declaration.

SUGGESTIONS

- Be sure to sign the declaration on your answer sheet.
- Unless this declaration is signed, your paper cannot be scored.

Tips for Teachers

CLASSROOM USE

All teachers will be able to use this book with their students as a companion to their regular textbooks and will find that their students gain considerable self-confidence and ability in test taking through its consistent use.

The Living Environment Core Curriculum (1999) defines the skills and abilities students should have at the point of commencement at the upper-secondary level. It is assumed that science concepts have been taught and assessed at an age-appropriate level throughout their career, so that little additional detail needs to be presented at the upper-secondary level.

An excellent companion to this book (and any comprehensive biology text) is Barron's *Let's Review: Biology—The Living Environment* (Hunter). The factual material and organization of this book lend themselves well to the development of Standards-Based Learning Units (SBLUs) and Essential Questions. The level of detail is consistent with what students really have to know in order to do well on the New York State Regents Examination on the Living Environment.

APPLICATION-BASED CURRICULUM

The curriculum focus can be characterized as application-based—one that is less concerned about content and more concerned about thinking. It is less about *how much* students know and more about *what they can do* with what they know. The latter, after all, is what real learning is all about; these are the abilities that will last a lifetime, not facts and scientific terminology.

This being said, it is acknowledged that students will have a difficult time expressing their views and making moral and ethical judgments about science if they lack a working knowledge of scientific principles and do not have at least a passing understanding of the terms used by biologists. For this reason, teachers and administrators will need to develop local curricula that complement the Core Curriculum. It is up to the teacher or administrator to decide what examples and factual knowledge best illustrate the concepts presented in the

Core Curriculum, what concepts need to be reinforced and enhanced, what experiences will add measurably to students' understanding of science, and what examples of local interest should be included.

The teacher will immediately recognize the need to go beyond this level in the classroom, with examples, specific content, and laboratory experiences that complement and illuminate these Major Understandings. It is at this level that the locally developed curriculum is essential. Each school system is challenged to develop an articulated K-12 curriculum in mathematics, science, and technology that will position students to achieve a passing standard at the elementary and intermediate levels, such that success is maximized at the commencement level.

The addition of factual content must be accomplished without contradicting the central philosophy of the learning standards. If local curricula merely revert to the fact-filled syllabi of the past, then little will have been accomplished in the standards movement other than to add yet another layer of content and requirements on the heads of students. A balance must be struck between the desire to build students' ability to think and analyze and the desire to add to the content they are expected to master.

LABORATORY EXPERIENCE

The reduction of factual detail in the Core Curriculum (1982–1999) should allow a more in-depth treatment of laboratory investigations to be planned and carried out than was possible under the previous syllabus. Laboratory experiences should be designed to address Standard 1 (inquiry techniques) but should also take into account Standards 2 (information systems), 6 (interconnectedness of content), and 7 (problem-solving approaches). They should also address the laboratory skills listed in Appendix A of the Core Curriculum.

Part D of the examination assesses student knowledge of and skills on any of four required laboratory experiences supplied to schools by the New York State Education Department. The specific laboratory experiences required in any year will vary according to a preset schedule (see chart on next page).

Questions on this section can be a combination of multiple-choice and constructed-response questions similar to those found in Parts A, B, and C of the Living Environment Regents Examination. The content of these questions reflect the four specific laboratory experiences required for a particular year. Teachers are strongly encouraged to include review of these laboratory experiences as part of their year-end Regents preparation activity.

The following chart summarizes the current laboratory requirement for New York State public schools:

Laboratory Title	Description
The Beaks of Finches	Explores the adaptive advantages of beaks with different physical characteristics
Relationships and Biodiversity	Explores the relationship between DNA structure and the biochemistry of inheritance
Making Connections	Explores the effects of physical activity on human metabolic activities
Diffusion Through a Membrane	Explores the nature of cross-membrane transport in living cells
Adaptations for Reproductive Success in Flowering Plants*	Not yet available
DNA Technology*	Not yet available
Environmental Conditions and Seed Germination*	Not yet available

*Not yet available as of this writing.

Study Questions and Answers

QUESTIONS ON STANDARD 1

Students will use mathematical analysis, scientific inquiry, and engineering design, as appropriate, to pose questions, seek answers, and develop solutions.

KEY IDEA 1—PURPOSE OF SCIENTIFIC INQUIRY

The central purpose of scientific inquiry is to develop explanations of natural phenomena in a continuing and creative process.

Performance Indicator	Description
1.1	The student should be able to elaborate on basic scientific and personal explanations of natural phenomena and develop extended visual models and mathematical formulations to represent one's thinking.
1.2	The student should be able to hone ideas through reasoning, library research, and discussion with others, including experts.
1.3	The student should be able to work toward reconciling competing explanations and clarify points of agreement and disagreement.
1.4	The student should be able to coordinate explanations at different levels of scale, points of focus, and degrees of complexity and specificity, and recognize the need for such alternative representations of the natural world.

Base your answers to questions 1 through 4 on the passage below and on your knowledge of biology.

To Tan or Not to Tan

Around 1870, scientists discovered that sunshine could kill bacteria. In 1903, Niels Finsen, an Icelandic researcher, won the Nobel Prize for his use of sunlight therapy against infectious diseases. Sunbathing then came into wide use as a treatment for tuberculosis, Hodgkin's disease (a form of cancer), and common wounds. The discovery of vitamin D, the "sunshine vitamin," reinforced the healthful image of the Sun. People learned that it was better to live in a sun-filled home than a dark dwelling. At that time, the relationship between skin cancer and exposure to the Sun was not known.

In the early twentieth century, many light-skinned people believed that a deep tan was a sign of good health. However, in the 1940s, the rate of skin cancer began to increase and reached significant proportions by the 1970s. At this time, scientists began to realize how damaging deep tans could really be.

Tanning occurs when ultraviolet radiation is absorbed by the skin, causing an increase in the activity of melanocytes, cells that produce the pigment melanin. As melanin is produced, it is absorbed by cells in the upper region of the skin, resulting in the formation of a tan. In reality, the skin is building up protection against damage caused by the ultraviolet radiation. It is interesting to note that people with naturally dark skin also produce additional melanin when their skin is exposed to sunlight.

Exposure to more sunlight means more damage to the cells of the skin. Research has shown that, although people usually do not get skin cancer as children, each time a child is exposed to the Sun without protection, the chance of that child getting skin cancer as an adult increases.

Knowledge connecting the Sun to skin cancer has greatly increased since the late 1800s. Currently, it is estimated that ultraviolet radiation is responsible for more than 90% of skin cancers. Yet, even with this knowledge, about 2 million Americans use tanning parlors that expose patrons to high doses of ultraviolet radiation. A recent survey showed that at least 10% of these people would continue to do so even if they knew for certain that it would give them skin cancer.

Many of the deaths due to this type of cancer can be prevented. The cure rate for skin cancer is almost 100% when it is treated early. Reducing exposure to harmful ultraviolet radiation helps to prevent it. During the past 15 years, scientists have tried to undo the tanning myth. If the word "healthy" is separated from the word "tan," maybe the occurrence of skin cancer will be reduced.

1. State *one* known benefit of daily exposure to the Sun. [1]

   _____

2. Explain what is meant by the phrase "the tanning myth." [1]

   _____

3. Which statement concerning tanning is correct?
   (1) Tanning causes a decrease in the ability of the skin to regulate body temperature.
   (2) Radiation from the Sun is the only radiation that causes tanning.
   (3) The production of melanin, which causes tanning, increases when skin cells are exposed to the Sun.
   (4) Melanocytes decrease their activity as exposure to the Sun increases, causing a protective coloration on the skin.   3 _____

4. Which statement concerning ultraviolet radiation is *not* correct?
   (1) It may damage the skin.
   (2) It stimulates the skin to produce antibodies.
   (3) It is absorbed by the skin.
   (4) It may stimulate the skin to produce excess pigment.   4 _____

5. Current knowledge concerning cells is a result of the investigations and observations of many scientists. The work of these scientists forms a well-accepted body of knowledge about cells. This body of knowledge is an example of a
   (1) hypothesis   (3) theory
   (2) controlled experiment   (4) research plan   5 _____

6. In his theory of evolution, Lamarck suggested that organisms will develop and pass on to offspring variations that they need in order to survive in a particular environment. In a later theory of evolution, Darwin proposed that changing environmental conditions favor certain variations that promote the survival of organisms. Which statement is best illustrated by this information?
   (1) Scientific theories that have been changed are the only ones supported by scientists.
   (2) All scientific theories are subject to change and improvement.
   (3) Most scientific theories are the outcome of a single hypothesis.
   (4) Scientific theories are not subject to change.   6 _____

Base your answers to questions 7 and 8 on the passage below and on your knowledge of biology.

The number in the parentheses ( ) at the end of a sentence is used to identify that sentence.

They Sure Do Look Like Dinosaurs

When making movies about dinosaurs, film producers often use ordinary lizards and enlarge their images many times (1). We all know, however, that although they look like dinosaurs and are related to dinosaurs, lizards are not actually dinosaurs (2).

Recently, some scientists have developed a hypothesis that challenges this view (3). These scientists believe that some dinosaurs were actually the same species as some modern lizard that had grown to unbelievable sizes (4). They think that such growth might be due to a special type of DNA called repetitive DNA, often referred to as "junk" DNA because scientists do not understand its functions (5). These scientists studied pumpkins that can reach sizes of nearly 1,000 pounds and found them to contain large amounts of repetitive DNA (6). Other pumpkins that grow to only a few ounces in weight have very little of this kind of DNA (7). In addition, cells that reproduce uncontrollably have almost always been found to contain large amounts of this DNA (8).

7. State *one* reason why scientists formerly thought of repetitive DNA as "junk." [1]

_____

_____

8. Write the number of the sentence that provides evidence supporting the hypothesis that increasing amounts of repetitive DNA are responsible for increased sizes of organisms. [1]

_____

Answers Explained

KEY IDEA 1—PURPOSE OF SCIENTIFIC INQUIRY

1. One response is required. Acceptable responses include:

    * *Kills bacteria*
    * *Produces vitamin D*
    * *Treats diseases and/or wounds*

2. One response is required. Acceptable responses include:

    * *The "tanning myth" involves people believing that a tan is a sign of good health.*
    * *The "tanning myth" says that a good tan is good for people.*

3. **3** *The production of melanin, which causes tanning, increases when skin cells are exposed to the Sun* is the correct statement concerning tanning. Melanin is a dark pigment that is produced in specialized skin cells in response to ultraviolet radiation in sunlight or an artificial source. This information is found in the third paragraph of the passage.

WRONG CHOICES EXPLAINED:

(1) *Tanning causes a decrease in the ability of the skin to regulate body temperature* is not a correct statement concerning tanning. There is no known relationship between tanning and body temperature regulation.

(2) *Radiation from the Sun is the only radiation that causes tanning* is not a correct statement concerning tanning. Tanning can also occur when the skin is exposed to artificial sources of ultraviolet radiation. A reference is made in the fourth paragraph of the passage to "tanning parlors" where people can be exposed to artificial doses of ultraviolet radiation.

(4) *Melanocytes decrease their activity as exposure to the Sun increases, causing a protective coloration on the skin* is not a correct statement concerning tanning. Melanin is produced as a protective pigment that helps prevent deep penetration of ultraviolet radiation into the deep layers of the skin. When ultraviolet radiation is absorbed by melanocytes, their activity increases, not decreases.

4. **2** *It stimulates the skin to produce antibodies* is not a correct statement concerning ultraviolet radiation. There is no information in the passage relating to the production of antibodies as a result of absorption of ultraviolet radiation, and no known research indicates this type of relationship.

WRONG CHOICES EXPLAINED:

(1), (3), (4) *It may damage the skin, it is absorbed by the skin,* and *it may stimulate the skin to produce excess pigment* are all correct statements concerning ultraviolet radiation. Ultraviolet radiation is an invisible but extremely powerful form of electromagnetic radiation. It can penetrate unshielded living tissues and alter the genetic makeup of the cells it encounters. In humans, this radiation can cause the production of melanin from melanocytes; in extreme cases, it can stimulate the growth of skin cancer.

5. **3** The body of knowledge described in this question is an example of a *theory*. When scientists begin to study a phenomenon in nature, their first step is normally to investigate it through repeated observation and experimentation. As a result of the analysis of the large quantity of data gathered during this process, the scientists then formulate a theory ("well-accepted body of knowledge") that describes the phenomenon in a way that is consistent with the data.

WRONG CHOICES EXPLAINED:

(1) A *hypothesis* is not the body of knowledge described in this question. Scientists develop a hypothesis ("educated guess") around their preliminary observations concerning a natural phenomenon. The hypothesis may be proven accurate or inaccurate as a result of the experimentation used to test it. For this reason, a hypothesis cannot be considered a "well-accepted body of knowledge."

(2) A *controlled experiment* is not the body of knowledge described in this question. A controlled experiment is a scientific method used to test an experimental hypothesis. The data that results from a controlled experiment can be used to support the development of a "well-accepted body of knowledge," but it does not constitute that body of knowledge.

(4) A *research plan* is not the body of knowledge described in this question. A research plan may be a series of controlled experiments designed to test various aspects of a natural phenomenon. The data that results from the research plan can be used to support the development of a "well-accepted body of knowledge," but it does not constitute that body of knowledge.

6. **2** *All scientific theories are subject to change and improvement* is the statement best illustrated by the information given. Both Lamarck and Darwin developed their theories of evolution based on observations made and inferences drawn before there was a good understanding of the genetic basis of variation. Lamarck's earlier theory of "use and disuse" was disproven by later experiments of other scientists. Darwin's later theory of "natural selection," though much closer to the currently accepted scientific theory of evolution, has been modified and improved on by the work of later scientists who have had the benefit of modern-day research in genetics, paleontology, and other sciences.

WRONG CHOICES EXPLAINED:

(1) *Scientific theories that have been changed are the only ones supported by scientists* is not the statement best illustrated by the information given. Scientists generally support theories that have stood the test of good scientific research. A theory that has not changed, as long as it is still supported by such research, is generally supported by most scientists.

(3) *Most scientific theories are the outcome of a single hypothesis* is not the statement best illustrated by the information given. In fact, scientific theories are based on the results of many experiments that each contain their own independent hypotheses.

(4) *Scientific theories are not subject to change* is not the statement best illustrated by the information given. Scientists are constantly questioning and reevaluating scientific theories. It is likely that a vast majority of all scientific theories undergo at least some modification.

7.   One response is required that indicates a reason why scientists formerly thought of repetitive DNA as "junk." Acceptable responses include: [1]

- *Scientists did not understand the function of repetitive DNA.*
- *They didn't know what it did, and so they thought it was junk.*

8.   One credit is allowed for indicating that either sentence 6 or sentence 7 provides evidence supporting the hypothesis that increased amounts of repetitive DNA are responsible for increased sizes of organisms. These sentences give information about the results of scientific investigations that measured the amount of repetitive DNA in the cells of giant pumpkins and miniature pumpkins and found that giant pumpkins contain more of this kind of DNA than miniature pumpkins.

KEY IDEA 2—METHODS OF SCIENTIFIC INQUIRY

Beyond the use of reasoning and consensus, scientific inquiry involves the testing of proposed explanations involving the use of conventional techniques and procedures and usually requiring considerable ingenuity.

Performance Indicator	Description
2.1	The student should be able to devise ways of making observations to test proposed explanations.
2.2	The student should be able to refine research ideas through library investigations, including electronic information retrieval and reviews of literature, and through peer feedback obtained from review and discussion.
2.3	The student should be able to develop and present proposals including formal hypotheses to test explanations (i.e., predict what should be observed under specific conditions if the experiment is true).
2.4	The student should be able to carry out research for testing explanations, including selecting and developing techniques, acquiring and building apparatus, and recording observations as necessary.

Base your answers to questions 9 and 10 on the diagram below of the field of view of a light compound microscope and on your knowledge of microscopes.

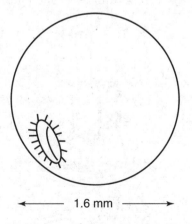

← 1.6 mm →

9. In order to center the organism in the field of view, the slide should be moved
   (1) down and to the right   (3) up and to the right
   (2) down and to the left   (4) up and to the left   9 ____

10. The approximate length of the organism is
   (1) 500 μm   (3) 50 μm
   (2) 1,600 μm   (4) 1.6 μm   10 ____

11. After viewing an organism under low power, a student switches to high power. The student should first
   (1) adjust the mirror
   (2) center the organism
   (3) raise the objective and switch to high power
   (4) close the diaphragm   11 ____

12. Using one or more complete sentences, explain why a specimen viewed under the high-power objective of a microscope appears darker than when it is viewed under low power.

   _____

   _____

Base your answers to questions 13 and 14 on the diagram below of a compound light microscope.

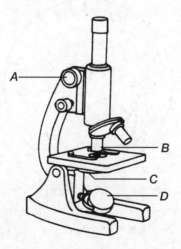

13. The letter *C* represents

(1) the mirror
(2) the diaphragm
(3) the eyepiece
(4) the high-power objective   13 _____

14. Select and name *one* of the labeled parts, and in one or more complete sentences describe its function.

_____

_____

15. The letter "p" as it normally appears in print is placed on the stage of a compound light microscope. Which best represents the image observed when a student looks through the microscope?

(1) p   (3) b
(2) q   (4) d   15 _____

16. To separate the parts of a cell by differences in density, a biologist would probably use

(1) a microdissection instrument
(2) an ultracentrifuge
(3) a compound light microscope
(4) an electron microscope   16 _____

17. The diagram below represents the field of view of a microscope. What is the approximate diameter, in micrometers, of the cell shown in the field?

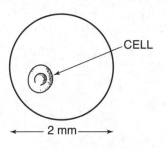

(1) 50 μm   (3) 1,000 μm
(2) 500 μm   (4) 2,000 μm   17 _____

Base your answer to the question on the diagram below.

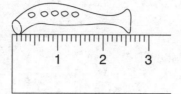

18. How many millimeters long is the organism resting on the metric
ruler?

_____

Questions 19 through 21 are based on the experiment described below.

A test tube was filled with a molasses solution, sealed with a membrane, and
inverted into a beaker containing 200 mL of distilled water. A second test tube
was filled with a starch solution, sealed with a membrane, and inverted into a
beaker containing 200 mL of distilled water. After several hours, the water in
each beaker was tested for the presence of molasses and starch.

The diagrams show the setup of the experiment.

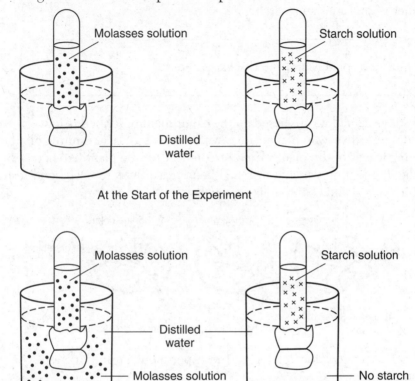

At the Start of the Experiment

After Several Hours

Answer each question related to the experiment in one or more complete sentences.

19. What principle was being tested in the experiment?

_____

20. What reagents were used in the experiment to test for the presence of molasses and starch?

_____

21. Draw one conclusion from this experiment.

_____

Questions 22 and 23 are based on the experiment described below.

An opaque disk was placed on several leaves of a geranium plant. The remaining leaves of the plant were untreated. After the plant had been exposed to sunlight, a leaf on which a disk had been placed was removed and tested as shown in parts _B_ and _C_ of the diagram below.

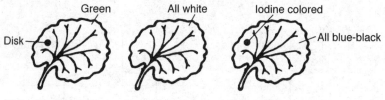

_A._ Leaf in Sunlight   _B._ Leaf after   _C._ Leaf after
                            Boiling in Alcohol   Testing with Iodine

Answer each question related to the experiment in one or more complete sentences.

22. What conclusion can be drawn from the result of the experiment?

_____

23. What process was being investigated by the experiment?

_____

Questions 24 and 25 are based on the experiment described below.

A student added 15 mL of water to each of three test tubes, labeled A, B, and C. A 1-cc piece of raw potato was added to tube B. A 1-cc piece of cooked potato was added to tube C. Five drops of hydrogen peroxide (H_2O_2) were added to each test tube. The results are shown in the following diagram.

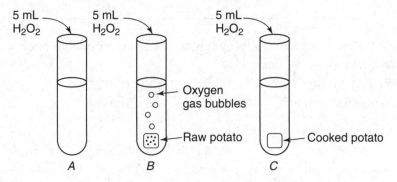

24. What conclusion can be drawn from the experiment?

_____

25. Which test tube is the control? Explain the reason for your choice.

_____

Base your answers to questions 26 through 28 on the diagram of the measuring device shown below.

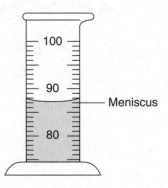

26. What is the name of this measuring device?

_____

27. In one complete sentence describe the procedure that you would follow to read the meniscus.

_____

28. What must a student do to obtain a volume of 85 milliliters of liquid in this measuring device?

   (1) Add 2.0 mL.   (3) Add 2.5 mL.

   (2) Remove 2.0 mL.   (4) Remove 8.7 mL.   28 _____

Answers Explained

KEY IDEA 2—METHODS OF SCIENTIFIC INQUIRY

9. **2** Specimens viewed under the microscope appear upside-down, backward, and reversed.

WRONG CHOICES EXPLAINED:

(1), (3), (4) With any of these choices, the specimen would be moved out of the field of view.

10. **1** The field of view is given as 1.6 mm. 1 mm = 1000 µm. 1.6 mm × 1000 µm = 1600 µm. The diagram shows that three specimens would fit across the field of view. One-third of 1600 µm = 533 µm. Of the choices given, *500 µm* (choice 1) is closest to this value.

WRONG CHOICES EXPLAINED:

(2), (3), (4) Each of these choices is mathematically incorrect.

11. **2** The student should first *center the organism*. The field of view is smaller under high power; therefore, less of a specimen can be seen. If the organism is not centered, it may fall out of the field of view under high power.

WRONG CHOICES EXPLAINED:

(1) The *mirror is adjusted* for maximum light under low power. Because the diameter of the high-power objective is very small, it is impossible to adjust the light under high power.

(3) A compound light microscope is parfocal; that is, it is not necessary to *lift the high-power objective* to focus under high power. The specimen remains in focus when switching from low power to high power.

(4) *Closing the diaphragm* reduces the amount of light entering the objective. Therefore, the specimen would appear very dark and would be difficult to see.

12. *The diameter of the high-power objective is smaller than the diameter of the low-power objective. Less light enters through the high-power objective, and therefore the specimen appears darker.*

13. **2** The letter C represents *the diaphragm*.

WRONG CHOICES EXPLAINED:
(1) *The mirror* is represented by *D*.
(3), (4) *The eyepiece* and *the high-power objective* are not labeled on the diagram.

14.   *Coarse adjustment (A)—used to focus a specimen under the low-power objective.*

or

Low-power objective (B)—along with the standard eyepiece, magnifies a specimen 100×.

or

Diaphragm (C)—regulates the amount of light entering the objectives.

or

Mirror (D)—provides a source of light that illuminates the specimen.

15.   **4** The image of a specimen as seen under a microscope is upside-down (*d*). The right side is on the left side, and the top is on the bottom.

WRONG CHOICES EXPLAINED:
(1) In this choice (*p*) there is no change in the way the image of the letter appears.
(2) In this choice (*q*) the image of the letter is reversed in only one direction: The right and left sides are reversed.
(3) In this choice (*b*) the image of the letter is reversed in only one direction: The top and bottom are reversed.

16.   **2** The *ultracentrifuge* is a machine that spins at a very high speed. A test tube of a liquid containing the parts of ruptured cells is placed in the machine. Each cell part has its own density (mass per unit volume). When the machine rotates, the cell parts fall to different levels in the test tube depending on their density.

WRONG CHOICES EXPLAINED:
(1) *A microdissection instrument* enables a biologist to remove a cell part from a single living cell. A micromanipulator is an example of such an instrument.
(3) A cell is transparent under a light microscope. Its structures cannot be seen unless the cell is stained. *A compound light microscope* can be used to view, but not to separate, cell parts.
(4) *An electron microscope* uses beams of electrons to view freeze-dried specimens; it cannot be used to separate cell parts for study.

17. **2** Study the information given in the diagram. Notice that the diameter of the circle is 2 mm. Since 1 mm is equal to 1,000 µm, 2 mm are equal to 2,000 µm. In relation to the entire circle, how large is the cell? Is the cell one-half as large or one-fourth as large? Dividing the circle into four parts shows us that the diameter of the cell is about one-quarter the diameter of the circle. Dividing 4 into 2,000 results in *500 µm*.

WRONG CHOICES EXPLAINED:
(1) *50 µm* is too small. The cell is ten times larger than 50.
(3) *1,000 µm* is too large. The cell is not one-half the diameter.
(4) *2,000 µm* is the diameter of the circle. The cell is only one-fourth as large.

18. The organism is *26 millimeters* long.

19. The principle of *diffusion* was being tested in the experiment.

20. *Benedict's solution* was used to test for the presence of molasses in the beaker. *Iodine* was used to test for the presence of starch in the beaker.

21. *Molasses can diffuse through a membrane.*
or
Starch cannot diffuse through a membrane.

22. *No starch was produced in the area covered by the disk.*

23. *The process of photosynthesis was being investigated.*

24. *Raw potato contains an enzyme that breaks down hydrogen peroxide.*
or
Cooking a potato destroys the enzyme that breaks down hydrogen peroxide.

25. *Test tube A is the control.* A control is the part of the experiment that provides the basis of comparison for the variable being tested.

26. The device is known as a *graduated cylinder*.

27. *The meniscus should be read at eye level.*

28. To obtain a volume of 85 mL, *2.0 mL* must be removed. The graduated cylinder contains 87 mL of liquid.

KEY IDEA 3—ANALYSIS IN SCIENTIFIC INQUIRY

The observations made while testing proposed explanations, when analyzed using conventional and invented methods, provide new insights into natural phenomena.

Performance Indicator	Description
3.1	The student should be able to use various methods of representing and organizing observations (e.g., diagrams, tables, charts, graphs, equations, matrices) and insightfully interpret the organized data.
3.2	The student should be able to apply statistical analysis techniques, when appropriate, to test if chance alone explains the results.
3.3	The student should be able to assess correspondence between the predicted result contained in the hypothesis and actual result, and reach a conclusion as to whether the explanation on which the prediction was based is supported.
3.4	The student should be able to, based on the results of the test and through public discussion, revise the explanation and contemplate additional research.
3.5	The student should be able to develop a written report for public scrutiny that describes the proposed explanation, including a literature review, the research carried out, its result, and suggestions for further research.

29. If curve A in the diagram represents a population of hawks in a community, what would most likely be represented by curve B?

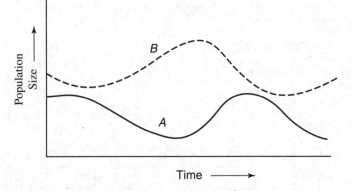

(1) the dominant trees in that community
(2) a population with which the hawks have a mutualistic relationship
(3) variations in the numbers of producers in that community
(4) a population on which the hawks prey

29 _____

Base your answers to questions 30 and 31 on this graph and on your knowledge of biology. The graph below depicts changes in the population growth rate of Kaibab deer.

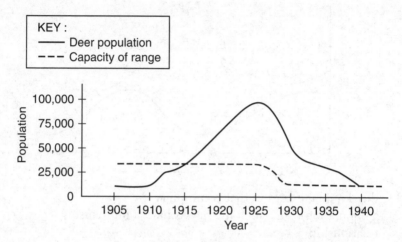

30. About how many deer could the range have supported in 1930 without some of them starving to death?

(1) 12,000

(2) 35,000

(3) 50,000

(4) 100,000

30 _____

31. In which year were the natural predators of the deer most likely being killed off faster than they could reproduce?

(1) 1905

(2) 1920

(3) 1930

(4) 1940

31 _____

32. Which process is illustrated by the diagram?

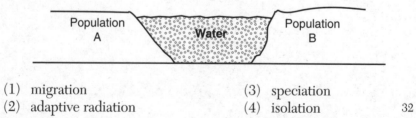

(1) migration

(2) adaptive radiation

(3) speciation

(4) isolation

32 _____

Base your answers to questions 33 through 36 on the information and data table below and on your knowledge of biology. The table shows the average systolic and diastolic blood pressure measured in millimeters of mercury (mm Hg) for humans between the ages of 2 and 14 years.

Data Table

Age	Average Blood Pressure (mm Hg)	
	Systolic	Diastolic
2	100	60
6	101	64
10	110	72
14	119	76

Directions (33–36): Using the information in the data table, construct a line graph on the grid provided, following the directions below.

33. Mark an appropriate scale on each labeled axis.

34. Plot the data for systolic blood pressure on your graph. Surround each point with a small triangle and connect the points.

35. Plot the data for diastolic blood pressure on your graph. Surround each point with a small circle and connect the points.

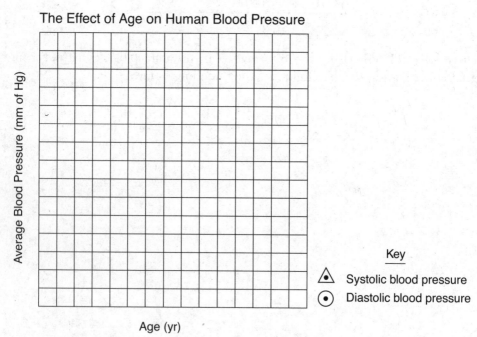

The Effect of Age on Human Blood Pressure

36. Using one or more complete sentences, state one conclusion that compares systolic blood pressure to diastolic blood pressure in humans between the ages of 2 and 14 years.

_____

_____

37. The graph below shows the results of an experiment.

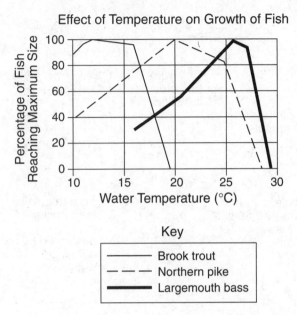

Effect of Temperature on Growth of Fish

Key

——— Brook trout
— — — Northern pike
━━━━ Largemouth bass

At 16°C, what percentage of the brook trout reached maximum size?

(1) 30%   (3) 75%
(2) 55%   (4) 95%   37 _____

38. An experiment is represented in the diagram below.

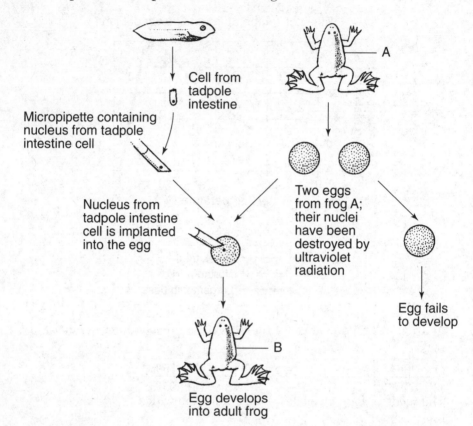

An inference that can be made from this experiment is that

(1) adult frog B will have the same genetic traits as the tadpole
(2) adult frog A can develop only from an egg and a sperm
(3) fertilization must occur in order for frog eggs to develop into adult frogs
(4) the nucleus of a body cell fails to function when transferred to other cell types

38 _____

39. The charts below show the relationship of recommended weight to height in men and women age 25–29.

Height-Weight Charts

MEN Age 25–29 Weight (lb)					WOMEN Age 25–29 Weight (lb)				
Height Feet	Inches	Small Frame	Medium Frame	Large Frame	Height Feet	Inches	Small Frame	Medium Frame	Large Frame
5	2	128–134	131–141	138–150	4	10	102–111	109–121	118–131
5	3	130–136	133–143	140–153	4	11	103–113	111–123	120–134
5	4	132–138	135–145	142–156	5	0	104–115	113–126	122–137
5	5	134–140	137–148	144–160	5	1	106–118	115–129	125–140
5	6	136–142	139–151	146–164	5	2	108–121	118–132	128–143
5	7	138–145	142–154	149–168	5	3	111–124	121–135	131–147
5	8	140–148	145–157	152–172	5	4	114–127	124–138	134–151
5	9	142–151	148–160	155–176	5	5	117–130	127–141	137–155
5	10	144–154	151–163	158–180	5	6	120–133	130–144	140–159
5	11	146–157	154–166	161–184	5	7	123–136	133–147	143–163
6	0	149–160	157–170	164–188	5	8	126–139	136–150	146–167
6	1	152–164	160–174	168–192	5	9	129–142	139–153	149–170
6	2	155–168	164–178	172–197	5	10	132–145	142–156	152–173
6	3	158–172	167–182	176–202	5	11	135–148	145–159	155–176
6	4	162–176	171–187	181–207	6	0	138–151	148–162	158–179

The recommended weight for a 6'0" tall man with a small frame is closest to that of a

(1) 5'10" man with a medium frame
(2) 5'9" woman with a large frame
(3) 6'0" man with a medium frame
(4) 6'0" woman with a medium frame

39 _____

Base your answers to questions 40 through 43 on the information below and on your knowledge of biology.

A group of biology students extracted the photosynthetic pigments from spinach leaves using the solvent acetone. A spectrophotometer was used to measure the percent absorption of six different wavelengths of light by the extracted pigments. The wavelengths of light were measured in units known as nanometers (nm). One nanometer is equal to one-billionth of a meter. The following data were collected:

yellow light (585 nm)—25.8% absorption

blue light (457 nm)—49.8% absorption

orange light (616 nm)—32.1% absorption

violet light (412 nm)—49.8% absorption

red light (674 nm)—41.0% absorption

green light (533 nm)—17.8% absorption

40. Complete all three columns in the data table below so that the wavelength of light either increases or decreases from the top to the bottom of the data table.

Color of Light	Wavelength of Light (nm)	Percent Absorption by Spinach Extract

Directions (41–42): Using the information in the data table, construct a line graph on the grid provided, following the directions below.

41. Mark an appropriate scale on the axis labeled "Percent Absorption."

42. Plot the data from the data table on your graph. Surround each point with a small circle and connect the points.

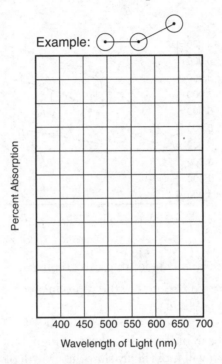

43. Which statement is a valid conclusion that can be drawn from the data obtained in this investigation?

(1) Photosynthetic pigments in spinach plants absorb blue light and violet light more efficiently than red light.

(2) The data would be the same for all pigments in spinach plants.

(3) Green light and yellow light are not absorbed by spinach plants.

(4) All plants are efficient at absorbing violet light and red light.

43 _____

44. The graph below represents the results of an investigation of the growth of three identical bacterial cultures incubated at different temperatures.

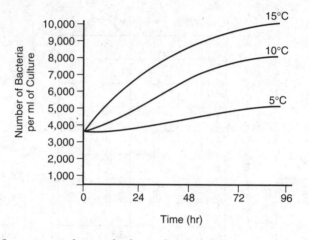

Which inference can be made from this graph?

(1) Temperature is unrelated to the reproductive rate of bacteria.
(2) Bacteria cannot grow at a temperature of 5°C.
(3) Life activities in bacteria slow down at high temperatures.
(4) Refrigeration will most likely slow the growth of these bacteria.

44 _____

45. A study was conducted using two groups of ten plants of the same species. During the study, the plants were kept under identical environmental conditions. The plants in one group were given a growth solution every 3 days. The heights of the plants in both groups were recorded at the beginning of the study and at the end of a 3-week period. The data showed that the plants given the growth solution grew faster than those not given the solution.

When other researchers conduct this study to test the accuracy of the results, they should

(1) give growth solution to both groups
(2) make sure that the conditions are identical to those in the first study
(3) give an increased amount of light to both groups of plants
(4) double the amount of growth solution given to the first group

45 _____

46. Worker bees acting as scouts are able to communicate the distance of a food supply from the hive by performing a "waggle dance." The graph below shows the relationship between the distance of a food supply from the hive and the number of turns in the waggle dance every 15 seconds.

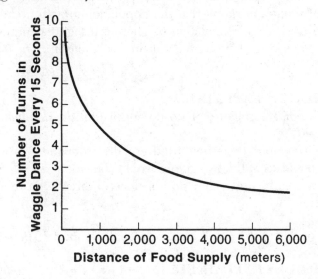

Using one or more complete sentences, state the relationship between the distance of the food supply from the hive and the number of turns a bee performs in the waggle dance every 15 seconds.

_____

47. Based on experimental results, a biologist in a laboratory reports a new discovery. If the experimental results are valid, biologists in other laboratories should be able to perform

(1) an experiment with a different variable and obtain the same results
(2) the same experiment and obtain different results
(3) the same experiment and obtain the same results
(4) an experiment under different conditions and obtain the same results

47 ____

Answers Explained

29.   **4** The diagram shows the population growth cycle for two organisms. An examination of the graph shows that the population growth cycle of the hawks closely follows the cycle of population B. There is a slight lag in the cycles. This type of graph is used to show a predator-prey relationship. *The hawks prey on population B.*

WRONG CHOICES EXPLAINED:

(1) Hawks are *not* herbivores. They do not live off the *dominant trees* in the community.

(2) If the two populations benefited equally from the relationship (were *mutualistic*), the peaks of the two graphs would coincide.

(3) Hawks are carnivores. They do not depend directly on the *producers* in the community.

30.   **1** According to the graph, the range could support *12,000* deer in 1930.

WRONG CHOICES EXPLAINED:

(2) The carrying capacity of the range was *35,000*. The *carrying capacity* is the maximum number of individuals that can be supported by the area. The number is usually constant unless severe environmental changes occur.

(3) The actual number of deer occupying the range in 1930 was *50,000*.

(4) A population of *100,000* deer was reached in 1925.

31.   **2** In nature, a predator-prey relationship keeps the prey population in check. In *1920*, the population of deer increased. It was at this time that the predators were removed by human hunting.

WRONG CHOICES EXPLAINED:

(1) The deer probably entered the region in *1905*. It takes time for an organism to adjust to a new environment.

(3) In *1930*, the deer population, which had exceeded the carrying capacity, was declining. The decline was caused by starvation. The deer had consumed almost all of the available vegetation in the area.

(4) By *1940*, the deer population reached the new carrying capacity of the range. The carrying capacity had been greatly reduced by overgrazing by the previously unchecked deer population.

32.   **4** As the result of *isolation*, the members of populations *A* and *B* are separated and are prevented from interbreeding. Variations that occur in one area are not transmitted to the individuals in the other area. Consequently, over a long period of time, the genetic differences become accentuated, and the new variations are maintained.

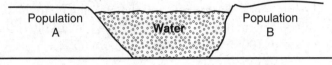

WRONG CHOICES EXPLAINED:

(1) *Migration* is not possible when populations are separated by a geographical barrier such as water.

(2) *Adaptive radiation* refers to a branching evolution and is not depicted in the diagram.

(3) *Speciation* indicates formation of new species from a parent population. This is not indicated in the diagram.

33–35.

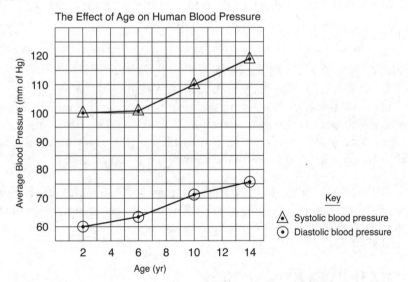

36. *Systolic pressure is higher than diastolic pressure.* or *Both systolic pressure and diastolic pressure increase between the ages of 2 and 14.*

[*Note:* Other correct complete-sentence responses are acceptable.]

37. **4** Brook trout growth is represented by the lighter solid line on the graph. Tracing along the horizontal axis to 16°C and then tracing up to the point at which the brook trout line is encountered, we see that the growth rate is between 80% and 100% but closer to 100%; we estimate it is about 95%.

WRONG CHOICES EXPLAINED:
(1) Brook trout growth rate is at 30% when the water temperature is about 18°C. Largemouth bass growth rate is about 30% at 16°C.
(2) None of the species indicated have a growth rate of 55% at 16°C. Brook trout growth rate is at 55% at about 17.5°C.
(3) Brook trout growth rate falls to 75% at about 17°C. The growth rate of northern pike is at 75% at about 16°C.

38. **1** The diagram represents an experiment in cloning. Because the nucleus (which contains the genetic material) used in this experiment comes from a tadpole, the egg will produce new cells that have *genetic characteristics identical to those of the tadpole.*

WRONG CHOICES EXPLAINED:
(2) It is unclear from this diagram how frog *A* came into being.
(3) This conclusion is refuted by the results of this experiment; a new frog was created without the use of fertilization.
(4) This conclusion is refuted by the results of this experiment; the tadpole nucleus was taken from an intestinal cell, which is a "body" cell.

39. **4** The chart on the left shows that a man 6′0″ tall with a small frame has an ideal weight range of 149–160 pounds. Of the choices given, the closest comparison can be made to the ideal weight range of a *6′0″ woman with a medium frame* at 148–162 pounds.

WRONG CHOICES EXPLAINED:
(1) The weight range shown for a *5′10″ man with a medium frame* is 151–163 pounds.
(2) The weight range shown for a *5′9″ woman with a large frame* is 149–170 pounds.
(3) The weight range shown for a *6′0″ man with a medium frame* is 157–170 pounds.

40.

Color of Light	Wavelength of Light (nm)	Percent Absorption by Spinach Extract
violet	412	49.8
blue	457	49.8
green	533	17.8
yellow	585	25.8
orange	616	32.1
red	674	41.0

41–42.

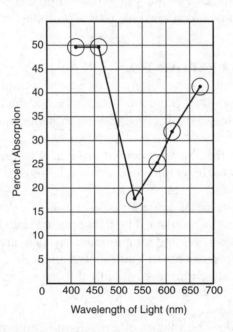

43.   **1** The statement *photosynthetic pigments in spinach plants absorb blue light and violet light more efficiently than red light* is a valid conclusion that can be drawn from the data. The high point of the chart/graph data is clearly shown to be above the blue and violet wavelengths of light.

WRONG CHOICES EXPLAINED:

(2) The statement *the data would be the same for all pigments in spinach plants* is not supported by the results of this experiment. The chart/graph data show considerable variation in the experimental results as the wavelength of light varies.

(3) The statement *green light and yellow light are not absorbed by spinach plants* is not supported by the results of this experiment. Although the chart/graph data show a lower absorption rate at these wavelengths, there is still some absorption in this range.

(4) The statement *all plants are efficient at absorbing violet light and red light* is not supported by the results of this experiment. The experimental data are limited to the absorption of light by pigments found in one type of plant. These data cannot be extended to all plants unless all other types of plants are tested under the same experimental conditions and the results are found to be similar.

44.   **4** Of those given, *refrigeration will most likely slow the growth of these bacteria* is the most reasonable inference that can be made from the graph data. The graph clearly shows a slower rate of growth (reproduction) at 5°C than at 10°C or 15°C.

WRONG CHOICES EXPLAINED:
(1) The inference that *temperature is unrelated to the reproductive rate of bacteria* is not supported by the data. Temperature is the independent (experimental) variable in this study. It clearly has an influence on the bacterial reproductive rate.

(2) The inference that *bacteria cannot grow at a temperature of 5°C* is not supported by the data. The graph clearly shows that growth at this temperature, while slow, occurs at a steady pace.

(3) The inference that *life activities in bacteria slow down at high temperatures* is not supported by the data. The data indicate that, if anything, bacterial activity increases with increasing temperature. No data are shown for bacterial growth at temperatures above 15°C, and so we cannot draw any inference about what happens to the rate of bacterial growth at these extremes.

45.   **2** These researchers should *make sure that the conditions are identical to those in the first study.* The validity of any scientific experiment can be verified only if the same results are obtained under the same experimental conditions. Any change in these conditions invalidates the results of the verification study.

WRONG CHOICES EXPLAINED:
(1) If the researchers *give growth solution to both groups,* there will be no control group against which to compare the experimental group. The results of the verification study will be invalid because the experimental conditions will have been changed.

(3) If the researchers *give an increased amount of light to both groups of plants,* the original experimental method will be altered. The results of the verification study will be invalid because the experimental conditions will have been changed.

(4) If the researchers *double the amount of growth solution given to the first group,* the original experimental method will not be followed. The results of the verification study will be invalid because the experimental conditions will have been changed.

46. *The number of turns in the waggle dance decreases as the distance of the food supply from the hive increases.* Or *The closer to the hive the food source is located, the more turns there are in the waggle dance.* [*Note:* Any correct, complete-sentence answer is acceptable.]

47. **3** Other biologists in other laboratories should be able to perform *the same experiment and obtain the same results* if the experimental results are valid. Any experimental results obtained by one scientist must be validated through independent research by other scientists following the same procedures.

WRONG CHOICES EXPLAINED:

(1) If different scientists perform *an experiment with a different variable and obtain the same results,* the original experimental results will be invalidated.

(2), (4) If different scientists perform *the same experiment and obtain different results,* or *an experiment under different conditions and obtain the same results,* they will neither validate nor invalidate the results of the original experiment. All variables and conditions must be kept the same if the experimental results are to be properly tested.

QUESTIONS ON STANDARD 4

Students will understand and apply scientific concepts, principles, and theories pertaining to the physical setting and living environment and recognize the historical development of ideas in science.

KEY IDEA 1—APPLICATION OF SCIENTIFIC PRINCIPLES

Living things are both similar to and different from each other and from nonliving things.

Performance Indicator	Description
1.1	The student should be able to explain how diversity of populations within ecosystems relates to the stability of ecosystems.
1.2	The student should be able to describe and explain the structures and functions of the human body at different organizational levels (e.g., systems, tissues, cells, organelles).
1.3	The student should be able to explain how a one-celled organism is able to function despite lacking the levels of organization present in more complex organisms.

48. In which life function is the potential energy of organic compounds converted to a form of stored energy that can be used by the cell?

(1) transport   (3) excretion

(2) respiration   (4) regulation   48 _____

49. Which life activity is *not* required for the survival of an individual organism?

(1) nutrition   (3) reproduction

(2) respiration   (4) synthesis   49 _____

50. Which function of human blood includes the other three?

(1) transporting nutrients

(2) transporting oxygen

(3) maintaining homeostasis

(4) collecting wastes   50 _____

51. In the human body, the blood with the greatest concentration of oxygen is found in the

  (1) left atrium of the heart
  (2) cerebrum of the brain
  (3) nephrons of the kidney
  (4) lining of the intestine   51 _____

52. Which type of vessel normally contains valves that prevent the backward flow of materials?

  (1) artery   (3) capillary
  (2) arteriole   (4) vein   52 _____

Directions (53–55): For each of questions 53 through 55, select the excretory structure, chosen from the list below, that best answers the question. Then record its number in the space provided at the right.

Excretory Structures
  (1) Alveolus
  (2) Nephron
  (3) Sweat gland
  (4) Liver

53. Which structure forms urine from water, urea, and salts?

_____

54. Which structure removes carbon dioxide and water from the blood?

_____

55. Which structure is involved in the breakdown of red blood cells?

_____

56. The bones of the lower arm are connected to the muscles of the upper arm by

  (1) ligaments   (3) cartilage
  (2) tendons   (4) skin   56 _____

57. The diagram below shows the same type of molecules in area *A* and area *B*. With the passage of time, some molecules move from area *A* to area *B*.

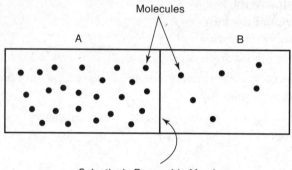

Selectively Permeable Membrane

The movement is the result of the process of

(1) phagocytosis   (3) diffusion
(2) pinocytosis   (4) cyclosis   57 _____

58. Which is the principal inorganic compound found in cytoplasm?

(1) lipid   (3) water
(2) carbohydrate   (4) nucleic acid   58 _____

59. A specific organic compound contains only the elements carbon, hydrogen, and oxygen in a ratio of 1:2:1. This compound is most probably a

(1) nucleic acid   (3) protein
(2) carbohydrate   (4) lipid   59 _____

60. Compared to ingested food molecules, end-product molecules of digestion are usually

(1) smaller and more soluble
(2) larger and more soluble
(3) smaller and less soluble
(4) larger and less soluble   60 _____

61. The cellular function of the endoplasmic reticulum is to

(1) provide channels for the transport of materials
(2) convert urea to a form usable by the cell
(3) regulate all cell activities
(4) change light energy into chemical bond energy   61 _____

62. In which organelles are polypeptide chains synthesized?

(1) nuclei   (3) ribosomes
(2) vacuoles   (4) cilia   62 _____

63. Which organelle contains hereditary factors and controls most cell activities?

(1) nucleus
(2) cell membrane
(3) vacuole
(4) endoplasmic reticulum   63 _____

64. Centrioles are cell structures involved primarily in

(1) cell division   (3) enzyme production
(2) storage of fats   (4) cellular respiration   64 _____

65. Which cell structure contains respiratory enzymes?

(1) cell wall   (3) mitochondrion
(2) nucleolus   (4) vacuole   65 _____

66. Which process is represented below?

simple organic molecules $\xrightarrow{\text{enzymes}}$ complex organic molecules + H_2O

(1) hydrolysis   (3) digestion
(2) synthesis   (4) respiration   66 _____

67. Amino acids derived from the digestion of a piece of meat are transported to living cells of an animal. In the cell they are

(1) converted to cellulose
(2) used to attack invading bacteria
(3) synthesized into specific proteins
(4) incorporated into glycogen molecules   67 _____

68. Which of the following variables has the *least* direct effect on the rate of a hydrolytic reaction regulated by enzymes?

(1) temperature
(2) pH
(3) carbon dioxide concentration
(4) enzyme concentration

68 _____

69. Which term refers to the chemical substance that aids in the transmission of the impulse through the area indicated by X?

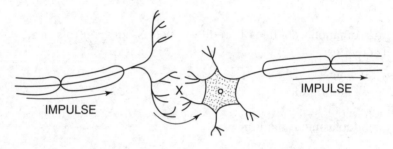

(1) neurotransmitter   (3) neuron
(2) synapse   (4) nerve

69 _____

70. Which lists human nervous-system structures in order of increasing size?

(1) neuron, nerve, ganglion, receptor
(2) nerve, ganglion, neuron, receptor
(3) neuron, receptor, ganglion, nerve
(4) ganglion, receptor, nerve, neuron

70 _____

71. Glands located within the digestive tube include

(1) gastric glands and thyroid glands
(2) gastric glands and intestinal glands
(3) thyroid glands and intestinal glands
(4) adrenal glands and intestinal glands

71 _____

72. In humans, which substance is directly responsible for controlling the calcium levels of the blood?

(1) adrenalin   (3) parathormone
(2) insulin   (4) thyroxin

72 _____

Base your answer to question 73 on the word equation below.

glucose → 2 pyruvic acid → 2 ethyl alcohol + 2 carbon dioxide + energy

73. The process represented by the word equation is known as

(1) aerobic respiration
(2) fermentation
(3) chemosynthesis
(4) dehydration synthesis   73 _____

74. The excretory organelles of some unicellular organisms are contractile vacuoles and

(1) cell membranes   (3) ribosomes
(2) cell walls   (4) centrioles   74 _____

75. Which is a type of asexual reproduction that commonly occurs in
many species of unicellular protists?

(1) external fertilization
(2) tissue regeneration
(3) binary fission
(4) vegetative propagation   75 _____

Answers Explained

48.   **2** *Respiration* is the life function by which ATP is made available to cells. Carbohydrate molecules are organic compounds. The breakdown of the carbohydrate molecules releases the energy stored in the bonds of the compounds. Potential energy is stored energy. The released potential energy is used to produce ATP.

WRONG CHOICES EXPLAINED:

(1) *Transport* is the life function by which materials are distributed throughout an organism.

(3) *Excretion* is the life function by which the wastes of metabolism are removed from an organism. Carbon dioxide, water, ammonia, and urea are metabolic wastes.

(4) *Regulation* is the life activity by which an organism responds to changes in its environment. The responses are controlled by the nervous system and the endocrine system.

49.   **3** *Reproduction*, the life function through which a parent organism gives rise to offspring, is not necessary for the survival of the parent. Although reproduction is not required for the survival of an individual, it is necessary for the survival of a species. If a given species loses its potential for reproduction, it will become extinct.

WRONG CHOICES EXPLAINED:

(1) *Nutrition* is a collective term that refers to the biochemical processes by which cells extract nutrient molecules from food substances. The nutrients are used to build tissues, provide energy, and regulate the many biochemical activities that occur in cells. Without nutrition, cells die and, consequently, so do organisms. Each organism is dependent on adequate nutrition for survival.

(2) *Respiration* refers to the series of chemical changes that fuel molecules undergo to release chemical energy for cells. Respiration is necessary for the survival of the individual. Tissue cells cannot live without a means of obtaining chemical energy to power cellular activities such as active transport and metabolism. Of course, death of tissue cells means death of the individual.

(4) *Synthesis* occurs when small molecules are joined chemically to form large molecules. Enzymes, hormones, and body tissues are the results of syntheses, without which an individual organism cannot survive. Synthesis is a building-up process in which molecules vital to the life of the organism are produced.

50. **3** *Maintaining homeostasis* is the function of human blood that includes the other three. By transporting nutrients, oxygen, wastes, and other materials around the body, the blood helps to make essential materials available to every living body cell while removing potentially harmful materials from these tissues. Equal distribution of these materials helps to promote a steady state in the tissues essential to homeostatic balance.

WRONG CHOICES EXPLAINED:

(1), (2), (4) *Transporting nutrients, transporting oxygen,* and *collecting wastes* are all functions of the blood that are involved in maintaining homeostasis. Nutrients provide cells with dissolved food molecules. Oxygen is used by cells in the release of energy from these food molecules. Wastes such as urea and carbon dioxide are carried away from the cells for excretion into the environment.

51. **1** Blood that has just returned from the lungs has the greatest concentration of oxygen. The *left atrium of the heart* receives blood directly from the lungs.

WRONG CHOICES EXPLAINED:

(2) Brain tissue is one of the largest consumers of oxygen. The blood circulating in the *cerebellum* gives up most of its oxygen to the nerve cells.

(3) The largest concentration of metabolic wastes is found in the *nephrons*. The nephrons are filtering units in the kidney.

(4) The largest concentration of digested nutrients is found in the *lining of the intestine*. Absorption of nutrients occurs through the villi in the small intestine.

52. **4** *Veins* are blood vessels that carry blood to the heart. They contain valves that prevent the backflow of blood. The blood in veins is usually deoxygenated; the exception is the pulmonary vein in which the blood is rich in oxygen.

WRONG CHOICES EXPLAINED:

(1) *Arteries* are blood vessels that transport blood away from the heart. Arteries are rather thick-walled and pump blood in rhythm with the heart. They have no valves.

(2) Small arteries are *arterioles*. This type of blood vessel functions similarly to arteries. Arterioles lead into capillaries.

(3) *Capillaries* are the smallest blood vessels. They are one cell thick and permit diffusion of water, nutrients, gases, and other substances into and out of the bloodstream. Capillaries have no valves; they are the connecting vessels between arterioles and venules.

53.   **2** The *nephron* is the unit of structure of the kidney. Each nephron has a glomerulus, Bowman's capsule, and kidney tubules. The kidney tubules filter out excess water, salts, and the wastes from protein metabolism. Urea and salts dissolved in water form urine.

54.   **1** The *alveolus* is an air sac in the lung. It not only permits the diffusion of oxygen from the lungs into the bloodstream but also aids in the diffusion of carbon dioxide and water vapor out of the blood into the lungs.

55.   **4** The *liver* is the largest gland in the body. One of its functions is to destroy old red blood cells and change the waste products into bile. The liver also synthesizes the anticoagulant known as heparin.

56.   **2** *Tendons* are tough connective tissues made strong by fibers. Tendons connect muscles to bones. The movable joints function when muscles pull on tendons.

WRONG CHOICES EXPLAINED:
   (1) *Ligaments* are strong connective tissues that contain elastic muscle fibers. Ligaments connect bone to bone.
   (3) *Cartilage* is a supporting tissue that provides strength to body structures without rigidity. Cartilage supports structures such as the ears and nose and covers the ends of bones that form joints. The ground substance, or matrix, of cartilage is made of protein.
   (4) *Skin* is composed of epithelial tissue. Skin serves as a body covering and has no function in the movement of bones or muscles.

57.   **3** *Diffusion* is the process that results in the movement of molecules from a region of higher concentration (area A) to a region of lower concentration (area B). This net movement occurs until the concentrations of molecules have reached equilibrium between area A and area B.

WRONG CHOICES EXPLAINED:
   (1), (2) *Phagocytosis* and *pinocytosis* are processes by which certain protists engulf their food and enclose it within a vacuole for digestion.
   (4) *Cyclosis* refers to the streaming of cytoplasm in the cell, a simple form of intracellular transport.

58.   **3** Inorganic compounds are compounds that do not contain carbon atoms. *Water*, the universal solvent, is the principal inorganic compound of cytoplasm. Water is the medium through which all chemical reactions take place in the cell.

WRONG CHOICES EXPLAINED:

(1), (2), (4) *Lipids, carbohydrates,* and *nucleic acids* are organic compounds. Organic compounds are carbon-containing compounds.

59. **2** Glucose is the building block of *carbohydrate* molecules. The ratio of carbon to hydrogen to oxygen is 1:2:1 in glucose and all reducing sugars. By dehydration synthesis, many glucose molecules form complex carbohydrates. However, the 1:2:1 ratio holds.

WRONG CHOICES EXPLAINED:

(1) A *nucleic acid* is composed of a phosphate group, a protein base, and a five-carbon sugar. DNA and RNA are nucleic acids. The CHO 1:2:1 ratio is not applicable.

(3) *Proteins* are built from amino acids, which, in addition to carbon, hydrogen, and oxygen, contain nitrogen. Some protein molecules also contain sulfur. The 1:2:1 ratio of elements does not apply to proteins because proteins are tissue builders whereas carbohydrates are fuel molecules.

(4) *Lipids* are fats and are composed of three fatty acid molecules and one glycerol molecule. The 1:2:1 ratio of carbon to hydrogen to oxygen does not apply to fats.

60. **1** The end products of digestion are usually *smaller and more soluble* than the ingested food molecules. Digestion makes available nutrient molecules that can diffuse across cell membranes and enter the cytoplasm of cells. Carbohydrates are broken down into glucose molecules. Fats are hydrolyzed into fatty acids and glycerol. Proteins are digested into their component amino acid molecules. Each of these end products of digestion is able to diffuse across cell membranes and enter into the biochemical activities of cells.

WRONG CHOICES EXPLAINED:

(2) Synthesis produces *larger molecules.* Larger molecules are more complex and are usually less soluble than smaller, simpler ones. Digestion results in smaller nutrient molecules.

(3) *Smaller molecules* are usually more soluble than larger ones. Digestion produces molecules that are more soluble than the complex nutrient molecules that were ingested.

(4) *Molecules derived from digestion* of ingested food are not larger than the molecules from which they came. Molecules produced by the digestion of complex carbohydrates, proteins, and fats are more soluble and are able to dissolve in water. Thus, these molecules can cross cell membranes.

61.   **1** The endoplasmic reticulum is a network of membranes that extends throughout the cell. The membranes form channels that *provide for the movement of materials through the cell*.

WRONG CHOICES EXPLAINED:

(2) *Urea* is a metabolic waste. It is a poisonous nitrogen compound. Urea must be removed from the cells if an organism is to survive.

(3) The *nucleus* is the organelle in the cell that regulates all cellular activities.

(4) The *chloroplasts* are organelles in plant cells that contain the green pigment chlorophyll. Chloroplasts are necessary for the process of photosynthesis.

62.   **3** Proteins are polypeptide chains. Proteins are synthesized in the *ribosomes*.

WRONG CHOICES EXPLAINED:

(1) The *nuclei* contain the genetic material carried in the chromosomes.

(2) *Vacuoles* are saclike organelles in the cytoplasm. Food vacuoles and contractile vacuoles are two common types of vacuoles.

(4) *Cilia* are microscopic hairs used for locomotion by some protozoans.

63.   **1** The *nucleus* contains the hereditary factors. Nuclei of plant and animal cells house the chromosomes, which are composed of deoxyribonucleic acid. Molecules of DNA function as genes. Points on the chromosomes are genes. Genetic information is passed from parent to offspring by way of the genes. Chromosomes are part of the fine structure of the nucleus. DNA molecules contribute to the chemical structure. Genes are sites or points that dot the length of the chromosome. Genes, DNA molecules, and chromosomes function in passing along hereditary factors.

WRONG CHOICES EXPLAINED:

(2) The *cell membrane* encloses the contents of the cell and directs the flow of materials into and out of the cell. The cell membrane does not contribute to the passing of genetic material from one generation to the next. The function of the membrane is to control cellular transport.

(3) A *vacuole* is a fluid-filled space in the cytoplasm. Vacuoles help to regulate the internal pressure of the cell. The vacuoles in fat cells are filled with oil.

(4) The membranes that line the cytoplasmic canals within cells are known collectively as the *endoplasmic reticulum*. This cytoplasmic fine structure aids in the transport of molecules from the cell membrane to various sites within the cell. Neither the endoplasmic reticulum nor the vacuoles of the cell membrane contain hereditary structures.

64.   **1** Centrioles are cell structures involved primarily in *cell division*. Centrioles are organelles that lie in the cytoplasm outside the nucleus; they are also found near the base of each flagellum and cilium. The centrioles of nonflagellated animal cells move to the spindle poles during cell division and seem to send out spindle fibers. The spindle fibers are attached to chromosomes and appear to pull the chromosomes from the center of the cell to the spindle poles.

WRONG CHOICES EXPLAINED:

(2) *Fats are stored* in cells. Fat in which the energy is channeled into heat production is stored in brown fat cells of hibernating mammals. At times, fat can be stored in arteries or accumulate around the heart. Fat cells are not involved in cell division.

(3) *Enzyme production* is controlled by the ribosomes that dot the membranes of the endoplasmic reticulum. Molecules of tRNA and mRNA regulate enzyme production.

(4) *Cellular respiration* is the process by which energy is released from glucose molecules. This process takes place in the mitochondria where oxygen is used as the final hydrogen carrier.

65.   **3** Cellular respiration occurs in the mitochondria (plural of *mitochondrion*). Each step in the process of cellular respiration is regulated by enzymes. Respiratory enzymes are located in the mitochondria.

WRONG CHOICES EXPLAINED:

(1) The *cell wall* is composed of cellulose. Cell walls give shape and protection to plant cells.

(2) The *nucleolus* contains the materials needed for the synthesis of RNA. It is located in the nucleus of the cell.

(4) A *vacuole* is a rounded sac that serves as a storage place for food and waste products. Some vacuoles, such as contractile vacuoles, maintain a stable internal environment.

66.   **2** *Synthesis* is the formation of complex molecules by combining simpler molecules. Water is removed from the simple molecules in this process.

WRONG CHOICES EXPLAINED:

(1) *Hydrolysis* is the addition of water to split complex molecules into simpler molecules. It is the opposite of synthesis.

(3) *Digestion* is another name for hydrolysis.

(4) *Respiration* is the process by which cells obtain energy. Glucose is converted to smaller molecules.

67.   **3** Amino acids are the building blocks of proteins. The dehydration synthesis of amino acids *produces protein molecules*.

WRONG CHOICES EXPLAINED:

(1) *Cellulose* is a polysaccharide composed of hundreds of simple sugar molecules. The sugars were joined together by dehydration synthesis.

(2) Antibodies attack *invading bacteria*. Antibodies are protein molecules produced by special white blood cells.

(4) *Glycogen*, a polysaccharide, is a product of the dehydration synthesis of many glucose units.

68.   **3** A *hydrolytic reaction* is a reaction in which a molecule is split. Enzymes are needed to speed up such a reaction. Any factor that affects the operation of the enzyme affects the speed at which the reaction takes place. The *concentration of carbon dioxide* has the least effect on enzyme activity.

WRONG CHOICES EXPLAINED:

(1) As the *temperature* is increased up to a point, the rate of the reaction increases. The increase in temperature increases the speed at which the enzyme and the substrate make contact with each other. The substrate is the molecule on which the enzyme acts. A very high temperature destroys the enzyme, and the reaction stops.

(2) Every enzyme works best at a particular *pH*. The enzymes in the stomach work in an acid environment, whereas the enzymes in the intestine work best in a basic medium.

(4) One molecule of an enzyme reacts with one molecule of a substrate. Increasing the *concentration of an enzyme* means that more substrate molecules will be acted on. The rate of the reaction will increase.

69.   **1** A *neurotransmitter* is a chemical substance that is released by an impulse arriving at the terminal end of a neuron. The neurotransmitter diffuses across the synapse and stimulates the second nerve cell. Acetylcholine is an example of a neurotransmitter.

WRONG CHOICES EXPLAINED:

(2) A *synapse* is the space between the terminal end of one nerve cell and the dendrites of a second nerve cell. The area marked by an X in the diagram is a synapse.

(3) A *neuron* is a nerve cell that is specially adapted for the conduction of impulses.

(4) A *nerve* is made up of many neurons.

70.   **1** A *neuron* is a single microscopic nerve cell. A *nerve* is composed of many nerve cells. A *ganglion* is a large mass of cell bodies of nerve cells; a ganglion functions as a coordinating center for impulses. A *receptor* is an organ specialized to receive environmental stimuli. The eye is an example of a receptor.

WRONG CHOICES EXPLAINED:
(2), (3), (4) In these three choices, either one or several structures are not arranged according to increasing size.

71.   **2** *Gastric glands* are embedded in the walls of the stomach. They are duct glands that secrete gastric juice, a mixture of water, hydrochloric acid, rennin, and pepsin. Gastric juice begins the digestion of protein in the stomach. *Intestinal glands* are duct glands that line the walls of the small intestine. They secrete intestinal juice, a mixture of water, proteases, amylases, and lipases. Both types of glands lie within the digestive tube.

WRONG CHOICES EXPLAINED:
(1) Gastric glands are described above. *Thyroid glands* lie outside the digestive tract at the base of the neck, straddled across the larynx. The thyroid is a ductless gland that secretes the hormone known as thyroxin. Thyroid glands do not function in the biochemical process of digestion.
(3) *Thyroid glands* and intestinal glands are described above. Thyroxin controls the metabolism of cells. The explanation above shows why this choice is wrong.
(4) *Adrenal glands* are dual endocrine glands that lie on top of each kidney. They are not within the digestive tract. The adrenal medulla, the inner gland, secretes the hormone adrenaline, also known as epinephrine. This hormone enables the body to function in emergencies. The adrenal cortex secretes about six active hormones, including cortisone, the antiarthritis hormone.

72.   **3** *Parathormone* is the hormone secreted by the parathyroid glands. The parathyroids are buried in the thyroids. Parathormone controls the level of calcium in the blood. Lack of blood calcium causes muscles to go into tetany. Tetany, or cramping, of the heart muscle causes death.

WRONG CHOICES EXPLAINED:
(1) *Adrenalin* is the hormone of the adrenal medulla, a ductless gland called the "gland of combat." Adrenaline stimulates the heart to beat faster, increases the rate of breathing, and controls the constriction and dilation of the arteriole walls.

(2) *Insulin* is secreted by the beta cells of the islets of Langerhans, which lie in the pancreas. Insulin controls sugar metabolism; specifically, it makes cell walls permeable to glucose and encourages the phosphorylation of fructose.

(4) *Thyroxin* is released by the thyroid gland. The rate of cellular metabolism is controlled by thyroxin. Iodine is used in the synthesis of thyroxin. People whose thyroid glands fail to develop become cretins; they are mentally retarded and physically undersized.

73. **2** Another name for anaerobic respiration is *fermentation*. In the process of fermentation, glucose is converted to energy, alcohol, and carbon dioxide.

WRONG CHOICES EXPLAINED:

(1) *Aerobic respiration* is another name for cellular respiration. This process requires oxygen. The following is an equation for aerobic respiration.

glucose + oxygen → pyruvic acid → carbon dioxide + water + energy

(3) *Chemosynthesis* is the synthesis of carbohydrates from inorganic compounds without the use of sunlight as a source of energy. Chemosynthesis is a form of autotrophic nutrition. It is carried out only by certain species of bacteria such as nitrifying bacteria.

(4) *Dehydration synthesis* is the method by which simple molecules are converted to complex molecules.

74. **1** *Cell membranes* and contractile vacuoles are excretory organelles of some unicellular organisms. The cell membrane is a selectively permeable membrane. It permits the diffusion of carbon dioxide and ammonia, two metabolic waste gases.

WRONG CHOICES EXPLAINED:

(2) *Cell walls* are composed of nonliving materials. Many canals penetrate through these walls, allowing the unrestricted passage of molecules.

(3) Proteins are synthesized in *ribosomes*.

(4) *Centrioles* are rodlike particles found in the centrosome. They function during the processes of mitosis and meiosis. Centrioles are found only in animal cells.

75.   **3** A unicellular protist (e.g., an ameba) is composed of a single cell. When this cell divides by mitosis, the process is known as *binary fission.*

WRONG CHOICES EXPLAINED:

(1) *External fertilization* is an element of sexual reproduction in many aquatic multicellular species. Both the sexual nature of this process and the fact that it is carried out by multicellular animals eliminate this as a correct choice.

(2) *Tissue regeneration* implies a process that occurs in multicellular organisms.

(4) *Vegetative propagation* is a form of asexual reproduction common to certain species of multicellular plants; it cannot be carried out by unicellular protists.

KEY IDEA 2—GENETIC CONTINUITY

Organisms inherit genetic information in a variety of ways that result in continuity of structure and function between parents and offspring.

Performance Indicator	Description
2.1	The student should be able to explain how the structure and replication of genetic material result in offspring that resemble their parents.
2.2	The student should be able to describe and explain how the technology of genetic engineering allows humans to alter genetic makeup of organisms.

76. Corn plants grown in the dark will be white and usually much taller than genetically identical corn plants grown in light, which will be green and shorter. The most probable explanation for this is that the

    (1) corn plants grown in the dark were all mutants for color and height
    (2) expression of a gene may be dependent on the environment
    (3) plants grown in the dark will always be genetically albino
    (4) phenotype of a plant is independent of its genotype   76 _____

77. In order for a substance to act as a carrier of hereditary information, it must be

    (1) easily destroyed by enzyme action
    (2) exactly the same in all organisms
    (3) present only in the nuclei of cells
    (4) copied during the process of mitosis   77 _____

78. During synapsis in meiosis, portions of one chromosome may be exchanged for corresponding portions of its homologous chromosome. This process is known as

    (1) nondisjunction   (3) crossing-over
    (2) polyploidy   (4) hybridization   78 _____

79. A DNA nucleotide is composed of three parts. These three parts may be

(1) phosphate, adenine, and thymine
(2) phosphate, deoxyribose, and thymine
(3) phosphate, glucose, and cytosine
(4) adenine, thymine, and cytosine

79 _____

80. A double-stranded DNA molecule replicates as it unwinds and "unzips" along weak

(1) hydrogen bonds
(2) carbon bonds
(3) phosphate groups
(4) ribose groups

80 _____

Base your answers to questions 81 through 84 on your knowledge of biology and the diagrams below. The diagram on the left represents a portion of a double-stranded DNA molecule. The diagrams at the right represent specific combinations of nitrogenous bases found in compounds transporting specific amino acids.

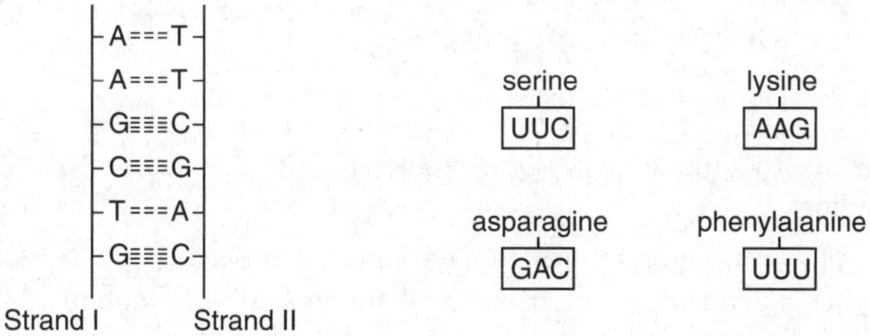

```
 -A===T-
 -A===T-          serine        lysine
 -G≡≡≡C-          UUC           AAG
 -C≡≡≡G-
 -T===A-          asparagine    phenylalanine
 -G≡≡≡C-          GAC           UUU
Strand I    Strand II
```

81. The amino acid whose genetic code is present in strand I is

(1) lysine
(2) serine
(3) asparagine
(4) phenylalanine

81 _____

82. The thymine (*T*) of strand I is accidentally replaced by adenine (*A*). This occurrence is called

(1) segregation
(2) disjunction
(3) cytoplasmic inheritance
(4) gene mutation

82 _____

83. The number of different amino acids coded by strand I is

(1) 1   (3) 8
(2) 2   (4) 12   83 _____

84. Which represents the sequence of nitrogenous bases in the molecule of messenger RNA synthesized by strand I?

(1) -T-T-C-G-U-C-   (3) -U-U-C-G-A-C-
(2) -A-A-C-G-T-C-   (4) -A-A-G-C-U-G-   84 _____

85. Molecules that transport amino acids to ribosomes are known as

(1) protein molecules   (3) mitochondria
(2) RNA molecules   (4) chromosomes   85 _____

86. A similarity between DNA molecules and RNA molecules is that they

(1) are built from nucleotides
(2) are double-stranded
(3) contain deoxyribose sugar
(4) contain uracil   86 _____

87. What is the function of DNA molecules in the synthesis of proteins?

(1) They catalyze the formation of peptide bonds.
(2) They determine the sequence of amino acids in a protein.
(3) They transfer amino acids from the cytoplasm to the nucleus.
(4) They supply energy for protein synthesis.   87 _____

88. In pea plants, the trait for smooth seeds is dominant over the trait for wrinkled seeds. When two hybrids are crossed, which results are most probable?

(1) 75% smooth and 25% wrinkled seeds
(2) 100% smooth seeds
(3) 50% smooth and 50% wrinkled seeds
(4) 100% wrinkled seeds   88 _____

89. A person who is homozygous for blood type A has a genotype that can be represented as

(1) $I^a I^b$ (3) $I^a i$

(2) $I^a I^a$ (4) ii 89 _____

90. Animal breeders often cross breed members of the same litter in order to maintain desirable traits. This procedure is known as

(1) hybridization

(2) inbreeding

(3) natural selection

(4) vegetative propagation   90 _____

Answers Explained

76.   **2** Corn plants grown in the dark will be white. The most probable explanation for this is that expression of the gene for color may *depend on the environment*. The plants have the genetic information for chlorophyll production. This can be assumed because they are genetically identical to the plants grown in the light. Light is needed to activate the chlorophyll gene.

WRONG CHOICES EXPLAINED:

(1) *Mutations* are sudden changes in the genetic material. Mutations are inherited. Because the plants grown in the dark were genetically identical to those grown in the light, neither group lacked the genetic information for chlorophyll production.

(3) *Albinism* is a condition resulting from the absence of a normal gene for color. Both groups of plants had the normal gene for color.

(4) The *phenotype* is the physical appearance of the organism. The phenotype depends on the genotype.

77.   **4** The hereditary information is contained in the chromosomes. During mitosis, the *chromosomes duplicate*. The duplication of chromosomes ensures the equal distribution of identical genetic material to the new cells.

WRONG CHOICES EXPLAINED:

(1) If the hereditary information is *destroyed*, the cells cannot function. The chromosomes contain the information necessary for carrying out all cellular activities.

(2) *No two organisms are exactly alike*. No two organisms have the same hereditary material. Identical twins are the only exception to these statements.

(3) Plasmagenes are *genes located outside the nucleus*. Drug resistance in some bacteria is transmitted through plasmagenes.

78.   **3** *Crossing-over* is the exchange of chromosomal material between homologous pairs of chromosomes. This process occurs during synapsis in meiosis.

WRONG CHOICES EXPLAINED:

(1) *Nondisjunction* is the failure of homologous chromosomes to separate from each other during meiosis. Cells with extra chromosomes and cells with too few chromosomes result from nondisjunction.

(2) *Polyploidy* is a condition in which the cells have extra sets of chromosomes beyond the normal 2*n* number.

(4) *Hybridization* is the crossing of two organisms that are distinctly different from each other. The purpose is to bring together new combinations of genes. Usually the individual with the new gene combinations is more sturdy than either parent. A tangelo is a cross between a tangerine and a grapefruit.

79. **2** A DNA nucleotide is composed of a *deoxyribose sugar molecule, a phosphoric acid molecule, and a nitrogen base.*

WRONG CHOICES EXPLAINED:
(1) *Adenine and thymine* are bases.
(3) *Glucose* is not the sugar molecule in DNA.
(4) *Adenine, thymine, and cytosine* are bases.

80. **1** The two strands of DNA are held together by *hydrogen bonds*. The hydrogen bonds form weak links between the base pairs of each strand.

WRONG CHOICES EXPLAINED:
(2) The sugars and bases of the nucleotides of DNA are organic compounds. Each individual compound is made up of *carbon bonds*.
(3) The nucleotides in each strand are joined together by *phosphate groups*.
(4) There are no *ribose groups* in DNA.

81. **1** *Lysine* is the amino acid whose genetic code is present on DNA strand I. Strand I of the DNA molecule contains two triplet codons (a triplet codon is a three-base sequence): AAG-CTG. Each triplet codon represents a specific amino acid. The amino acids are carried by tRNA molecules. The tRNA codon matches the DNA codon (except that U replaces T). There are two possible tRNA codons that could match the DNA strand I sequence; they are AAG and CUG. Of these, only the AAG tRNA appears in the diagram. The AAG tRNA carries the amino acid known as lysine.

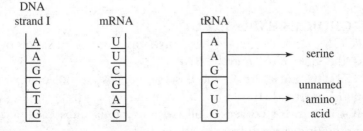

WRONG CHOICES EXPLAINED:

(2) The DNA triplet code for *serine* is TTC. This codon does not appear on strand I of the DNA molecule in the diagram.

(3) The DNA triplet code for *asparagine* is GAC. This codon does not appear on strand I of the DNA molecule in the diagram.

(4) The DNA triplet code for *phenylalenine* is TTT. This codon does not appear on strand I of the DNA molecule in the diagram.

82. **4** A gene controls the production of a protein. The substitution of one base for another changes the triplet code. One amino acid will be substituted for another. The result is a *mutation.* The replacement of glutamic acid by valine in a hemoglobin molecule causes sickle cell anemia.

WRONG CHOICES EXPLAINED:

(1) *Segregation* is the separation of alleles from each other during the formation of gametes.

(2) *Disjunction* is the separation of homologous chromosomes during the process of meiosis.

(3) *Cytoplasmic inheritance* is the inheritance of genes located in the cytoplasm, not in the nucleus. The cytoplasmic genes are called plasmagenes.

83. **2** *Two* different amino acids are coded by strand I. Strand I has two triplet codes, six bases.

WRONG CHOICES EXPLAINED:

(1) Only three bases would have to be shown in the diagram to code for *1* amino acid.

(3) 24 bases are needed for *8* amino acids.

(4) 48 bases are needed for *12* amino acids.

84. **3** *UUCGAC* is the correct sequence. Base pairing is an important concept in DNA duplication and RNA synthesis. Adenine pairs with thymine; cytosine pairs with guanine. There is no thymine in RNA. Uracil takes its place.

WRONG CHOICES EXPLAINED:

(1) *TTCGUC* is not correct. Because thymine is present in the base sequences, the molecule cannot be RNA.

(2) *AACGTC* is not correct. The base sequences are not complementary to either strand I or strand II.

(4) *AAGCUG* is not correct. The base sequences are complementary to strand II in the diagram not strand I.

85.   **2** Amino acids are transported to the ribosomes by *RNA molecules* known as transfer RNA, tRNA.

WRONG CHOICES EXPLAINED:
(1) *Protein molecules* are synthesized in the ribosomes. The code for the synthesis is contained in mRNA.
(3) Cellular respiration takes place in the *mitochondria*.
(4) *Chromosomes* are structures found in the nucleus. They are composed of DNA and protein. The genes are located on the chromosomes.

86.   **1** Both DNA and RNA *are built from nucleotides*.

WRONG CHOICES EXPLAINED:
(2) Only DNA is *double-stranded*. RNA is single-stranded.
(3) *Deoxyribose* is the sugar in the DNA nucleotides. Ribose is the sugar in the RNA nucleotides.
(4) The base thymine is replaced by *uracil* in RNA nucleotides.

87.   **2** The *sequence of amino acids* in a protein is determined by the triplet codes in DNA. The codes are carried to the ribosomes when mRNA is synthesized. A DNA strand is the template in mRNA synthesis.

WRONG CHOICES EXPLAINED:
(1) Enzymes *catalyze the formation of peptide bonds*. A peptide bond is a C–N bond formed by the dehydration synthesis of amino acids.
(3) *Amino acids are transferred* from the cytoplasm to the ribosomes, not to the nucleus. Transfer RNA is the carrier molecule.
(4) ATP molecules *supply the energy for protein synthesis*.

88.   **1** A *hybrid* is an individual that has two different alleles for a particular trait. The hybrid may be represented by the symbols *Ss*. The hybrid is smooth; the smooth trait is dominant over the wrinkled trait. When two hybrids are crossed, *75% of the offspring will have the smooth trait and 25% of the offspring will have the wrinkled trait.*

	S	*s*
S	*SS*	*Ss*
s	*Ss*	*ss*

WRONG CHOICES EXPLAINED:

(2), (3), (4) These choices are incorrect based on the information provided by the Punnett square.

89. **2** The term *homozygous* means *pure for the trait*. A homozygous individual has two identical alleles for a gene. There are three alleles for blood type: I^a, I^b, and i. The allele I^a produces a protein for blood type A; the allele I^b produces a protein for blood type B; the allele i does not produce either protein. The type of blood is determined by the combination of alleles. The genotype refers to the allelic combination. Because the person in the question is homozygous for type A blood, his genotype is I^aI^a.

WRONG CHOICES EXPLAINED:

(1) The I^a and I^b alleles are both dominant over the i allele. When both I^a and I^b alleles occur in the same person, the person has type AB blood.

(3) The type of blood represented by the genotype I^ai is type A.

(4) The genotype of a person with type O blood is ii.

90. **2** The mating of members of the same litter to maintain desirable traits is known as *inbreeding*. Because mating pairs come from the same litter, they are genetically similar to each other. Inbreeding is used to maintain pure breeds.

WRONG CHOICES EXPLAINED:

(1) *Hybridization,* or outbreeding, is the mating of organisms with contrasting traits. It is the opposite of inbreeding.

(3) Factors in the environment select the organisms that are best adapted to survive in the environment. This principle is known as *natural selection.* It is an essential feature in the theory of evolution.

(4) *Vegetative propagation* is asexual reproduction in plants.

KEY IDEA 3—ORGANIC EVOLUTION

Individual organisms and species change over time.

Performance Indicator	Description
3.1	The student should be able to explain the major patterns of evolution.

91. In modern classification, protozoa and algae are known as molds, and bacteria are known as

(1) bryophytes   (3) protists

(2) plants   (4) animals   91 ____

92. Most modern biologists agree that an ideal classification system should reflect

(1) nutritional similarities among organisms

(2) habitat requirements of like groups

(3) distinctions between organisms based on size

(4) evolutionary relationships among species   92 ____

93. Which term includes the other three?

(1) genus   (3) kingdom

(2) species   (4) phylum   93 ____

94. In one modern classification system, organisms are grouped into three

(1) kingdoms   (3) genera

(2) phyla   (4) species   94 ____

95. Which is one basic assumption of the heterotroph hypothesis?

    (1) More complex organisms appeared before less complex organisms.

    (2) Living organisms did not appear until there was oxygen in the atmosphere.

    (3) Large autotrophic organisms appeared before small photosynthesizing organisms.

    (4) Autotrophic activity added molecular oxygen to the environment.   95 _____

96. According to the heterotroph hypothesis, scientists believe that life arose in

    (1) a desert environment

    (2) a forest environment

    (3) a vacuum

    (4) an ocean environment   96 _____

97. From an evolutionary standpoint, the greatest advantage of sexual reproduction is the

    (1) variety of organisms produced

    (2) appearance of similar traits generation after generation

    (3) continuity within a species

    (4) small number of offspring produced   97 _____

98. According to modern theories of evolution, which of the following factors would be *least* effective in bringing about species changes?

    (1) geographic isolation

    (2) changing environments

    (3) genetic recombination

    (4) asexual reproduction   98 _____

99. A factor that tends to cause species to change is a

    (1) stable environment

    (2) lack of migration

    (3) recombination of genes

    (4) decrease in mutations   99 _____

100. If a fossil mammoth were discovered frozen in ice, its cells could be analyzed to determine whether its proteins were similar to those of the modern elephant. This type of investigation is known as comparative

(1) anatomy (3) biochemistry

(2) embryology (4) ecology 100 _____

101. If members of the same species have been geographically isolated from each other for an extended period of time, which will they most likely exhibit?

(1) mutations identical to each other

(2) random recombination occurring in the same manner

(3) evolution of traits of high adaptive value for their particular environments

(4) evolution into two new species which will have no problem interbreeding 101 _____

102. Skeletal similarities between two animals of different species are probably due to the fact that both species

(1) live in the same environment

(2) perform the same functions

(3) are genetically related to a common ancestor

(4) have survived until the present time 102 _____

103. The best means of discovering if there is a close evolutionary relationship between animals is to compare

(1) blood proteins (3) foods consumed

(2) use of forelimbs (4) habitats occupied 103 _____

104. In the process of evolution, the effect of the environment is to

(1) prevent the occurrence of mutations

(2) act as a selective force on variations in species

(3) provide conditions favorable for the formation of fossils

(4) provide stable conditions favorable to the survival of all species 104 _____

105. In a stable population in which the gene frequencies have been constant for a long time, the rate of evolution

(1) increases
(2) decreases
(3) remains the same
(4) increases, then decreases   105 _____

106. Certain strains of bacteria that were susceptible to penicillin in the past have now become resistant. The probable explanation for this is that

(1) the mutation rate must have increased naturally
(2) the strains have become resistant because they needed to do so for survival
(3) a mutation was retained and passed on to succeeding generations because it had high survival value
(4) the principal forces influencing the pattern of survival in a population are isolation and mating   106 _____

107. The frequency of traits that presently offer high adaptive value to a population may *decrease* markedly in future generations if

(1) conditions remain stable
(2) the environment changes
(3) all organisms with these traits survive
(4) mating remains random   107 _____

108. Since the publication of Darwin's theory, evolutionists have developed the concept that

(1) a species produces more offspring than can possibly survive
(2) the individuals that survive are those best fitted to the environment
(3) through time, favorable variations are retained in a species
(4) mutations are partially responsible for the variations within a species   108 _____

109. One factor that Darwin was unable to explain satisfactorily in his theory of evolution was

(1) natural selection
(2) overproduction
(3) survival of the fittest
(4) the source of variations

109 _____

Answers Explained

91.   **3** The *protists* include all unicellular organisms and organisms that have both plant and animal features within one cell. Protozoa and algae are protists.

WRONG CHOICES EXPLAINED:
(1) *Bryophytes* are multicellular green plants that do not have vascular tissue. Mosses are examples of bryophytes.
(2) Multicellular photosynthetic organisms make up the *plant* kingdom, which includes both vascular and nonvascular plants.
(4) The *animal* kingdom is composed of multicellular organisms that cannot manufacture their own food. The organisms within this kingdom lack cell walls, and most are capable of some type of locomotion.

92.   **4** A classification system should reflect *evolutionary relationships among species*. Evolutionary relationships are determined on the basis of the similarities in the anatomy, embryology, and biochemistry among organisms.

WRONG CHOICES EXPLAINED:
(1) All animals from protozoans to humans utilize the same nutrients in a similar manner. The process of photosynthesis is the same in tree cells and unicellular algae. *Nutritional similarities among organisms* are *not* useful in a system of classification.
(2) Both the whale and the fish live in an ocean environment. However, the whale is a mammal. Other than *sharing the same habitat*, the whale has no fish characteristics.
(3) Algae and protozoans are both microscopic organisms. However, algae have plant characteristics and protozoans have animal characteristics. *Size cannot be used* as the basis for a system of classification.

93.   **3** According to the classification system, the largest grouping of organisms is the *kingdom*. Following this, the other groups, in order, are phylum, class, order, family, genus, species. Depending on its chief characteristics, an organism is placed in one of five kingdoms, Monera, Protist, Fungi, Animal, or Plant.

94.   **1** In one modern classification system, organisms are grouped into three *kingdoms:* Animal, Plant, and Protist. Organisms that are not typical plants or animals are classified as protists. Examples are protozoa, slime molds, and bacteria.

WRONG CHOICES EXPLAINED:

(2) A *phylum* (plural phyla) is a large grouping that consists of classes, orders, families, genera, and species.

(3) A *genus* (plural genera) is a classification group composed of species. Members of a genus are more closely related than groups belonging to a given phylum.

(4) The *species* is the unit of classification. All members of a species are so closely related that they can mate and produce viable offspring.

95. **4** A heterotroph is an organism that must get its food from a source outside its own body cells; it cannot synthesize its food from inorganic materials. The heterotroph hypothesis proposes that the first living things on Earth were heterotrophs that obtained their food from the organic materials in the primitive seas. Autotrophs are organisms, such as green plants, that can synthesize their own food. At some stage in Earth's history, *autotrophic activity* used up the carbon dioxide in the air and, as a consequence of photosynthesis, *added molecular oxygen to the atmosphere*.

WRONG CHOICES EXPLAINED:

(1) Coacervates and then relatively *simple cells developed before more complex organisms*.

(2) Living heterotrophs *appeared before molecular oxygen was added to the atmosphere*.

(3) *Small cells that carried on photosynthesis appeared before the more complex vascular and seed plants.* The course of evolution is from the simple to the complex.

96. **4** According to the heterotroph hypothesis, life on earth evolved through a sequence of stages. The gases of the primitive atmosphere, such as methane, ammonia, and hydrogen, were washed down by heavy rains into the *early oceans*. They were acted on by ultraviolet radiation, cosmic rays, the earth's heat, and radioactivity. The bonding together of the molecules resulted in the formation of larger organic molecules.

WRONG CHOICES EXPLAINED:

(1) A *desert* environment could not support "first" life. The intense heat and the rapid evaporation of water are conditions that do not allow for the movement or maintenance of molecules in a fluid medium.

(2) A *forest* environment does not provide the pools of warm water on a continuous basis necessary for aggregate molecules to form.

(3) A *vacuum*, a place without air, cannot support life.

97.   **1** Sexual reproduction helps to maximize the number of different allelic combinations that occur in offspring, leading to a greater *variety of organisms produced* within the species' population as a whole. Individuals displaying favorable traits in a changing environment are more likely to survive and to pass these traits on to their offspring, a fact that helps to promote evolutionary change.

WRONG CHOICES EXPLAINED:

(2), (3) *Appearance of similar traits generation after generation* and *continuity within a species* describe conditions that promote stability and uniformity within species, both of which are maximized during asexual reproduction.

(4) Evolutionarily speaking, the larger the number of offspring produced during reproduction, the more successful a particular species variety tends to be in competing with other varieties of the same species. *Small number of offspring produced,* therefore, is not an evolutionary advantage of sexual or any other type of reproduction.

98.   **4** Variations among organisms are necessary for speciation. *Asexual reproduction* is least effective in bringing about changes in species. There are no variations among organisms that are reproduced asexually. These organisms are genetically like their parents.

WRONG CHOICES EXPLAINED:

(1) *Geographic isolation* increases the chance that a group of organisms will develop a new gene pool, which will give rise to a new species.

(2) *Changes in the environment* cause shifts in the gene pool. The genes that ensure the survival of organisms increase in the pool. Thus, the environment changes the characteristics of the original population, and a new species is formed.

(3) *Genetic recombination* occurs during meiosis and fertilization. The shuffling of genes results in the appearance of new characteristics in a population.

99.   **3** The *recombination of genes* is one factor that tends to cause species to change. Mutations are changes in genes. When like mutations combine in the fertilized egg, the new characteristic will be expressed in the offspring. If this mutation adds to the survival value of the organism, the gene change will be passed on to progeny because individuals having this mutation will live to reproduce.

WRONG CHOICES EXPLAINED:

(1) A *stable environment* will probably not cause species to change. Beneficial mutations become effective in changing environments. For example, the mutations that produced white fur in the polar bear were beneficial. At one time, the polar regions were tropical. A change to a glacial environment was accompanied by a change or changes in the animal species that inhabited the region. Bears with a dark coat color became immediate targets for natural enemies.

(2) *Migration* aids the recombination of genes because organisms have greater opportunities for interbreeding.

(4) A *decrease in mutations* does not aid speciation but slows it.

100. **3** *Biochemistry* is the study of the chemistry of living organisms. Proteins are compounds found only in living organisms.

WRONG CHOICES EXPLAINED:

(1) *Anatomy* is the study of the structure of organisms.

(2) *Embryology* is the study of the development of embryos.

(4) *Ecology* is the study of the relationship of living organisms to each other and to their environment.

101. **3** Geographic isolation involves the separation of organisms by natural barriers. Each group of isolated individuals develops its own gene pool because each group lives under different environmental conditions. In the case of members of the same species who have been geographically isolated from each other, the *selection for individuals with special survival traits* is different in each environment.

WRONG CHOICES EXPLAINED:

(1) *Mutations* are the raw materials for evolution. Although mutations might have been the same in each group, they do not have the same adaptive value in each group.

(2) *Random recombination* occurs in the same manner in each group. However, the recombination process operates on two distinctly different gene pools.

(4) Usually, *different species cannot mate*. If their mating happens to be successful, their offspring will not be fertile.

102. **3** Morphology is the study of the structure and form of living things. When the arm of a human and the wing of a bird are studied, they are seen to have similar bone structure. This indicates that both organisms descended from a *common ancestor*. They have both undergone many changes since then and are now very different from each other. However, they still retain some of the same genes and therefore show a similarity in many parts of their bodies, including the arrangement of the bones in their forelimbs.

103. **1** A close evolutionary relationship between animals can be shown by a study of their *blood proteins*. The precipitin test is used to show such a relationship. A rabbit can be sensitized to human blood by being injected with human serum. When the sensitized rabbit serum is mixed with human serum, a white precipitate forms. If the sensitized rabbit serum is mixed with serum from a chicken, there is no reaction. However, a precipitate does form when the sensitized rabbit serum is mixed with the serum of a chimpanzee. In a like manner, the serum of a dog and a wolf show precipitation with serum sensitized to dog serum.

WRONG CHOICES EXPLAINED:

(2) The *forelimbs* of many unrelated animals may be used for the same purpose.

(3) Many unrelated animals *consume similar food*.

(4) Many unrelated animals *live in the same habitat*.

104. **2** There are many variations among the organisms of a species. Some variations allow an organism to survive best in a particular environment. Such factors as climate, food supply, and type of predators determine which organisms are *best adapted to that environment*.

WRONG CHOICES EXPLAINED:

(1) *Mutations* occur naturally and randomly. Mutations increase the variations among organisms.

(3) *Fossils* are the remains of organisms that lived in the past. They present evidence that evolution has occurred.

(4) There are many factors in an environment that influence the survival of different species. A *stable environment* preserves the species that have already adapted to that environment.

105. **3** Stable gene pools are a hallmark of nonevolving populations. The rate of evolution in such populations neither increases nor decreases, but *remain the same*.

WRONG CHOICES EXPLAINED:

(1), (2), and (4) all refer to changing rates of evolution that do not occur.

106. **3** Bacteria resistant to penicillin developed as a result of mutation. Organisms that did not receive the mutated gene were killed by the antibiotic. Those in which gene mutation occurred survived and *passed the mutation on to succeeding generations*.

WRONG CHOICES EXPLAINED:

(1) *The mutation rate did not increase.* The survivors had the mutated gene that allowed the bacteria to resist the effects of penicillin. These resistant strains reproduced, creating populations that replaced the nonresistant strains.

(2) *Need does not determine mutation.* Mutations are chance occurrences.

(4) Survival of a species depends on the ability of its members to obtain food, carry out respiration, and *reproduce successfully. Isolation* does not increase species survival.

107. **2** As long as environmental conditions remain stable, the alleles controlling traits that promote individual survival in that environment tend to be maintained at a high level in the population's gene pool. When *the environment changes,* however, the factors that promoted survival may no longer be present. Selection pressure may then operate to reduce the frequency of the once-prevailing alleles in favor of alleles controlling other, contrasting traits that increase individuals' chances for survival in the new environment.

WRONG CHOICES EXPLAINED:

(1) As long as *conditions remain stable,* selection pressures on individuals displaying favorable traits remain low, promoting the maintenance of a high frequency of alleles controlling these traits.

(3) If *all organisms with these traits survive,* their genes will be passed on to future generations at a high rate. This will help to maintain a high frequency of alleles controlling these traits in the population.

(4) If *mating remains random,* the probability that alleles will pair in unrestricted combinations will remain high, helping to ensure that the laws of probability will operate freely and that gene frequencies for existing traits will remain stable.

108. **4** Darwin proposed his theory of evolution in 1856. His theory did not explain how variation arose in organisms. In 1901, Hugo De Vries discovered the existence of mutations. *Mutations accounted for the rise of variations in organisms.*

WRONG CHOICES EXPLAINED:
(1) One of the principles of Darwin's theory of evolution stated that a species produced *more offspring than could possibly survive.*
(2) Another principle stated that the *individuals that survived were those best suited to the environment.*
(3) *The variations favored by the environment are retained within a species.* It is the environment that determines which variations are favorable.

109. **4** Darwin's theory of evolution was completed and published before Mendel completed his study of inheritance in the garden pea. Darwin could not explain *how variations occurred* or how they were passed on from parent to offspring.

WRONG CHOICES EXPLAINED:
(1) Darwin's theory of *natural selection* was divided into five distinct principles that formulated his concept of evolution. These ideas were set forth in his book *The Origin of the Species by Natural Selection.*
(2) *Overproduction* was one of the principles of Darwin's theory. He explained that for a species to continue in existence, it must overproduce in order to maintain the species number. For example, one female codfish lays about 9 million eggs. Not all of these eggs are fertilized and not all codfish lay about 9 million eggs. Not all of these eggs are fertilized and not all codfish fry reach adulthood. If the 9 million eggs per female were fertilized and if these zygotes developed into adult fish, the seas would be overrun with codfish. However, if the number of gametes produced by codfish were greatly reduced, the species would die out. This overproduction of gametes is necessary to maintain codfish survival.
(3) Another of Darwin's principles was *survival of the fittest.* No two organisms are alike; each has variations. These variations may either help or hinder the organism in its struggle for existence. An organism with variations that help it reach food faster is more fit and has a better potential for survival than a slower, less fit member of the species.

KEY IDEA 4—REPRODUCTIVE CONTINUITY

The continuity of life is sustained through reproduction and development.

Performance Indicator	Description
4.1	The student should be able to explain how organisms, including humans, reproduce their own kind.

110. Which occurs in a plant cell but *not* in an animal cell during mitotic cell division?

(1) formation of spindle fibers
(2) chromosome duplication
(3) formation of a cell plate
(4) cytoplasmic division

110 _____

111. A plant cell with 12 chromosomes undergoes normal mitosis. What is the total number of chromosomes in each of the resulting daughter cells?

(1) 24
(2) 12
(3) 6
(4) 4

111 _____

112. Asexual reproduction *differs* from sexual reproduction in that, in asexual reproduction,

(1) new organisms are usually genetically identical to the parent
(2) the reproductive cycle involves the production of gametes
(3) nuclei of sex cells fuse to form a zygote
(4) offspring show much genetic variation

112 _____

113. In most multicellular animals, meiotic cell division occurs in specialized organs known as

(1) gonads
(2) gametes
(3) kidneys
(4) cytoplasmic organelles

113 _____

114. Which is an important adaptation for reproduction among land animals?

    (1) fertilization of gametes outside the body of the female
    (2) fertilization of gametes within the body of the female
    (3) production of sperm cells with thick cell walls
    (4) production of sperm cells with thin cell walls   114 _____

115. In humans, a single primary sex cell may produce four gametes. These gametes are known as

    (1) diploid egg cells   (3) polar bodies
    (2) monoploid egg cells   (4) sperm cells   115 _____

116. In sexual reproduction, the $2n$ chromosome number is restored as a direct result of

    (1) fertilization   (3) cleavage
    (2) gamete formation   (4) meiosis   116 _____

117. In human females, the main function of the follicle-stimulating hormone (FSH) secreted by the pituitary gland is to

    (1) stimulate the adrenal glands to produce cortisone
    (2) stimulate activity in the ovaries
    (3) control the metabolism of calcium
    (4) regulate the rate of oxidation in the body   117 _____

118. If the first stage of an uninterrupted human menstrual cycle is the follicle stage, the last stage includes the

    (1) formation of sperm cells in the testis
    (2) release of a mature egg
    (3) buildup of the uterine lining
    (4) shedding of the uterine lining   118 _____

119. Which statement best describes internal fertilization?

    (1) It does not require motile gametes.
    (2) It helps to make terrestrial life possible.
    (3) It requires the presence of many eggs.
    (4) It normally occurs in the male.   119 _____

120. What are the normal chromosome numbers of a sperm, egg, and zygote, respectively?

(1) monoploid, monoploid, and monoploid
(2) monoploid, diploid, and diploid
(3) diploid, diploid, and diploid
(4) monoploid, monoploid, and diploid   120 _____

121. When compared with the number of gametes produced from a single primary sex cell during oogenesis, the number of gametes produced from a single human primary sex cell during spermatogenesis is usually

(1) four times as great   (3) half as great
(2) twice as great   (4) the same   121 _____

122. In human males, sperm cells are suspended in a fluid medium. The main advantage gained from this adaptation is that the fluid

(1) removes polar bodies from the surface of the sperm
(2) activates the egg nucleus so that it begins to divide
(3) acts as a transport medium for sperm
(4) provides currents that propel the egg down the oviduct   122 _____

Base your answers to questions 123 and 124 on the diagrams and the information below.

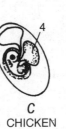

| A | B | C | D |
| BEAN | CHIMPANZEE | CHICKEN | AMEBA |

123. Which organisms were produced as a result of fertilization?

(1) *A*, *B*, and *C*, only   (3) *C* and *D*, only
(2) *B* and *C*, only   (4) *B*, *C*, and *D*, only   123 _____

124. Structures that function in the storage of food to be used by growing embryonic cells are indicated by

(1) 1 and 3   (3) 2 and 4
(2) 2 and 3   (4) 3 and 4   124 _____

Base your answers to questions 125 through 127 on your knowledge of biology and the information below.

A biologist cut a flap of ectoderm from the top of a developing embryo. He did not remove the piece of ectoderm but just folded it back. Then he cut out the mesoderm underneath and completely removed it. He folded the flap of ectoderm back in place. The ectoderm healed; however, a complete nervous system did not develop.

125. This experiment was most likely performed immediately after

(1) cleavage   (3) fertilization

(2) gestation   (4) gastrulation   125 _____

126. This experiment interfered with the process of

(1) differentiation   (3) cleavage

(2) zygote formation   (4) ovulation   126 _____

127. This experiment demonstrates that the

(1) ectoderm is solely responsible for development of the nervous system

(2) nervous system is destroyed during surgical operations

(3) mesoderm influences the development of the nervous system

(4) digestive enzymes have a major role in the development of embryonic layers   127 _____

128. In a developing embryo, the mesoderm layer normally gives rise to

(1) epidermal tissue

(2) skeletal tissue

(3) digestive tract lining

(4) respiratory tract lining   128 _____

129. What is the function of the placenta in a mammal?

(1) It surrounds the embryo and protects it from shock.

(2) It allows mixing of the maternal and fetal blood.

(3) It permits the passage of nutrients and oxygen from the mother to the fetus.

(4) It replaces the heart of the fetus until the fetus is born.   129 _____

Answers Explained

110. **3** Mitosis is the process by which two identical nuclei are formed. Mitotic cell division is usually followed by cytoplasmic division. A plant cell has a rigid cell wall. Division of the cytoplasm begins with the appearance of a *cell plate* between the two nuclei. The cell plate is composed of membrane fragments from the endoplasmic reticulum.

WRONG CHOICES EXPLAINED:

(1) *Spindle fibers* are elastic-like protein fibers. Chromosome movement is controlled by spindle fibers.

(2) In order for two nuclei to be identical, they must have the same number and kind of chromosomes. The *chromosomes duplicate* before the nucleus divides. The mitotic process is the same in both plant and animal cells.

(4) *Cytoplasmic division*, or cytokinesis, usually follows nuclear division in both plant and animal cells.

111. **2** Chromosomes are structures in the nucleus. During mitosis, two cells with identical chromosomes are formed. Because the cell had 12 chromosomes, the daughter cells must also have *12* chromosomes.

WRONG CHOICES EXPLAINED:

(1) A cell with *24* chromosomes has twice the diploid number. The condition in which there are extra sets of chromosomes is known as polyploidy.

(3) A cell with *6* chromosomes has one-half the diploid number. Monoploid cells arise through meiosis.

(4) A cell with *4* chromosomes can only arise through a complete breakdown of the mitotic or meiotic process.

112. **1** In asexual reproduction, new organisms are produced by a single parent. Asexual reproduction involves the mitotic process. The *genetic material of the offspring is identical to that of the parent*.

WRONG CHOICES EXPLAINED:

(2) *Gametes* are produced by sexually reproducing organisms.

(3) The fusion of sex cells (gametes) to *form a zygote* is characteristic of sexually reproducing organisms.

(4) *Genetic variation* among offspring is characteristic of sexually reproducing organisms. The process of meiosis through synapsis and segregation ensures new combinations of genetic material.

113. **1** *Gonads* are sex glands. In these glands, gametes, or sex cells, are produced from primary sex cells that undergo meiosis, also known as reduction division. Male gonads are called testes, and female gonads, ovaries.

WRONG CHOICES EXPLAINED:

(2) *Gametes* are sex cells and not organs. Sex cells are specialized for fertilization.

(3) *Kidneys* are organs of excretion and are specialized for filtering metabolic wastes out of the blood. The nephron is the unit of structure and function in the kidney. Meiosis does not take place in kidney cells.

(4) *Cytoplasmic organelles* such as mitochondria, ribosomes, lysosomes, and endoplasmic reticula are not organs.

114. **2** A gamete is a reproductive cell that must fuse with another gamete to produce a new individual. Sperm cells and egg cells are gametes. Fertilization is the fusion of an egg cell and a sperm cell. In land animals, *fertilization occurs within the body of the female* and is known as internal fertilization.

WRONG CHOICES EXPLAINED:

(1) External fertilization, the *union of gametes outside the female's body*, occurs in animals that live in a watery environment. Fish and amphibians reproduce by external fertilization.

(3), (4) The question refers to reproduction in animals. Animal cells, including gametes, do not have *cell walls*.

115. **4** Primary sex cells give rise to gametes. Gametes are formed by the process of meiosis. In meiosis, a diploid cell divides twice to form four monoploid cells. In humans the four gametes, which are identical in size, are known as *sperm cells*.

WRONG CHOICES EXPLAINED:

(1) Chromosomes occur in pairs. The diploid number of chromosomes is the full number of chromosomes of all the pairs. Meiosis is cell division in which the nucleus receives one member of each pair of chromosomes. The nucleus of an *egg cell* thus contains half the diploid chromosome number, or the monoploid number.

(2) In formation of the egg cell, the cytoplasm does not divide equally. One large monoploid cell, the egg cell, and three very small cells (polar bodies) are produced from one primary sex cell.

(3) The three small monoploid cells accompanying the egg cell are known as *polar bodies*. Polar bodies degenerate and do *not* function in fertilization.

116. **1** The diploid chromosome number is represented as $2n$, and the monoploid number as n. When two gametes in the n condition combine, a $2n$ cell is produced. *Fertilization* is the union of two gametes.

WRONG CHOICES EXPLAINED:

(2) *Gamete formation* reduces the chromosome number from $2n$ to n.

(3) *Cleavage* is mitotic cell division without growth. It is the process by which a fertilized egg cell becomes a multicellular embryo.

(4) *Meiosis*, or reduction division, reduces the chromosome number of diploid cells.

117. **2** Follicles contain immature egg cells. The follicles are found in the ovary. FSH *stimulates ovarian* follicle development.

WRONG CHOICES EXPLAINED:

(1) ACTH stimulates the *adrenal glands* to produce cortisone. ACTH is secreted by the pituitary gland.

(3) *Calcium metabolism* is controlled by parathormone. The parathyroid gland secretes parathormone.

(4) Thyroxin secreted by the thyroid gland is the major regulator of the rate of *oxidation*. The hormones from the adrenal glands and the pancreas also play a role in the oxidation of glucose.

118. **4** The menstrual cycle is a series of changes that occur within the female reproductive system. The events of the cycle prepare the uterus to receive an embryo. The lining of the uterus is built up. If the cycle is not interrupted, the egg is not fertilized and no embryo is formed. In the last stage of the cycle, the *lining of the uterus disintegrates and is shed*.

WRONG CHOICES EXPLAINED:

(1) *Sperm cells* are produced by the male.

(2) Ovulation (*release of a mature egg*) occurs midway through the menstrual cycle.

(3) Once a month the *uterus* is prepared to receive an embryo. What happens to the lining of the uterus depends upon presence or absence of an embryo. If an embryo is present, the uterus continues to develop and the menstrual cycle is interrupted.

119. **2** *It helps to make terrestrial life possible* is the correct response. Internal fertilization, as its name implies, occurs within the body of the parent (usually female). The conditions in the female reproductive tract provide an ideal environment for the survival and pairing of gametes, helping to ensure that fertilization occurs successfully. This method of reproduction is especially helpful in the survival of terrestrial animal species, who live where harsh conditions (such as drying, heat, and cold) can easily damage or kill gametes released into the environment for external fertilization.

WRONG CHOICES EXPLAINED:
(1) *Motile gametes* (such as human sperm cells) are common in species employing both external and internal fertilization. Motility (ability to move) enables the sperm cells to swim toward the egg cell in either environment.

(3) Because of the dangers posed to fragile gametes in any environment, the *presence of many eggs* is characteristic of species employing external fertilization. Species using internal fertilization produce relatively few eggs in the reproductive process.

(4) In most species internal *fertilization occurs within the body of the female*, not that of the *male*.

120. **4** *Monoploid, monoploid, and diploid* is the correct combination. Sperm cells and egg cells are monoploid (n) gametes formed during the process of meiotic cell division. A zygote results from the fusion of two monoploid nuclei in fertilization and so must be diploid ($2n$) in chromosome number.

WRONG CHOICES EXPLAINED:
(1), (2), (3) Each of these distracters contains an incorrect combination of choices (see above).

121. **1** In the process of oogenesis, a single primary sex cell gives rise to a single monoploid egg cell and three nonfunctional monoploid polar bodies. The process of spermatogenesis yields four functional monoploid sperm cells for each primary sex cell. Therefore, a comparison of these two processes leads to the conclusion that, per primary sex cell, spermatogenesis yields *four times* as many gametes as oogenesis does.

WRONG CHOICES EXPLAINED:
(2), (3), (4) Each of these distracters contains a mathematical comparison that is not consistent with the explanation above.

122. **3** The fluid surrounding human sperm cells *acts as a transport medium for sperm*. This fluid is known as semen. Its primary function is to provide a protective watery medium for sperm cells as they enter the female reproductive tract.

WRONG CHOICES EXPLAINED:
(1) *Removes polar bodies from the surface of the sperm* is a "nonsense" distracter. Polar bodies are not associated with sperm production.
(2) *Activates the egg nucleus so that it begins to divide* is not a function of semen. The egg is stimulated to divide by the act of fertilization. Semen is not directly involved in this process.
(4) *Provides currents that propel the egg down the oviduct* is not a function of semen. Cilia that line the oviduct are responsible for establishing fluid currents that both carry the egg downward toward the uterus and carry sperm upward toward the ovary. Semen is not directly involved in this process.

123. **1** Fertilization is one of the processes in sexual reproduction. Sexual reproduction is the method of reproduction in the bean plant, chimpanzee, and chicken (*A, B, and C, only*). The ameba reproduces asexually by binary fission.

WRONG CHOICES EXPLAINED:
(2) The bean plant (*B*) was omitted in this choice.
(3) The ameba (*D*) is an incorrect answer. Both the bean (*A*) and the chimpanzee (*B*) were omitted in this choice.
(4) The bean plant (*A*) was omitted and the ameba (*D*) is an incorrect answer.

124. **3** The structures that function in the storage of food for the embryonic cells are labeled 2 and 4. Structure 2 is the cotyledon of the seed. The yolk *sac* is structure 4.

WRONG CHOICES EXPLAINED:
(1) Structure 3 refers to the wall of the uterus. The embryo of a chimpanzee is nourished through the placenta, not the uterine wall. Structure 1 is the leaf of the embryo bean plant.
(2) Although structure 2 is a correct answer, structure 3 is incorrect.
(4) Although structure 4 is a correct answer, structure 3 is incorrect.

125. **4** The experiment was performed after *gastrulation*. Gastrulation is a stage in embryonic development that gives rise to three germ layers of cells. The three germ layers are the ectoderm, mesoderm, and endoderm.

WRONG CHOICES EXPLAINED:

(1) *Cleavage* is a stage of embryonic development in which the zygote undergoes rapid mitotic divisions. The final result is a ball of cells.

(2) *Gestation* is a prebirth period. It is the time a developing embryo spends in the uterus.

(3) The union of a sperm cell nucleus with an egg cell nucleus is called *fertilization*. The result of the process is a zygote.

126. **1** The experiment interfered with the development of a nervous system. The development of special tissues and organisms is known as *differentiation*.

WRONG CHOICES EXPLAINED:

(2) *Zygote formation* must occur before an embryo can develop.

(3) The process of *cleavage* provides the embryo with hundreds of undifferentiated cells.

(4) *Ovulation* is the release of an egg from the ovary.

127. **3** The experiment demonstrates that the development of the *nervous system is influenced by the presence of the mesoderm*. The nervous system does not develop when the mesoderm is removed.

WRONG CHOICES EXPLAINED:

(1) If the *ectoderm was solely responsible for the development of the nervous system*, the nervous system would have developed after the mesoderm was removed.

(2) There was no *nervous system* present when the surgery was performed.

(4) The experiment was not concerned with the reasons for the *development of the embryonic layers*.

128. **2** Each of the three germ layers of the embryo is responsible for the development of the systems of the body. The *skeletal system* develops from the mesoderm. The muscle system, circulatory system, and excretory system also evolve from the mesoderm.

WRONG CHOICES EXPLAINED:

(1) *Epidermal cells* form the outer covering or skin of the body. The skin and nervous system develop from the ectoderm.

(3), (4) The linings of the *digestive* and *respiratory tracts* develop from the endoderm.

129. **3** The placenta is an area of spongy tissue in the uterus. It is very rich in blood vessels. The placenta functions as a respiratory and excretory organ of the fetus. *Vital materials are exchanged between the capillaries of the fetus and the capillaries of the mother.*

WRONG CHOICES EXPLAINED:

(1) The amnion is a fluid-filled sac surrounding the embryo. The fluid bathes the cells of the fetus and *protects it against shock.*

(2) The circulatory systems of the mother and the fetus are separate from each other. *Blood does not flow from one system into the other.*

(4) The *embryo develops its own heart.* The placenta provides an area for the diffusion of materials into and out of the fetus.

KEY IDEA 5—DYNAMIC EQUILIBRIUM AND HOMEOSTASIS

Organisms maintain a dynamic equilibrium that sustains life.

Performance Indicator	Description
5.1	The student should be able to explain the basic biochemical processes in living organisms and their importance in maintaining dynamic equilibrium.
5.2	The student should be able to explain disease as a failure of homeostasis.
5.3	The student should be able to relate processes at the system level to the cellular level in order to explain dynamic equilibrium.

130. Which energy conversion occurs in the process of photosynthesis?

   (1) Light energy is converted to nuclear energy.
   (2) Chemical bond energy is converted to nuclear energy.
   (3) Light energy is converted to chemical bond energy.
   (4) Mechanical energy is converted to light energy.   130 _____

131. An environmental change that would most likely increase the rate of photosynthesis in a bean plant would be an increase in the

   (1) intensity of green light
   (2) concentration of nitrogen in the air
   (3) concentration of oxygen in the air
   (4) concentration of carbon dioxide in the air   131 _____

132. During photosynthesis, molecules of oxygen are released as a result of the "splitting" of water molecules. This is a direct result of the

   (1) dark reaction   (4) formation of CO_2 132 _____
   (2) light reaction
   (3) formation of PGAL

133. An organism that makes its own food without the direct need for any light energy is known as a

(1) chemosynthetic heterotroph
(2) chemosynthetic autotroph
(3) photosynthetic heterotroph
(4) photosynthetic autotroph   133 _____

134. While looking through a microscope at a section of a leaf from a freshwater plant, a student observed some cells in which chloroplasts were moving around with the cytoplasm. This type of movement is known as

(1) pinocytosis   (3) osmosis
(2) synapsis   (4) cyclosis   134 _____

135. By what process does carbon dioxide pass through the stomates into the leaf ?

(1) diffusion   (3) respiration
(2) osmosis   (4) pinocytosis   135 _____

136. Two end products of aerobic respiration are

(1) oxygen and alcohol
(2) oxygen and water
(3) carbon dioxide and water
(4) carbon dioxide and oxygen   136 _____

137. Homeostatic regulation of the body is made possible through the coordination of all body systems. This coordination is achieved mainly by

(1) respiratory and reproductive systems
(2) skeletal and excretory systems
(3) nervous and endocrine systems
(4) circulatory and digestive systems   137 _____

138. Phenylketonuria (PKU) is an inherited condition characterized by mental retardation. The symptoms of the disorder result from an inability to synthesize a single type of

(1) enzyme   (3) blood cell

(2) nutrient   (4) brain cell   138 _____

139. All the children of a hemophiliac male and a normal female are normal with respect to blood clotting. However, some of their grandsons are hemophiliacs. This is an example of the pattern of hereditary known as

(1) sex determination

(2) sex linkage

(3) incomplete dominance

(4) multiple alleles   139 _____

140. What is the total number of chromosomes in a typical body cell of a person with Down's syndrome?

(1) 22   (3) 44

(2) 23   (4) 47   140 _____

Answers Explained

130. **3** In the process of photosynthesis, carbon dioxide and water molecules are converted to glucose. Light is the energy source for the reaction. *Light energy is transformed into the chemical bond energy* of the glucose molecules.

WRONG CHOICES EXPLAINED:

(1), (2), (4) The Law of Conservation of Energy states that energy cannot be created or destroyed, but it can be changed from one form to another. All the choices refer to this law, but only choice (3) occurs in living organisms.

131. **4** Experiments have shown that the rate of photosynthesis depends on the availability of carbon dioxide. The greater the *concentration of carbon dioxide*, the greater the rate of photosynthesis.

WRONG CHOICES EXPLAINED:

(1) Chlorophyll is a light-absorbing pigment found in chloroplasts. However, it does not absorb much of the green wavelength of light. An increase in the *intensity of green light* has no effect on the photosynthetic rate.

(2) Carbohydrates are the products of photosynthesis. They do not contain atoms of nitrogen. The nitrogen needed by plants comes from nitrates in the soil, not from *the concentration of nitrogen in the air*.

(3) Oxygen is released during photosynthesis. Therefore, *the concentration of oxygen in the air* has no direct effect on photosynthesis.

132. **2** Sunlight provides the energy needed to split water molecules into hydrogen and oxygen. Because light is required, this part of photosynthesis is known as the *light reaction*.

WRONG CHOICES EXPLAINED:

(1) The *dark reaction* does not use light energy. In this reaction, the hydrogen released from the light reaction is combined with carbon dioxide.

(3) *PGAL*, phosphoglyceric aldehyde, is the first stable compound formed during the dark reaction. This compound is later converted to glucose.

(4) *Carbon dioxide is not formed* but is used during photosynthesis.

133. **2** An autotroph manufactures its own food. A *chemosynthetic autotroph* produces its own food without the use of light energy. It obtains its energy from certain chemical reactions that take place in the cell.

WRONG CHOICES EXPLAINED:

(1) A *chemosynthetic organism* cannot be a heterotroph. Heterotrophs do not have the ability to manufacture their own food. All animals are heterotrophs.

(3) *Photosynthetic organisms* manufacture their own food. They cannot be heterotrophs.

(4) A *photosynthetic autotroph* utilizes light energy. All green plants are photosynthetic autotrophs.

134. **4** The movement of chloroplasts in the plant cell was due to the movement of cytoplasm in the cell. *Cyclosis* is the streaming of cytoplasm in a cell.

WRONG CHOICES EXPLAINED:

(1) *Pinocytosis* is the formation of a pocket by an infolding of the cell membrane. Large molecules are brought into the cell by this process.

(2) *Synapsis* is the pairing of homologous chromosomes during meiosis.

(3) *Osmosis* is the movement of water across a selectively permeable membrane. Osmosis is the diffusion of water.

135. **1** Carbon dioxide passes through the stomates of a leaf by *diffusion*. Diffusion is passive transport. Molecules move along a concentration gradient from an area of high density to an area of lower density.

WRONG CHOICES EXPLAINED:

(2) *Osmosis* is the diffusion of water.

(3) *Respiration* is an energy-releasing process.

(4) *Pinocytosis* is active transport. Cells use energy to draw in large molecules by the infolding of their cell membranes.

136. **3** Respiration is a process by which cells release energy from glucose molecules. Aerobic respiration requires the presence of oxygen. In the process of aerobic respiration, glucose is oxidized to *carbon dioxide and water*. Both compounds are end products, the results of a chemical reaction.

WRONG CHOICES EXPLAINED:
(1), (2), (4) All three choices are incorrect because oxygen is consumed in aerobic respiration. Oxygen is not the end product of the reaction.

137. **3** Homeostasis refers to the steady state of control of the cell and, in turn, the entire body. The biochemical processes that take place in body cells occur in even and regular sequences. The cells, tissues, and organs in all body systems must function cooperatively so that the organism can carry out its life functions effectively. The coordination of all these biochemical activities is made possible by the work of the *nervous and endocrine systems*. The nervous system carries impulses from sense organs to the brain or spinal cord and then to effector organs such as muscles or glands. The endocrine system secretes hormones that control the functions of certain glands, tissues, and organs. Together the nervous and endocrine systems maintain the homeostasis of the body.

WRONG CHOICES EXPLAINED:
(1) The *respiratory system* is specialized for the intake and distribution of oxygen. It also expels waste gases. The *reproductive system* is specialized for the developing of embryos. Both systems are controlled by the nervous and endocrine systems. In mammals, hormones control gestation and birth.

(2) The *skeletal system* gives support to the body. Hormones control the growth of long bones. Nerve cell fibers help the muscles to function. The *excretory system* coordinates waste removal from the body. Hormones control water loss and reabsorption by the kidney tubules.

(4) Blood and lymph circulate by way of the *circulatory system*. Hydrolysis of food takes place in the *digestive system*. Both systems depend on the nervous and endocrine systems for the coordination of mechanical and biochemical activities.

138. **1** According to the one gene–one enzyme theory, a single gene is responsible for the production of a single enzyme. Because PKU is an inherited defect, the gene for an *enzyme* is absent in the victim.

WRONG CHOICES EXPLAINED:
(2) *Nutrients* are *not* synthesized by animals. Nutrients must be taken in from organic sources.

(3) *Blood cells* are *not* affected nor involved in the PKU disorder.

(4) The development of *brain cells* is affected by the PKU condition. Because of the lack of an enzyme, phenylalanine is converted to phenylpyruvic acid, which accumulates in brain tissue. The result is mental retardation.

139. **2** A trait that appears more often in one sex than in the other sex is said to be *sex-linked*. The gene for the trait is located on a sex chromosome. Hemophilia and color blindness are examples of sex-linked traits.

WRONG CHOICES EXPLAINED:

(1) *Sex determination* is controlled by a pair of sex chromosomes. There are two kinds of sex chromosomes, an X and a Y chromosome. A female has two X chromosomes (XX). A male has one X chromosome and one Y chromosome (XY).

(3) *Incomplete dominance* is a type of inheritance in which neither allele in a hybrid is dominant. The hybrid shows a trait completely different from either parent. The inheritance of color in a Japanese four-o'clock flower is an example of incomplete dominance, or blending.

(4) Alleles are different forms of the same gene. The term *multiple alleles* implies that a gene has more than two forms. Inheritance of human blood type involves multiple alleles.

140. **4** The normal number of chromosomes in human body cells is 46. Down's syndrome results from the presence of an extra chromosome. The extra or 47th chromosome is due to meiotic nondisjunction.

WRONG CHOICES EXPLAINED:

(1), (3) The numbers 22 and 44 do not apply to any known normal human chromosome number.

(2) The number 23 is the monoploid number of chromosomes in human gametes.

KEY IDEA 6—INTERDEPENDENCE OF LIVING THINGS

Plants and animals depend on each other and their physical environment.

Performance Indicator	Description
6.1	The student should be able to explain factors that limit growth of individuals and populations.
6.2	The student should be able to explain the importance of preserving diversity of species and habitats.
6.3	The student should be able to explain how the living and nonliving environments change over time and respond to disturbances.

141. Animals *cannot* synthesize nutrients from inorganic raw materials. Therefore, animals obtain their nutrients by

(1) combining carbon dioxide with water
(2) consuming preformed organic compounds
(3) hydrolyzing large quantities of simple sugars
(4) oxidizing inorganic molecules for energy   141 _____

142. Which organisms carry out heterotrophic nutrition?

(1) ferns   (3) fungi
(2) grasses   (4) mosses   142 _____

143. Which activity is an example of intracellular digestion?

(1) a grasshopper chewing blades of grass
(2) a maple tree converting starch to sugar in its roots
(3) an earthworm digesting proteins in its intestine
(4) a fungus digesting dead leaves   143 _____

144. A hydra ingests a daphnia, digests it, and later egests some materials. All these events are most closely associated with the life process known as

(1) transport   (3) growth
(2) synthesis   (4) nutrition   144 _____

145. Some bacteria are classified as saprophytes because they are organisms that
    (1) feed on other living things
    (2) feed on dead organic matter
    (3) manufacture food by photosynthesis
    (4) contain vascular bundles   145 _____

146. Aerobic organisms are dependent on autotrophs. One reason for this dependency is that most autotrophs provide the aerobic organisms with

    (1) oxygen   (3) nitrogen gas
    (2) carbon dioxide   (4) hydrogen   146 _____

147. Of the following, the greatest amount of the Earth's food production is thought to occur in

    (1) coastal ocean waters   (3) taiga forests
    (2) desert biomes   (4) tundra biomes   147 _____

148. Most of the minerals within an ecosystem are recycled and returned to the environment by the direct activities of organisms known as

    (1) producers   (3) decomposers
    (2) secondary consumers   (4) primary consumers 148 _____

149. Which type of organism is *not* shown in the following representation of a food chain?

    grass → mouse → snake → hawk

    (1) herbivore   (3) producer
    (2) decomposer   (4) carnivore   149 _____

150. In the food chain shown below, which organism represents a primary consumer?

(1) grasshopper   (3) frog
(2) grass   (4) snake   150 _____

151. A lake contains minnows, mosquito larvae, sunfish, algae, and pike. Which of these organisms would probably be present in the largest number?

(1) minnows   (3) sunfish
(2) larvae   (4) algae   151 _____

152. An abiotic factor that affects the ability of pioneer organisms such as lichens to survive is the

(1) type of climax vegetation   (3) type of substratum
(2) species of algae   (4) species of bacteria   152 _____

153. In order to avoid predators, the clown fish hides unharmed in the stinging tentacles of the sea anemone. The clown fish attracts food to the sea anemone. This is an example of a type of relationship known as

(1) mutualism   (3) predator–prey
(2) commensalism   (4) parasitism   153 _____

154. Which world biome has the greatest number of organisms?

(1) tundra   (3) temperate deciduous forest
(2) tropical forest   (4) marine   154 _____

155. In a particular area, living organisms and the nonliving environment function together as

(1) a population   (3) an ecosystem
(2) a community   (4) a species   155 _____

Answers Explained

141. **2** The portions of food that are usable to an animal are known as nutrients. Carbohydrates, lipids, and proteins are the organic nutrients needed by all organisms. Animals cannot synthesize their own nutrients. They must eat other organisms that contain the *preformed organic nutrients*.

WRONG CHOICES EXPLAINED:

(1) *Carbon dioxide and water* are inorganic compounds. These compounds are converted to nutrients by plants only. The process is called photosynthesis.

(3) Hydrolysis is the breakdown of compounds to simpler molecules through the action of enzymes in the presence of water. Glucose is a *simple sugar*. The hydrolyzing of glucose results in the release of energy in a cell.

(4) The use of *inorganic molecules* for the production of energy occurs only in certain species of bacteria. Chemosynthesis does not occur in members of the animal kingdom.

142. **3** *Fungi* are nongreen plants. They are heterotrophs, which means that they cannot manufacture their own food. Organisms that carry out heterotrophic nutrition must take in preformed organic molecules.

WRONG CHOICES EXPLAINED:

(1), (2), (4) *Ferns*, *grasses*, and *mosses* are green plants or bryophytes. Green plants are *autotrophs*, which means that they manufacture their own food.

143. **2** Digestion that occurs within a cell is known as intracellular digestion. Plants do not have special digestive systems. Digestion, or the *conversion of starch to sugar*, occurs within the individual cells of a plant, including those of the root.

WRONG CHOICES EXPLAINED:

(1) *Grasshoppers* have a digestive system. Digestion is extracellular and takes place in a digestive tube outside the body cells. Chewing a blade of grass is an example of mechanical digestion taking place in the mouth.

(3) *Earthworms* also have a digestive system. Proteins are digested outside the body cells in a portion of the digestive system known as the intestine.

(4) *Fungi* demonstrate a special form of extracellular digestion. Digestive enzymes are secreted into the external environment. The nutrients from the digested food diffuse into the cells.

144. **4** *Nutrition* is the life process most closely associated with a hydra ingesting, digesting, and egesting a daphnia. Ingestion is the process by which food materials are taken into the body of an organism such as a hydra. Digestion is the process by which the complex food molecules within the daphnia are hydrolyzed to soluble end products. Egestion is the process by which the undigestible materials of the daphnia's body are expelled from the body of the hydra.

WRONG CHOICES EXPLAINED:

(1) *Transport* is the life process by which soluble foods and other materials are circulated through the body of an organism such that they reach all parts of the organism's body.

(2) *Synthesis* is a process by which complex materials are constructed from simpler chemical components. The processes described in the question represent hydrolysis, the opposite of synthesis.

(3) *Growth* is a process that involves an increase in cell number and cell size, leading to an increase in the size of the organism. The end products of digestion can be used to supply raw materials for such growth.

145. **2** *Saprophytes* are organisms that *feed on dead organic matter*. Fungi and the bacteria of decay are examples of saprophytes.

WRONG CHOICES EXPLAINED:

(1) Heterotrophs *live on or off other living organisms*. A dog flea is an example of a heterotroph also known as a parasite.

(3) Autotrophs *manufacture their own food by photosynthesis*. Algae, mosses, and grasses are examples of autotrophs.

(4) Higher plants *contain vascular bundles*. These plants, called *tracheophytes,* include ferns, conifers, and flowering plants.

146. **1** Aerobic organisms need *oxygen* for cellular respiration. Some autotrophs are photosynthetic organisms. Oxygen is released by photosynthesis. Aerobic organisms depend on the autotrophs to release oxygen into the environment.

WRONG CHOICES EXPLAINED:

(2) *Carbon dioxide* is a waste product from the cellular respiration of aerobic organisms.

(3) *Nitrogen gas* makes up 78% of the atmosphere. However, the nitrogen cannot be used in the gaseous form by aerobic organisms and most autotrophs.

(4) *Hydrogen* does *not* exist as a gas on our planet. It is combined with other elements. The hydrogen needed by organisms comes mostly from water and organic compounds.

147. **1**. The area of greatest food production is in the region where the greatest rate of photosynthesis occurs. The area must be rich in minerals, water, gases, and light. The *coastal ocean waters* meet these requirements.

WRONG CHOICES EXPLAINED:

(2) There is very little precipitation in the *desert*. Water is the factor that limits plant growth.

(3) The water in the *taiga* is frozen part of the year.

(4) The *tundra* is a frozen plain. Water is frozen almost all year long on the tundra.

148. **3** Bacteria of decay are *decomposers* that release minerals from decaying plant and animal bodies and return them to the environment.

WRONG CHOICES EXPLAINED:

(1) *Producers* are autotrophs, that is, green plants that synthesize food by photosynthesis from carbon dioxide and water in the presence of sunlight.

(2) *Secondary consumers* are animals that eat other animals. For example, a frog feeds on flies. Thus, a frog is a secondary consumer.

(4) *Primary consumers* are organisms that feed on plants only. Herbivores are primary consumers.

149. **2** *Decomposers* are not shown in the food chain. They are organisms that live on dead things. Fungi and bacteria are decomposers.

WRONG CHOICES EXPLAINED:

(1) An *herbivore* is a primary consumer; that is, it eats vegetation. The mouse is the herbivore in the food chain.

(3) A *producer* is a green plant. It depends on sunlight to synthesize its own food. The grass is the producer in the food chain.

(4) A *carnivore* is an animal that eats the flesh of other animals. Both the snake and the hawk are carnivores.

150. **1** The *grasshopper* is a primary consumer because it feeds on vegetation.

WRONG CHOICES EXPLAINED:

(2) *Grass* is an autotroph, that is, a producer or self-feeder. Grass is a green plant that can make its own food.

(3) A *frog* is a secondary consumer; it eats insects that are plant eaters.

(4) A *snake* is a secondary consumer; it eats animals that are primary consumers.

151. **4** The organisms in the question make up a food chain. The number of organisms at each level of the food chain decreases as one moves down the chain. The pyramid of energy shown below represents this fact. The producers, which form the base of the pyramid, are the most numerous. *Algae* are the producers in this food chain.

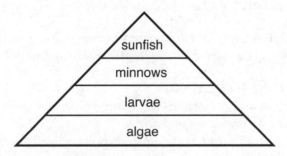

WRONG CHOICES EXPLAINED:

(1) The *minnows* are carnivores. They are the secondary consumers.

(2) The *larvae* are herbivores. They are primary consumers.

(3) The *sunfish* are tertiary consumers and occupy the top level of the pyramid. The organisms at the top of the pyramid are the least numerous.

152. **3** A *substratum* is the surface on which organisms grow. Lichens grow on rocks. Rocks are nonliving. The nonliving parts of the environment make up the abiotic factors.

WRONG CHOICES EXPLAINED:

(1) *Climax vegetation* is the type of vegetation that occupies an area in its final stage of succession. Plants make up the biotic, or living, environment.

(2) *Algae* are living organisms. They are part of the biotic (living) environment.

(4) *Bacteria* also make up the biotic environment.

153. **1** A relationship between two organisms in which both benefit from the association is known as *mutualism*. The clown fish is protected by the sea anemone. The sea anemone is sessile. The clown fish draws food to the sea anemone.

WRONG CHOICES EXPLAINED:

(2) *Commensalism* is a relationship between two organisms in which one organism is benefited by the association. The second organism is neither harmed nor benefited by the association. Barnacles attached to a whale are an example of a commensal relationship.

(3) A predator is a carnivore that hunts, kills, and eats its prey. The prey is the hunted organism. A *predator-prey* relationship is important in controlling the population of both.

(4) *Parasitism* is a relationship between two organisms in which one organism is benefited by the association whereas the second organism is harmed by the association. The parasitized organism is called the host.

154. **4** A biome is a large area dominated by one major type of vegetation and one type of climate. The *marine* biome has the greatest number of organisms.

WRONG CHOICES EXPLAINED:

(1) The *tundra* is a region where the ground is frozen all year long. Mosses and lichens are the dominant vegetation.

(2) The *tropical rain forest* is dominated by broadleaf plants. The region is always warm, and the rainfall is abundant and continuous.

(3) The *temperate deciduous forest* is made up of trees that shed their leaves once a year. The winters are cold, and the summers are warm. The rainfall is distributed throughout the year.

155. **3** An *ecosystem* is an area in which communities of living organisms interact with the nonliving environment.

WRONG CHOICES EXPLAINED:

(1) A *population* is all the organisms of a particular species living in a given area.

(2) A *community* is made up of populations of different species that interact with each other.

(4) A *species* is a group of organisms whose members are able to interbreed with each other. The offspring resulting from the matings are fertile and can reproduce.

KEY IDEA 7—HUMAN IMPACT ON THE ENVIRONMENT

Human decisions and activities have a profound impact on the physical and living environment.

Performance Indicator	Description
7.1	The student should be able to describe the range of interrelationships of humans with the living and nonliving environment.
7.2	The student should be able to explain the impact of technological development and growth in the human population on the living and nonliving environment.
7.3	The student should be able to explain how individual choices and societal actions can contribute to improving the environment.

156. Human impact on the environment is most often more dramatic than the impact of most other living things because humans have a greater

(1) need for water
(2) need for food
(3) ability to adapt to change
(4) ability to alter the environment   156 _____

157. Which human activity would have the most direct impact on the oxygen–carbon dioxide cycle?

(1) reducing the rate of ecological succession
(2) decreasing the use of water
(3) destroying large forest areas
(4) enforcing laws that prevent the use of leaded gasoline   157 _____

158. Fertilizers used to improve lawns and gardens may interfere with the equilibrium of an ecosystem because they

(1) cause mutations in all plants
(2) cannot be absorbed by roots
(3) can be carried into local water supplies
(4) cause atmospheric pollution   158 _____

159. The tall wetland plant purple loosestrife was brought from Europe to the United States in the early 1800s as a garden plant. The plant's growth is now so widespread across the United States that it is crowding out a number of native plants. This situation is an example of

(1) the results of the use of pesticides
(2) the recycling of nutrients
(3) the flow of energy present in all ecosystems
(4) an unintended effect of adding a species to an ecosystem   159 _____

160. Choose *one* ecological problem from the list below.

Ecological Problems
Global warming
Destruction of the ozone shield
Loss of biodiversity

Discuss the ecological problem you chose. In your answer be sure to state:

- the problem you selected and *one* human action that may have caused the problem [1]
- *one* way in which the problem can negatively affect humans [1]
- *one* positive action that could be taken to reduce the problem [1]

_____

_____

_____

Base your answers to questions 161 through 163 on the information below and on your knowledge of biology.

The planning board of a community held a public hearing in response to complaints by residents concerning a waste-recycling plant. The residents claimed that the waste-hauling trucks were polluting air, land, and water and that the garbage has brought an increase in rats, mice, and pathogenic bacteria to the area. The residents insisted that the waste-recycling plant be closed permanently.

Other residents recognized the health risks but felt that the benefits of waste recycling outweighed the health issues.

161. Identify two specific health problems that could result from living near the waste-recycling plant. [2]

_____

162. Identify one specific contaminant that might be released into the environment from operation of the waste-recycling plant. [1]

_____

163. State one ecological benefit of recycling wastes. [1]

_____

164. Which organism is a near-extinct species?

(1) Japanese beetle   (3) blue whale
(2) dodo bird   (4) passenger pigeon   164 _____

165. Which human activity has probably contributed most to the acidification of lakes in the Adirondack region?

(1) passing environmental protection laws
(2) establishing reforestation projects in lumbered areas
(3) burning fossil fuels that produce air pollutants containing sulfur and nitrogen
(4) using pesticides for the control of insects that feed on trees   165 _____

166. Compared to a natural forest, the wheat field of a farmer lacks

(1) heterotrophs   (3) autotrophs
(2) significant biodiversity   (4) stored energy   166 _____

167. Which factor is not considered by ecologists when they evaluate the impact of human activities on an ecosystem?

(1) amount of energy released from the Sun
(2) quality of the atmosphere
(3) degree of biodiversity
(4) location of power plants   167 _____

168. A new type of fuel gives off excessive amounts of smoke. Before this type of fuel is widely used, an ecologist would most likely want to know

   (1) what effect the smoke will have on the environment
   (2) how much it will cost to produce the fuel
   (3) how long it will take to produce the fuel
   (4) if the fuel will be widely accepted by consumers   168 _____

169. Which of the following is the most ecologically promising method of insect control?

   (1) interference with insect reproductive processes
   (2) stronger insecticides designed to kill higher percentages of insects
   (3) physical barriers to insect pests
   (4) draining marshes and other insect habitats   169 _____

170. Which is an example of biological control of a pest species?

   (1) DDT was used to destroy the red mite.
   (2) Most of the predators of a deer population were destroyed by humans.
   (3) Gypsy moth larvae (tree defoliators) are destroyed by beetle predators that were cultured and released.
   (4) Drugs were used in the control of certain pathogenic bacteria.   170 _____

171. To ensure environmental quality for the future, each individual should

   (1) acquire and apply knowledge of ecological principles
   (2) continue to take part in deforestation
   (3) use Earth's finite resources
   (4) add and take away organisms from ecosystems   171 _____

172. Ladybugs were introduced as predators into an agricultural area of the United States to reduce the number of aphids (pests that feed on grain crops). Describe the positive and negative effects of this method of pest control. Your response must include at least:

- two advantages of this method of pest control [2]
- two possible dangers of using this method of pest control [2]

_____

_____

_____

_____

173. Some people claim that certain carnivores should be destroyed because they kill beneficial animals. Explain why these carnivores should be protected. Your answer must include information concerning:

- prey population growth [1]
- extinction [1]
- the importance of carnivores in an ecosystem [1]

_____

_____

_____

_____

Answers Explained

156. **4** The fact that humans have a greater *ability to alter the environment* means that human impact on the environment is often more dramatic than that of most other living things. In addition to our ability to make physical changes in the environment, humans have the unique ability to alter the environment chemically, introducing many materials that are not found in nature and that cannot be converted to useful products by nature.

WRONG CHOICES EXPLAINED:

(1), (2) On an individual basis, humans' *need for water* and *need for food* are not significantly greater than those of other living things. However, the fact is that our large population places incredible demands on the environment to supply these basic resources. As a result, our tendency to destroy natural habitats to create additional water and agricultural resources is a significant factor affecting the natural world.

(3) On an individual basis, humans' *ability to adapt to change* is not significantly greater than that of other living things. However, as a species, we have created artificial environments to protect ourselves from harsh environmental conditions. To the extent that these artificial environments are dependent on energy and other natural resources, their construction and maintenance have resulted in significant alterations of the natural world.

157. **3** *Destroying large forest areas* is the human activity that would have the most direct impact on the oxygen-carbon cycle. Reducing the number of trees over a large area would decrease the forest's ability to absorb carbon dioxide and water and convert them to atmospheric oxygen and glucose. The millions of leaves in a forest are capable of releasing many tons of oxygen gas to the atmosphere. The massive bodies of forest trees can likewise store tons of carbon in the form of complex carbohydrates such as cellulose.

WRONG CHOICES EXPLAINED:

(1) *Reducing the rate of ecological succession* is not the human activity that would have the most direct impact on the oxygen-carbon cycle. Ecological succession is a process by which one plant-animal community is replaced over time by other plant-animal communities until a stable climax community is established. Reducing its rate would only have the effect of prolonging each successive community longer than might otherwise be expected but would not directly alter the cycling of carbon and oxygen.

(2) *Decreasing the use of water* is not the human activity that would have the most direct impact on the oxygen-carbon cycle. Water is a precious resource in many parts of the world. Reducing water use so as to conserve it would represent a positive impact of human activity on the environment but would not directly alter the cycling of carbon and oxygen.

(4) *Enforcing laws that prevent the use of leaded gasoline* is not the human activity that would have the most direct impact on the oxygen-carbon cycle. Lead is a dangerous heavy metal pollutant released when leaded gasoline is burned. Enforcing laws that limit its use would represent a positive impact of human activity on the environment, but would not directly alter the cycling of carbon and oxygen.

158. **3** Fertilizers used to improve lawns and gardens may interfere with the equilibrium of an ecosystem because they *can be carried into local water supplies*. Once dissolved fertilizers enter streams, ponds, wetlands, or lakes, they provide an abundant nutrient source for the growth of algae. As masses of algae die off in the water environment, their decomposition can rob the water of oxygen needed for the survival of fish and other water-dwelling populations, causing their elimination from the habitat. When these species disappear, other species that depend on them for food must migrate or starve. Because the changes caused by the entry of fertilizers into water environments are so significant, it can be said that ecosystem equilibrium is destroyed.

WRONG CHOICES EXPLAINED:

(1) It is not true that fertilizers *cause mutations in all plants*. Some compounds with chemical structures similar to that of fertilizers are known to stimulate rapid gene mutation in plant cells that may lead to the death of the plant. However, the class of chemical compounds known as fertilizers do not have this effect on all plants.

(2) It is not true that fertilizers *cannot be absorbed by all plants*. When dissolved in water, fertilizers can easily enter plants by being absorbed via simple diffusion into root hairs.

(4) It is not normally true that fertilizers *cause atmospheric pollution*. Most fertilizers are relatively stable chemical compounds that are solids at normal temperatures. For this reason fertilizers are not normally responsible for atmospheric pollution unless they are applied in a gaseous form (such as ammonia) or become airborne (when attached to dry soil particles).

159. **4** The situation described in the question is an example of *an unintended effect of adding a species to an ecosystem.* Although purple loosestrife has adapted well to North American habitats, its rapid growth in wetland environments has stressed or eliminated populations of cattail, pickerelweed, and other native plant species. The introduction of nonnative purple loosestrife to the North American continent has had an unintended negative effect on these native species and on the balance of nature established over many centuries.

WRONG CHOICES EXPLAINED:
(1), (2), (3) The situation described in the question is not an example of *the results of the use of pesticides, the recycling of nutrients,* or *the flow of energy present in all ecosystems.* The introduction of a nonnative plant (a living thing) is not the same as the introduction of a chemical pesticide, the recycling of nutrients, or the flow of energy (nonliving things).

160. A three-part response is required that must include the following points:
- One human activity that may have caused the ecological problem selected from the list [1]
- One way the problem may negatively affect humans [1]
- One positive action that could be taken to reduce the problem [1]

Note: No credit is awarded for discussing an ecological problem not on the list.

Acceptable responses include: [3]
- *Global warming is a worldwide ecological problem that may be caused by the release of carbon dioxide and other gases in automobile exhaust. [1] This problem may negatively affect humans if the warming conditions disrupt weather patterns and lead to droughts, floods, or other natural disasters. [1] One positive action that could be taken to help the problem would be to find an energy source for automobiles that would not release carbon dioxide into the atmosphere. [1]*
- *An ecological problem affecting humans is destruction of the ozone layer, which is caused by the use of chemicals known as CFCs as propellants in aerosol sprays. [1] This is a problem for humans because the ozone layer protects us from ultraviolet radiation from the sun; without this protection we would have an increased chance of getting skin cancer. [1] A way to help solve this problem would be to ban the use of CFCs in aerosols. [1]*
- *Loss of biodiversity is an ecological problem that negatively impacts humans. This problem is caused whenever humans destroy a natural habitat and convert it to other uses. [1] The overall health of our environment depends on the diversity of species that fill different roles in nature. When*

species diversity and environmental health are reduced, our health is threatened as well. [1] This problem can be reversed only if we use education to learn that protecting natural species is just as important as protecting our own. [1]

161. Two responses are required. Acceptable responses include:

- *Asthma*
- *Respiratory infections*
- *Allergic reactions*
- *Cancer*
- *Bacterial infections*
- *Viral infections*
- *Disease linked to a pathogen*
- *Poisoning linked to toxic contamination of groundwater*

162. One response is required. Acceptable responses include:

- *Particles in the air*
- *Presence of viruses or bacteria on trucks*
- *Chemicals in air or water*
- *Carcinogens*
- *Mold and fungus spores*

163. One response is required. Acceptable responses include:

- *Conservation of natural resources*
- *Protection of finite resources*
- *Energy conservation*
- *Reduction in pollution*
- *Landfills last longer*
- *Preservation of open space resources*

164. **3** The blue whale is near extinction because of uncontrolled hunting by humans.

WRONG CHOICES EXPLAINED:
(1) The Japanese beetle is a plant pest that was accidentally introduced into the United States. Its population is kept in check by the praying mantis, its predator.
(2) The dodo bird became extinct because of hunting by humans.
(4) The passenger pigeon became extinct in the 1900s due to hunting by humans.

165. **3** *Burning fossil fuels that produce air pollutants containing sulfur and nitrogen* is the human activity that has probably contributed the most to the acidification of lakes in the Adirondack region. These pollutants combine with water in the atmosphere to form sulfuric and nitric acids. These acids then enter lakes in rainfall and runoff, adding to the acidic quality of the lake water and killing many susceptible species.

WRONG CHOICES EXPLAINED:

(1) *Passing environmental protection laws* is not an activity that results in the acidification of lakes. In fact, it is a positive human activity that can help to limit the production and release of such gases into the atmosphere.

(2) *Establishing reforestation projects in lumbered areas* is not an activity that results in the acidification of lakes. In fact, it is a positive human activity that can help to replace trees lost because of the acidification of soils by acid rain.

(4) *Using pesticides for the control of insects that feed on trees* is not an activity that results in the acidification of lakes. It is a negative human activity carried out to protect commercial crops from destruction and does not normally result in the production of sulfur and nitrogen gases.

166. **2** *Significant biodiversity* is the factor lacking in a wheat field as compared to a natural forest. *Biodiversity* is a term relating to the variety of life forms in an environment. Natural environments, including forests, are typically made up of thousands of species that interact to provide a balanced, ecologically responsive community. By contrast, farm fields are often limited to a small number of different species, and predominantly a single species. Communities lacking in biodiversity are unstable and prone to collapse when environmental conditions change.

WRONG CHOICES EXPLAINED:

(1) *Heterotrophs* are not lacking in a farm field compared to a forest. Heterotrophs are found within a wheat field, although their number and variety are normally limited to those that use wheat or its by-products as food.

(3) *Autotrophs* are not lacking in a farm field compared to a forest. Wheat is a type of autotroph, as are the various weed species that may be interspersed among the wheat plants in the field.

(4) *Stored energy* is not lacking in a farm field compared to a forest. As the wheat grows in the field, it absorbs the Sun's energy and stores it as the chemical bond energy of carbohydrates and other organic compounds.

167. **1** The *amount of energy released from the Sun* is not normally considered by an ecologist when evaluating the impact of human activities on an ecosystem. The amount of solar energy emitted by the Sun is generally constant and out of our direct control. Because it is not a variable that can be directly affected by human activities, it is usually not a consideration in decisions of this kind.

WRONG CHOICES EXPLAINED:

(2) The *quality of the atmosphere* is often a factor considered by ecologists in evaluating the impact of human activities on an ecosystem. Many human activities introduce chemical contaminants into the atmosphere. These chemicals may have a negative impact on the health and survival of humans and other species.

(3) The *degree of biodiversity* is often a factor considered by ecologists in evaluating the impact of human activities on an ecosystem. Human activities often put pressure on natural species, eliminating those unable to migrate or adapt. As biodiversity in an area declines, so does environmental stability. This situation threatens the health and survival of humans and other species.

(4) The *location of power plants* is often a factor considered by ecologists in evaluating the impact of human activities on an ecosystem. Fossil fuel plants can pollute the atmosphere and consume valuable petroleum products. Nuclear plants can release radiation and heat into the environment. Hydroelectric, solar, wind, and geothermal plants can destroy natural habitats because of space considerations. Each of these consequences can affect the health and survival of humans and other species.

168. **1** An ecologist would want to know *what effect the smoke will have on the environment* before a new type of fuel is widely used. By understanding this effect, the ecologist can make more informed judgments about whether the smoke will harm the environment and human health.

WRONG CHOICES EXPLAINED:

(2), (3), (4) An ecologist is less likely to want to know *how much it will cost to produce the fuel, how long it will take to produce the fuel,* and *if the fuel will be widely accepted by consumers.* Although these are important questions for the manufacturer, they do not provide critical information for the ecologist, whose main concern is the protection of environmental quality for humans and other organisms.

169. **1** Interference with insect reproductive processes is known as biological control. It is the most promising method of controlling insects because it is the least ecologically damaging.

WRONG CHOICES EXPLAINED:

(2) The use of insecticides is a chemical control of insects. Insecticides kill both harmful and helpful insects. The chemicals accumulate in the bodies of birds, fish, and mammals and interfere with their normal life activities.

(3) It is impossible to set up physical barriers for insects because they are motile and are also carried from place to place by animals and humans.

(4) Draining marshes and other insect habitats has helped to control many insects such as mosquitoes. However, this method interferes with the life cycles of useful organisms living in the area.

170. **3** Insecticides are chemical pest controls. Biological controls are other insect species that feed on or in some way prey on an insect pest species. The example given here is control of the gypsy moth larvae by a certain species of beetle.

WRONG CHOICES EXPLAINED:

(1) DDT is an insecticide and represents chemical control. DDT is no longer used because it destroyed the insect food of birds and other wildlife.

(2) Humans upset the balance of nature (the balance of natural communities) by killing off deer predators. The deer population then increased so dramatically that deer starved to death because there was not enough food to support them.

(4) The use of drugs to cure disease is an example of chemical control of pathogens.

171. **1** Each individual should *acquire and apply knowledge of ecological principles* in order to ensure environmental quality for the future. By understanding how environmental principles operate, we can make more informed judgments about activities that may harm the environment and human health.

WRONG CHOICES EXPLAINED:

(2) If each individual were to *continue to take part in deforestation*, environmental quality would be degraded. Because forests are a natural part of the environment, eliminating them disturbs the balance of nature and can have significant negative consequences for environmental quality.

(3) If each individual were to *use Earth's finite resources*, environmental quality would be threatened. As these resources are used up, fewer remain for future generations. In addition, processing these resources consumes energy, produces pollutants, and adds to the solid waste problem.

(4) If each individual were to *add and take away organisms from ecosystems*, environmental quality would be diminished. Each natural community has established itself based on the particular niches filled by each type of organism. Adding to or taking away from this community upsets the balance of nature and would likely cause negative consequences.

172. Write one or more paragraphs describing positive and negative effects of this method of pest control. Include the following points:

- Two advantages of this method of pest control [2]
  - Chemicals are not added to the environment.
  - Biological controls are more specific than chemical controls.
  - Ladybugs are less likely to kill beneficial organisms.
  - Desirable garden plants are protected from aphid attacks.
  - Birds and other unintended victims of pesticide use are spared.
  - Human health is protected against the toxic effects of pesticides.

- Two possible dangers of using this method of pest control [2]
  - The control insects may eat the food of other organisms.
  - The population of natural predators of the aphids may be eliminated or greatly reduced.
  - The control organism may become overpopulated.
  - The control organisms may themselves become pests.

Sample paragraph: The method of pest control described is known as "biological control." This method of insect control has some distinct advantages over chemical controls: First, biological controls don't release toxic chemicals into the air and water, a fact that helps to protect wildlife and humans from being unintended victims of chemical pesticides. Second, biological controls are usually specific, which means that beneficial insects such as ladybugs and preying mantises aren't harmed. [2] There are also some things we should be careful of in the use of biological controls: First, we should know a lot about the control organism to be sure that it doesn't crowd out our native beneficial organisms. Second, we should remember that the control organism could become a pest, too, if it gets too numerous in the environment. [2]

173. Write one or more paragraphs explaining why carnivores should be protected. Include the following points:

- Information concerning prey population growth [1]
  - If predators are destroyed, the prey population will increase.
  - If unchecked by predation or disease, a natural population will tend to increase in number geometrically.

- Information concerning extinction [1]
  - If too many carnivores of a particular species are killed, the species may become extinct.
  - Extinction is a definite possibility when any species has too few members alive to carry out effective breeding.
  - Complete elimination of any species from its natural range can destabilize the ecosystem.

- Information concerning the importance of carnivores in an environment [1]
  - By feeding on herbivores, carnivores help keep certain species of plants from being eliminated because of overgrazing in a particular area.
  - Without predators to limit its number, a prey population could exceed the capacity of its range, resulting in widespread starvation and death of the prey population.
  - Carnivorous animals are part of the natural scheme that promotes ecological equilibrium.

Sample paragraph: Carnivores are important in an ecosystem because by reducing the number of prey organisms, the food organisms of the prey are kept from being eliminated from the environment. [1] If the predators were destroyed, the prey population would increase [1], perhaps to the point of consuming so many of the plants that the prey feed on that these plants would become extinct. [1]

Standard/Key Idea	Question Numbers	Number of Correct Responses	Number of Incorrect Responses
1.1 Purpose of Scientific Inquiry	1–8		
1.2 Methods of Scientific Inquiry	9–28		
1.3 Analysis in Scientific Inquiry	29–47		
4.1 Application of Scientific Principles	48–75		
4.2 Genetic Continuity	76–90		
4.3 Organic Evolution	91–109		
4.4 Reproductive Continuity	110–129		
4.5 Dynamic Equilibrium and Homeostasis	130–140		
4.6 Interdependence of Living Things	141–155		
4.7 Human Impact on the Environment	156–173		

Glossary

PROMINENT SCIENTISTS

Crick, Francis A 20th-century British scientist who, with James Watson, developed the first workable model of DNA structure and function.

Darwin, Charles A 19th-century British naturalist whose theory of organic evolution by natural selection forms the basis for the modern scientific theory of evolution.

Fox, Sidney A 20th-century American scientist whose experiments showed that Stanley Miller's simple chemical precursors could be joined to form more complex biochemicals.

Hardy, G. H. A 20th-century British mathematician who, with W. Weinberg, developed the Hardy-Weinberg principle of gene frequencies.

Lamarck, Jean An 18th-century French scientist who devised an early theory of organic evolution based on the concept of "use and disuse."

Linnaeus, Carl An 18th-century Dutch scientist who developed the first scientific system of classification, based on similarity of structure.

Mendel, Gregor A 19th-century Austrian monk and teacher who was the first to describe many of the fundamental concepts of genetic inheritance through his work with garden peas.

Miller, Stanley A 20th-century American scientist whose experiments showed that the simple chemical precursors of life could be produced in the laboratory.

Morgan, Thomas Hunt A 20th-century American geneticist whose pioneering work with Drosophila led to the discovery of several genetic principles, including sex linkage.

Watson, James A 20th-century American scientist who, with Francis Crick, developed the first workable model of DNA structure and function.

Weinberg, W. A 20th-century German physician who, with G. H. Hardy, developed the Hardy-Weinberg principle of gene frequencies.

Weismann, August A 19th-century German biologist who tested Lamarck's theory of use and disuse and found it to be unsupportable by scientific methods.

BIOLOGICAL TERMS

Abiotic factor Any of several nonliving, physical conditions that affect the survival of an organism in its environment.

Absorption The process by which water and dissolved solids, liquids, and gases are taken in by the cell through the cell membrane.

Accessory organ In human beings, any organ that has a digestive function but is not part of the food tube. (See **liver; gallbladder; pancreas.**)

Acid A chemical that releases hydrogen ion (H+) in solution with water.

Acid precipitation A phenomenon in which there is thought to be an interaction between atmospheric moisture and the oxides of sulfur and nitrogen that results in rainfall with low pH values.

Active immunity The immunity that develops when the body's immune system is stimulated by a disease organism or a vaccination.

Active site The specific area of an enzyme molecule that links to the substrate molecule and catalyzes its metabolism.

Active transport A process by which materials are absorbed or released by cells against the concentration gradient (from low to high concentration) with the expenditure of cell energy.

Adaptation Any structural, biochemical, or behavioral characteristic of an organism that helps it to survive potentially harsh environmental conditions.

Addition A type of chromosome mutation in which a section of a chromosome is transferred to a homologous chromosome.

Adenine A nitrogenous base found in DNA and RNA molecules.

Adenosine triphosphate (ATP) An organic compound that stores respiratory energy in the form of chemical-bond energy for transport from one part of the cell to another.

Adrenal cortex A portion of the adrenal gland that secretes steroid hormones which regulate various aspects of blood composition.

Adrenal gland An endocrine gland that produces several hormones, including **adrenaline**. (See **adrenal cortex; adrenal medulla**.)

Adrenal medulla A portion of the adrenal gland that secretes the hormone adrenaline, which regulates various aspects of the body's metabolic rate.

Adrenaline A hormone of the adrenal medulla that regulates general metabolic rate, the rates of heartbeat and breathing, and the conversion of glycogen to glucose.

Aerobic phase of respiration The reactions of aerobic respiration in which two pyruvic acid molecules are converted to six molecules of water and six molecules of carbon dioxide.

Aerobic respiration A type of respiration in which energy is released from organic molecules with the aid of oxygen.

Aging A stage of postnatal development that involves differentiation, maturation, and eventual deterioration of the body's tissues.

Air pollution The addition, due to technological oversight, of some unwanted factor (e.g., chemical oxides, hydrocarbons, particulates) to our air resources.

Albinism A condition, controlled by a single mutant gene, in which the skin lacks the ability to produce skin pigments.

Alcoholic fermentation A type of anaerobic respiration in which glucose is converted to ethyl alcohol and carbon dioxide.

Allantois A membrane that serves as a reservoir for wastes and as a respiratory surface for the embryos of many animal species.

Allele One of a pair of genes that exist at the same location on a pair of homologous chromosomes and exert parallel control over the same genetic trait.

Allergy A reaction of the body's immune system to the chemical composition of various substances.

Alveolus One of many "air sacs" within the lung that function to absorb atmospheric gases and pass them on to the bloodstream.

Amino acid An organic compound that is the component unit of proteins.

Amino group A chemical group having the formula —NH_2 that is found as a part of all amino acid molecules.

Ammonia A type of nitrogenous waste with high solubility and high toxicity.

Amniocentesis A technique for the detection of genetic disorders in human beings in which a small amount of amniotic fluid is removed and the chromosome content of its cells analyzed. (See **karyotyping**.)

Amnion A membrane that surrounds the embryo in many animal species and contains a fluid to protect the developing embryo from mechanical shock.

Amniotic fluid The fluid within the amnion membrane that bathes the developing embryo.

Amylase An enzyme specific for the hydrolysis of starch.

Anaerobic phase of respiration The reactions of aerobic respiration in which glucose is converted to two pyruvic acid molecules.

Anaerobic respiration A type of respiration in which energy is released from organic molecules without the aid of oxygen.

Anal pore The egestive organ of the paramecium.

Anemia A disorder of the human transport system in which the ability of the blood to carry oxygen is impaired, usually because of reduced numbers of red blood cells.

Angina pectoris A disorder of the human transport system in which chest pain signals potential damage to the heart muscle due to narrowing of the opening of the coronary artery.

Animal One of the five biological kingdoms; it includes multicellular organisms whose cells are not bounded by cell walls and which are incapable of photosynthesis (e.g., human being).

Annelida A phylum of the Animal Kingdom whose members (annelids) include the segmented worms (e.g., earthworm).

Antenna A receptor organ found in many arthropods (e.g., grasshopper), which is specialized for detecting chemical stimuli.

Anther The portion of the stamen that produces pollen.

Antibody A chemical substance, produced in response to the presence of a specific antigen, which neutralizes that antigen in the immune response.

Antigen A chemical substance, usually a protein, that is recognized by the immune system as a foreign "invader" and is neutralized by a specific antibody.

Anus The organ of egestion of the digestive tract.

Aorta The principal artery carrying blood from the heart to the body tissues.

Aortic arches A specialized part of the earthworm's transport system that serves as a pumping mechanism for the blood fluid.

Apical meristem A plant growth region located at the tip of the root or tip of the stem.

Appendicitis A disorder of the human digestive tract in which the appendix becomes inflamed as a result of bacterial infection.

Aquatic biome An ecological biome composed of many different water environments.

Artery A thick-walled blood vessel that carries blood away from the heart under pressure.

Arthritis A disorder of the human locomotor system in which skeletal joints become inflamed, swollen, and painful.

Arthropoda A phylum of the Animal Kingdom whose members (arthropods) have bodies with chitinous exoskeletons and jointed appendages (e.g., grasshopper).

Artificial selection A technique of plant/animal breeding in which individual organisms displaying desirable characteristics are chosen for breeding purposes.

Asexual reproduction A type of reproduction in which new organisms are formed from a single parent organism.

Asthma A disorder of the human respiratory system in which the respiratory tube becomes constricted by swelling brought on by some irritant.

Atrium In human beings, one of the two thin-walled upper chambers of the heart that receive blood.

Autonomic nervous system A subdivision of the peripheral nervous system consisting of nerves associated with automatic functions (e.g., heartbeat, breathing).

Autosome One of several chromosomes present in the cell that carry genes controlling "body" traits not associated with primary and secondary sex characteristics.

Autotroph An organism capable of carrying on autotrophic nutrition. Self feeder.

Autotrophic nutrition A type of nutrition in which organisms manufacture their own organic foods from inorganic raw materials.

Auxin A biochemical substance, plant hormone, produced by plants that regulates growth patterns.

Axon An elongated portion of a neuron that conducts nerve impulses, usually away from the cell body of the neuron.

Base A chemical that releases hydroxyl ion (OH^-) in solution with water.

Bicarbonate ion The chemical formed in the blood plasma when carbon dioxide is absorbed from body tissues.

Bile In human beings, a secretion of the liver that is stored in the gallbladder and that emulsifies fats.

Binary fission A type of cell division in which mitosis is followed by equal cytoplasmic division.

Binomial nomenclature A system of naming, used in biological classification, that consists of the genus and species names (e.g., *Homo sapiens*).

Biocide use The use of pesticides that eliminate one undesirable organism but that have, due to technological oversight, unanticipated effects on beneficial species as well.

Biological controls The use of natural enemies of various agricultural pests for pest control, thereby eliminating the need for biocide use—a positive aspect of human involvement with the environment.

Biomass The total mass of living material present at the various trophic levels in a food chain.

Biome A major geographical grouping of similar ecosystems, usually named for the climax flora in the region (e.g., Northeast Deciduous Forest).

Biosphere The portion of the earth in which living things exist, including all land and water environments.

Biotic factor Any of several conditions associated with life and living things that affect the survival of living things in the environment.

Birth In placental mammals, a stage of embryonic development in which the baby passes through the vaginal canal to the outside of the mother's body.

Blastula In certain animals, a stage of embryonic development in which the embryo resembles a hollow ball of undifferentiated cells.

Blood The complex fluid tissue that functions to transport nutrients and respiratory gases to all parts of the body.

Blood typing An application of the study of immunity in which the blood of a person is characterized by its antigen composition.

Bone A tissue that provides mechanical support and protection for bodily organs, and levers for the body's locomotive activities.

Bowman's capsule A cup-shaped portion of the nephron responsible for the filtration of soluble blood components.

Brain An organ of the central nervous system that is responsible for regulating conscious and much unconscious activity in the body.

Breathing A mechanical process by which air is forced into the lungs by means of muscular contraction of the diaphragm and rib muscles.

Bronchiole One of several subdivisions of the bronchi that penetrate the lung interior and terminate in alveoli.

Bronchitis A disorder of the human respiratory system in which the bronchi become inflamed.

Bronchus One of the two major subdivisions of the breathing tube; the bronchi are ringed with cartilage and conduct air from the trachea to the lung interior.

Bryophyta A phylum of the Plant Kingdom that consists of organisms lacking vascular tissues (e.g., moss).

Budding A type of asexual reproduction in which mitosis is followed by unequal cytoplasmic division.

Bulb A type of vegetative propagation in which a plant bulb produces new bulbs that may be established as independent organisms with identical characteristics.

Cambium The lateral meristem tissue in woody plants responsible for annual growth in stem diameter.

Cancer Any of a number of conditions characterized by rapid, abnormal, and uncontrolled division of affected cells.

Capillary A very small, thin-walled blood vessel that connects an artery to a vein and through which all absorption into the blood fluid occurs.

Carbohydrate An organic compound composed of carbon, hydrogen, and oxygen in a 1:2:1 ratio (e.g., $C_6H_{12}O_6$).

Carbon-14 A radioactive isotope of carbon used to trace the movement of carbon in various biochemical reactions, and also used in the "carbon dating" of fossils.

Carbon-fixation reactions A set of biochemical reactions in photosynthesis in which hydrogen atoms are combined with carbon and oxygen atoms to form PGAL and glucose.

Carbon-hydrogen-oxygen cycle A process by which these three elements are made available for use by other organisms through the chemical reactions of respiration and photosynthesis.

Carboxyl group A chemical group having the formula—COOH and found as part of all amino acid and fatty acid molecules.

Cardiac muscle A type of muscle tissue in the heart and arteries that is associated with the rhythmic nature of the pulse and heartbeat.

Cardiovascular disease In human beings, any disease of the circulatory organs.

Carnivore A heterotrophic organism that consumes animal tissue as its primary source of nutrition. (See **secondary consumer**.)

Carrier An individual who, though not expressing a particular recessive trait, carries this gene as part of his/her heterozygous genotype.

Carrier protein A specialized molecule embedded in the cell membrane that aids the movement of materials across the membrane.

Cartilage A flexible connective tissue found in many flexible parts of the body (e.g., knee); common in the embryonic stages of development.

Catalyst Any substance that speeds up or slows down the rate of a chemical reaction. (See **enzyme.**)

Cell plate A structure that forms during cytoplasmic division in plant cells and serves to separate the cytoplasm into two roughly equal parts.

Cell theory A scientific theory that states, "All cells arise from previously existing cells" and "Cells are the unit of structure and function of living things."

Cell wall A cell organelle that surrounds and gives structural support to plant cells; cell walls are composed of cellulose.

Central nervous system The portion of the vertebrate nervous system that consists of the brain and the spinal cord.

Centriole A cell organelle found in animal cells that functions in the process of cell division.

Centromere The area of attachment of two chromatids in a double-stranded chromosome.

Cerebellum The portion of the human brain responsible for the coordination of muscular activity.

Cerebral hemorrhage A disorder of the human regulatory system in which a broken blood vessel in the brain may result in severe dysfunction or death.

Cerebral palsy A disorder of the human regulatory system in which the motor and speech centers of the brain are impaired.

Cerebrum The portion of the human brain responsible for thought, reasoning, sense interpretation, learning, and other conscious activities.

Cervix A structure that bounds the lower end of the uterus and through which sperm must pass in order to fertilize the egg.

Chemical digestion The process by which nutrient molecules are converted by chemical means into a form usable by the cells.

Chemosynthesis A type of autotrophic nutrition in which certain bacteria use the energy of chemical oxidation to convert inorganic raw materials to organic food molecules.

Chitin A polysaccharide substance that forms the exoskeleton of the grasshopper and other arthropods.

Chlorophyll A green pigment in plant cells that absorbs sunlight and makes possible certain aspects of the photosynthetic process.

Chloroplast A cell organelle found in plant cells that contains chlorophyll and functions in photosynthesis.

Chordata A phylum of the Animal Kingdom whose members (chordates) have internal skeletons made of cartilage and/or bone (e.g., human being).

Chorion A membrane that surrounds all other embryonic membranes in many animal species, protecting them from mechanical damage.

Chromatid One strand of a double-stranded chromosome.

Chromosome mutation An alteration in the structure of a chromosome involving many genes. (See **nondisjunction; translocation; addition; deletion**.)

Cilia Small, hairlike structures in paramecia and other unicellular organisms that aid in nutrition and locomotion.

Classification A technique by which scientists sort, group, and name organisms for easier study.

Cleavage A series of rapid mitotic divisions that increase cell number in a developing embryo without corresponding increase in cell size.

Climax community A stable, self-perpetuating community that results from an ecological succession.

Cloning A technique of genetic investigation in which undifferentiated cells of an organism are used to produce new organisms with the same set of traits as the original cells.

Closed transport system A type of circulatory system in which the transport fluid is always enclosed within blood vessels (e.g., earthworm, human).

Clot A structure that forms as a result of enzyme-controlled reactions following the rupturing of a blood vessel and serves as a plug to prevent blood loss.

Codominance A type of intermediate inheritance that results from the simultaneous expression of two dominant alleles with contrasting effects.

Codon See **triplet codon**.

Coelenterata A phylum of the Animal Kingdom whose members (coelenterates) have bodies that resemble a sack (e.g., hydra, jellyfish).

Coenzyme A chemical substance or chemical subunit that functions to aid the action of a particular enzyme. (See **vitamin.**)

Cohesion A force binding water molecules together that aids in the upward conduction of materials in the xylem.

Commensalism A type of symbiosis in which one organism in the relationship benefits and the other is neither helped nor harmed.

Common ancestry A concept central to the science of evolution which postulates that all organisms share a common ancestry whose closeness varies with the degree of shared similarity.

Community A level of biological organization that includes all of the species populations inhabiting a particular geographic area.

Comparative anatomy The study of similarities in the anatomical structures of organisms, and their use as an indicator of common ancestry and as evidence of organic evolution.

Comparative biochemistry The study of similarities in the biochemical makeups of organisms, and their use as an indicator of common ancestry and as evidence of organic evolution.

Comparative cytology The study of similarities in the cell structures of organisms, and their use as an indicator of common ancestry and as evidence of organic evolution.

Comparative embryology The study of similarities in the patterns of embryological development of organisms, and their use as an indicator of common ancestry and as evidence of organic evolution.

Competition A condition that arises when different species in the same habitat attempt to use the same limited resources.

Complete protein A protein that contains all eight essential amino acids.

Compound A substance composed of two or more different kinds of atom (e.g., water: H_2O).

Compound light microscope A tool of biological study capable of producing a magnified image of a biological specimen by using a focused beam of light.

Conditioned behavior A type of response that is learned, but that becomes automatic with repetition.

Conservation of resources The development and application of practices to protect valuable and irreplaceable soil and mineral resources—a positive aspect of human involvement with the environment.

Constipation A disorder of the human digestive tract in which fecal matter solidifies and becomes difficult to egest.

Consumer Any heterotrophic animal organism (e.g., human being).

Coronary artery An artery that branches off the aorta to feed the heart muscle.

Coronary thrombosis A disorder of the human transport system in which the heart muscle becomes damaged as a result of blockage of the coronary artery.

Corpus luteum A structure resulting from the hormone-controlled transformation of the ovarian follicle that produces the hormone progesterone.

Corpus luteum stage A stage of the menstrual cycle in which the cells of the follicle are transformed into the corpus luteum under the influence of the hormone LH.

Cotyledon A portion of the plant embryo that serves as a source of nutrition for the young plant before photosynthesis begins.

Cover-cropping A proper agricultural practice in which a temporary planting (cover crop) is used to limit soil erosion between seasonal plantings of main crops.

Crop A portion of the digestive tract of certain animals that stores food temporarily before digestion.

Cross-pollination A type of pollination in which pollen from one flower pollinates flowers of a different plant of the same species.

Crossing-over A pattern of inheritance in which linked genes may be separated during synapsis in the first meiotic division, when sections of homologous chromosomes may be exchanged.

Cuticle A waxy coating that covers the upper epidermis of most leaves and acts to help the leaf retain water.

Cutting A technique of plant propagation in which vegetative parts of the parent plant are cut and rooted to establish new plant organisms with identical characteristics.

Cyclosis The circulation of the cell fluid (cytoplasm) within the cell interior.

Cyton The "cell body" of the neuron, which generates the nerve impulse.

Cytoplasm The watery fluid that provides a medium for the suspension of organelles within the cell.

Cytoplasmic division The separation of daughter nuclei into two new daughter cells.

Cytosine A nitrogenous base found in both DNA and RNA molecules.

Daughter cell A cell that results from mitotic cell division.

Daughter nucleus One of two nuclei that form as a result of mitosis.

Deamination A process by which amino acids are broken down into their component parts for conversion into urea.

Death The irreversible cessation of bodily functions and cellular activities.

Deciduous A term relating to broadleaf trees which shed their leaves in the fall.

Decomposer Any saprophytic organism that derives its energy from the decay of plant and animal tissues (e.g., bacteria of decay, fungus); the final stage of a food chain.

Decomposition bacteria In the nitrogen cycle, bacteria that break down plant and animal protein and produce ammonia as a by-product.

Dehydration synthesis A chemical process in which two organic molecules may be joined after removing the atoms needed to form a molecule of water as a by-product.

Deletion A type of chromosome mutation in which a section of a chromosome is separated and lost.

Dendrite A cytoplasmic extension of a neuron that serves to detect an environmental stimulus and carry an impulse to the cell body of the neuron.

Denitrifying bacteria In the nitrogen cycle, bacteria that convert excess nitrate salts into gaseous nitrogen.

Deoxygenated blood Blood that has released its transported oxygen to the body tissues.

Deoxyribonucleic acid (DNA) A nucleic acid molecule known to be the chemically active agent of the gene; the fundamental hereditary material of living organisms.

Deoxyribose A five-carbon sugar that is a component part of the nucleotide unit in DNA only.

Desert A terrestrial biome characterized by sparse rainfall, extreme temperature variation, and a climax flora that includes cactus.

Diabetes A disorder of the human regulatory system in which insufficient insulin production leads to elevated blood sugar concentrations.

Diarrhea A disorder of the human digestive tract in which the large intestine fails to absorb water from the waste matter, resulting in watery feces.

Diastole The lower pressure registered during blood pressure testing. (See **systole**.)

Differentiation The process by which embryonic cells become specialized to perform the various tasks of particular tissues throughout the body.

Diffusion A form of passive transport by which soluble substances are absorbed or released by cells.

Digestion The process by which complex foods are broken down by mechanical or chemical means for use by the body.

Dipeptide A chemical unit composed of two amino acid units linked by a peptide bond.

Diploid chromosome number The number of chromosomes found characteristically in the cells (except gametes) of sexually reproducing species.

Disaccharidase Any disaccharide-hydrolyzing enzyme.

Disaccharide A type of carbohydrate known also as a "double sugar"; all disaccharides have the molecular formula $C_{12}H_{22}O_{11}$.

Disjunction The separation of homologous chromosome pairs at the end of the first meiotic division.

Disposal problems Problems, due to technological oversight, that result when commercial and technological activities produce solid and/or chemical wastes that must be disposed of.

Dissecting microscope A tool of biological study that magnifies the image of a biological specimen up to 20 times normal size for purposes of gross dissection.

Dominance A pattern of genetic inheritance in which the effects of a dominant allele mask those of a recessive allele.

Dominant allele (gene) An allele (gene) whose effect masks that of its recessive allele.

Double-stranded chromosome The two-stranded structure that results from chromosomal replication.

Down's syndrome In human beings, a condition, characterized by mental and physical retardation, that may be caused by the nondisjunction of chromosome number 21.

Drosophila The common fruit fly, an organism that has served as an object of genetic research in the development of the gene-chromosome theory.

Ductless gland See **endocrine gland**.

Ecology The science that studies the interactions of living things with each other and with the nonliving environment.

Ecosystem The basic unit of study in ecology, including the plant and animal community in interaction with the nonliving environment.

Ectoderm An embryonic tissue that differentiates into skin and nerve tissue in the adult animal.

Effector An organ specialized to produce a response to an environmental stimulus: effectors may be muscles or glands.

Egestion The process by which undigested food materials are eliminated from the body.

Electron microscope A tool of biological study that uses a focused beam of electrons to produce an image of a biological specimen magnified up to 25,000 times normal size.

Element The simplest form of matter; an element is a substance (e.g., nitrogen) made up of a single type of atom.

Embryo An organism in the early stages of development following fertilization.

Embryonic development A series of complex processes by which animal and plant embryos develop into adult organisms.

Emphysema A disorder of the human respiratory system in which lung tissue deteriorates, leaving the lung with diminished capacity and efficiency.

Emulsification A process by which fat globules are surrounded by bile to form fat droplets.

Endocrine ("ductless") gland A gland (e.g., thyroid, pituitary) specialized for the production of hormones and their secretion directly into the bloodstream; such glands lack ducts.

Endoderm An embryonic tissue that differentiates into the digestive and respiratory tract lining in the adult animal.

Endoplasmic reticulum (ER) A cell organelle known to function in the transport of cell products from place to place within the cell.

Environmental laws Federal, state, and local legislation enacted in an attempt to protect environmental resources—a positive aspect of human involvement with the environment.

Enzymatic hydrolysis An enzyme-controlled reaction by which complex food molecules are broken down chemically into simpler subunits.

Enzyme An organic catalyst that controls the rate of metabolism of a single type of substrate; enzymes are protein in nature.

Enzyme-substrate complex A physical association between an enzyme molecule and its substrate within which the substrate is metabolized.

Epicotyl A portion of the plant embryo that specializes to become the upper stem, leaves, and flowers of the adult plant.

Epidermis The outermost cell layer in a plant or animal.

Epiglottis In a human being, a flap of tissue that covers the upper end of the trachea during swallowing and prevents inhalation of food.

Esophagus A structure in the upper portion of the digestive tract that conducts the food from the pharynx to the midgut.

Essential amino acid An amino acid that cannot be synthesized by the human body, but must be obtained by means of the diet.

Estrogen A hormone, secreted by the ovary, that regulates the production of female secondary sex characteristics.

Evolution Any process of gradual change through time.

Excretion The life function by which living things eliminate metabolic wastes from their cells.

Exoskeleton A chitinous material that covers the outside of the bodies of most arthropods and provides protection for internal organs and anchorage for muscles.

Exploitation of organisms Systematic removal of animals and plants with commercial value from their environments, for sale—a negative aspect of human involvement with the environment.

Extensor A skeletal muscle that extends (opens) a joint.

External development Embryonic development that occurs outside the body of the female parent (e.g., birds).

External fertilization Fertilization that occurs outside the body of the female parent (e.g., fish).

Extracellular digestion Digestion that occurs outside the cell.

Fallopian tube See **oviduct.**

Fatty acid An organic molecule that is a component of certain lipids.

Fauna The animal species comprising an ecological community.

Feces The semisolid material that results from the solidification of undigested foods in the large intestine.

Fertilization The fusion of gametic nuclei in the process of sexual reproduction.

Filament The portion of the stamen that supports the anther.

Flagella Microscopic, whiplike structures found on certain cells that aid in locomotion and circulation.

Flexor A skeletal muscle that flexes (closes) a joint.

Flora The plant species comprising an ecological community.

Flower The portion of a flowering plant that is specialized for sexual reproduction.

Fluid-mosaic model A model of the structure of the cell membrane in which large protein molecules are thought to be embedded in a bilipid layer.

Follicle One of many areas within the ovary that serve as sites for the periodic maturation of ova.

Follicle stage The stage of the menstrual cycle in which an ovum reaches its final maturity under the influence of the hormone FSH.

Follicle-stimulating hormone (FSH) A pituitary hormone that regulates the maturation of, and the secretion of estrogen by, the ovarian follicle.

Food chain A series of nutritional relationships in which food energy is passed from producer to herbivore to carnivore to decomposer; a segment of a food web.

Food web A construct showing a series of interrelated food chains and illustrating the complex nutritional interrelationships that exist in an ecosystem.

Fossil The preserved direct or indirect remains of an organism that lived in the past, as found in the geologic record.

Fraternal twins In human beings, twin offspring that result from the simultaneous fertilization of two ova by two sperm; such twins are not genetically identical.

Freshwater biome An aquatic biome made up of many separate freshwater systems that vary in size and stability and may be closely associated with terrestrial biomes.

Fruit Any plant structure that contains seeds; a mechanism of seed dispersal.

Fungi One of the five biological kingdoms; it includes organisms unable to manufacture their own organic foods (e.g., mushroom).

Gallbladder An accessory organ that stores bile.

Gallstones A disorder of the human digestive tract in which deposits of hardened cholesterol lodge in the gallbladder.

Gamete A specialized reproductive cell produced by organisms of sexually reproducing species. (See **sperm; ovum; pollen; ovule**.)

Gametogenesis The process of cell division by which gametes are produced. (See **meiosis; spermatogenesis; oogenesis**.)

Ganglion An area of bunched nerve cells that acts as a switching point for nerve impulses traveling from receptors and to effectors.

Garden pea The research organism used by Mendel in his early scientific work in genetic inheritance.

Gastric cecum A gland in the grasshopper that secretes digestive enzymes.

Gastrula A stage of embryonic development in animals in which the embryo assumes a tube-within-a-tube structure and distinct embryonic tissues (ectoderm, mesoderm, endoderm) begin to differentiate.

Gastrulation The process by which a blastula becomes progressively more indented, forming a gastrula.

Gene A unit of heredity; a discrete portion of a chromosome thought to be responsible for the production of a single type of polypeptide; the "factor" responsible for the inheritance of a genetic trait.

Gene frequency The proportion (percentage) of each allele for a particular trait that is present in the gene pool of a population.

Gene linkage A pattern of inheritance in which genes located along the same chromosome are prevented from assorting independently, but are linked together in their inheritance.

Gene mutation An alteration of the chemical nature of a gene that changes its ability to control the production of a polypeptide chain.

Gene pool The sum total of all the inheritable genes for the traits in a given sexually reproducing population.

Gene-chromosome theory A theory of genetic inheritance that is based on current understanding of the relationships between the biochemical control of traits and the process of cell division.

Genetic counseling Clinical discussions concerning inheritance patterns that are designed to inform prospective parents of the potential for expression of a genetic disorder in their offspring.

Genetic engineering The use of various techniques to move genes from one organism to another.

Genetic screening A technique for the detection of human genetic disorders in which bodily fluids are analyzed for the presence of certain marker chemicals.

Genotype The particular combination of genes in an allele pair.

Genus A level of biological classification that represents a subdivision of the phylum level; having fewer organisms with great similarity (e.g., *Drosophila, paramecium*).

Geographic isolation The separation of species populations by geographical barriers, facilitating the evolutionary process.

Geologic record A supporting item of evidence of organic evolution, supplied within the earth's rock and other geological deposits.

Germination The growth of the pollen tube from a pollen grain; the growth of the embryonic root and stem from a seed.

Gestation The period of prenatal development of a placental mammal; human gestation requires approximately 9 months.

Gizzard A portion of the digestive tract of certain organisms, including the earthworm and the grasshopper, in which food is ground into smaller fragments.

Glomerulus A capillary network lying within Bowman's capsule of the nephron.

Glucagon A hormone, secreted by the islets of Langerhans, that regulates the release of blood sugar from stored glycogen.

Glucose A monosaccharide produced commonly in photosynthesis and used by both plants and animals as a "fuel" in the process of respiration.

Glycerol An organic compound that is a component of certain lipids.

Glycogen A polysaccharide synthesized in animals as a means of storing glucose; glycogen is stored in the liver and in the muscles.

Goiter A disorder of the human regulatory system in which the thyroid gland enlarges because of a deficiency of dietary iodine.

Golgi complex Cell organelles that package cell products and move them to the plasma membrane for secretion.

Gonad An endocrine gland that produces the hormones responsible for the production of various secondary sex characteristics. (See **ovary; testis**.)

Gout A disorder of the human excretory system in which uric acid accumulates in the joints, causing severe pain.

Gradualism A theory of the time frame required for organic evolution which assumes that evolutionary change is slow, gradual, and continuous.

Grafting A technique of plant propagation in which the stems of desirable plants are attached (grafted) to rootstocks of related varieties to produce new plants for commercial purposes.

Grana The portion of the chloroplast within which chlorophyll molecules are concentrated.

Grassland A terrestrial biome characterized by wide variation in temperature and a climax flora that includes grasses.

Growth A process by which cells increase in number and size, resulting in an increase in size of the organism.

Growth-stimulating hormone (GSH) A pituitary hormone regulating the elongation of the long bones of the body.

Guanine A nitrogenous base found in both DNA and RNA molecules.

Guard cell One of a pair of cells that surround the leaf stomate and regulate its size.

Habitat The environment or set of ecological conditions within which an organism lives.

Hardy-Weinberg principle A hypothesis, advanced by G. H. Hardy and W. Weinberg, which states that the gene pool of a population should remain stable as long as a set of "ideal" conditions is met.

Heart In human beings, a four-chambered muscular pump that facilitates the movement of blood throughout the body.

Helix Literally a spiral; a term used to describe the "twisted ladder" shape of the DNA molecule.

Hemoglobin A type of protein specialized for the transport of respiratory oxygen in certain organisms, including earthworms and human beings.

Herbivore A heterotrophic organism that consumes plant matter as its primary source of nutrition. (See **primary consumer**.)

Hermaphrodite An animal organism that produces both male and female gametes.

Heterotroph An organism that typically carries on heterotrophic nutrition.

Heterotroph hypothesis A scientific hypothesis devised to explain the probable origin and early evolution of life on earth.

Heterotrophic nutrition A type of nutrition in which organisms must obtain their foods from outside sources of organic nutrients.

Heterozygous A term used to refer to an allele pair in which the alleles have different contrasting effects (e.g., *Aa, RW*).

High blood pressure A disorder of the human transport system in which systolic and diastolic pressures register higher than normal because of narrowing of the artery opening.

Histamine A chemical product of the body that causes irritation and swelling of the mucous membranes.

Homeostasis The condition of balance and dynamic stability that characterizes living systems under normal conditions.

Homologous chromosomes A pair of chromosomes that carry corresponding genes for the same traits.

Homologous structures Structures present within different species that can be shown to have had a common origin, but that may or may not share a common function.

Homozygous A term used to refer to an allele pair in which the alleles are identical in terms of effect (e.g., *AA, aa*).

Hormone A chemical product of an endocrine gland which has a regulatory effect on the cell's metabolism.

Host The organism that is harmed in a parasitic relationship.

Hybrid A term used to describe a heterozygous genotype. (See **heterozygous**.)

Hybridization A technique of plant/animal breeding in which two varieties of the same species are crossbred in the hope of producing offspring with the favorable traits of both varieties.

Hydrogen bond A weak electrostatic bond that holds together the twisted strands of DNA and RNA molecules.

Hydrolysis The chemical process by which a complex food molecule is split into simpler components through the addition of a molecule of water to the bonds holding it together.

Hypocotyl A portion of the plant embryo that specializes to become the root and lower stem of the adult plant.

Hypothalamus An endocrine gland whose secretions affect the pituitary gland.

Identical twins In human beings, twin offspring resulting from the separation of the embryonic cell mass of a single fertilization into two separate masses; such twins are genetically identical.

Importation of organisms The introduction of nonactive plants and animals into new areas where they compete strongly with native species—a negative aspect of human involvement with the environment.

In vitro fertilization A laboratory technique in which fertilization is accomplished outside the mother's body using mature ova and sperm extracted from the parents' bodies.

Inbreeding A technique of plant/animal breeding in which a "purebred" variety is bred only with its own members, so as to maintain a set of desired characteristics.

Independent assortment A pattern of inheritance in which genes on different, nonhomologous chromosomes are free to be inherited randomly and regardless of the inheritance of the others.

Ingestion The mechanism by which an organism takes in food from its environment.

Inorganic compound A chemical compound that lacks the element carbon or hydrogen (e.g., table salt: $NaCl$).

Insulin A hormone, secreted by the islets of Langerhans, that regulates the storage of blood sugar as glycogen.

Intercellular fluid (ICF) The fluid that bathes cells and fills intercellular spaces.

Interferon A substance, important in the fight against human cancer, that may now be produced in large quantities through techniques of genetic engineering.

Intermediate inheritance Any pattern of inheritance in which the offspring expresses a phenotype different from the phenotypes of its parents and usually representing a form intermediate between them.

Internal development Embryonic development that occurs within the body of the female parent.

Internal fertilization Fertilization that occurs inside the body of the female parent.

Interneuron A type of neuron, located in the central nervous system, that is responsible for the interpretation of impulses received from sensory neurons.

Intestine A portion of the digestive tract in which chemical digestion and absorption of digestive end-products occur.

Intracellular digestion A type of chemical digestion carried out within the cell.

Iodine A chemical stain used in cell study; an indicator used to detect the presence of starch. (See **staining**.)

Islets of Langerhans An endocrine gland, located within the pancreas, that produces the hormones insulin and glucagon.

Karyotype An enlarged photograph of the paired homologous chromosomes of an individual cell that is used in the detection of certain genetic disorders involving chromosome mutation.

Karyotyping A technique for the detection of human genetic disorders in which a karyotype is analyzed for abnormalities in chromosome structure or number.

Kidney The excretory organ responsible for maintaining the chemical composition of the blood. (See **nephron**.)

Kidney failure A disorder of the human excretory system in which there is a general breakdown of the kidney's ability to filter blood components.

Kingdom A level of biological classification that includes a broad grouping of organisms displaying general structural similarity; five kingdoms have been named by scientists.

Lacteal A small extension of the lymphatic system, found inside the villus, that absorbs fatty acids and glycerol resulting from lipid hydrolysis.

Lactic acid fermentation A type of anaerobic respiration in which glucose is converted to two lactic acid molecules.

Large intestine A portion of the digestive tract in which undigested foods are solidified by means of water absorption to form feces.

Lateral meristem A plant growth region located under the epidermis or bark of a stem. (See **cambium**.)

Latin The language used in biological classification for naming organisms by means of binomial nomenclature.

Lenticel A small pore in the stem surface that permits the absorption and release of respiratory gases within stem tissues.

Leukemia A disorder of the human transport system in which the bone marrow produces large numbers of abnormal white blood cells. (See **cancer.**)

Lichen A symbiosis of alga and fungus that frequently acts as a pioneer species on bare rock.

Limiting factor Any abiotic or biotic condition that places limits on the survival of organisms and on the growth of species populations in the environment.

Lipase Any lipid-hydrolyzing enzyme.

Lipid An organic compound composed of carbon, hydrogen, and oxygen in which hydrogen and oxygen are *not* in a 2:1 ratio (e.g., a wax, plant oil); many lipids are constructed of a glycerol and three fatty acids.

Liver An accessory organ that stores glycogen, produces bile, destroys old red blood cells, deaminates amino acids, and produces urea.

Lock-and-key model A theoretical model of enzyme action that attempts to explain the concept of enzyme specificity.

Lung The major organ of respiratory gas exchange.

Luteinizing hormone (LH) A pituitary hormone that regulates the conversion of the ovarian follicle into the corpus luteum.

Lymph Intercellular fluid (ICF) that has passed into the lymph vessels.

Lymph node One of a series of structures in the body that act as reservoirs of lymph and also contain white blood cells as part of the body's immune system.

Lymph vessel One of a branching series of tubes that collect ICF from the tissues and redistribute it as lymph.

Lymphatic circulation The movement of lymph throughout the body.

Lymphocyte A type of white blood cell that produces antibodies.

Lysosome A cell organelle that houses hydrolytic enzymes used by the cell in the process of chemical digestion.

Malpighian tubules In arthropods (e.g., grasshopper), an organ specialized for the removal of metabolic wastes.

Maltase A specific enzyme that catalyzes the hydrolysis (and dehydration synthesis) of maltose.

Maltose A type of disaccharide; a maltose molecule is composed of two units of glucose joined together by dehydration synthesis.

Marine biome An aquatic biome characterized by relatively stable conditions of moisture, salinity, and temperature.

Marsupial mammal See **nonplacental mammal**.

Mechanical digestion Any of the processes by which foods are broken apart physically into smaller particles.

Medulla The portion of the human brain responsible for regulating the automatic processes of the body.

Meiosis The process by which four monoploid nuclei are formed from a single diploid nucleus.

Meningitis A disorder of the human regulatory system in which the membranes of the brain or spinal cord become inflamed.

Menstrual cycle A hormone-controlled process responsible for the monthly release of mature ova.

Menstruation The stage of the menstrual cycle in which the lining of the uterus breaks down and is expelled from the body via the vaginal canal.

Meristem A plant tissue specialized for embryonic development. (See **apical meristem; lateral meristem; cambium.**)

Mesoderm An embryonic tissue that differentiates into muscle, bone, the excretory system, and most of the reproductive system in the adult animal.

Messenger RNA (m-RNA) A type of RNA that carries the genetic code from the nuclear DNA to the ribosome for transcription.

Metabolism All of the chemical processes of life considered together; the sum total of all the cell's chemical activity.

Methylene blue A chemical stain used in cell study. (See **staining**.)

Microdissection instruments Tools of biological study that are used to remove certain cell organelles from within cells for examination.

Micrometer (μm) A unit of linear measurement equal in length to 0.001 millimeter (0.000001 meter), used for expressing the dimensions of cells and cell organelles.

Mitochondrion A cell organelle that contains the enzymes necessary for aerobic respiration.

Mitosis A precise duplication of the contents of a parent cell nucleus, followed by an orderly separation of these contents into two new, identical daughter nuclei.

Mitotic cell division A type of cell division that results in the production of two daughter cells identical to each other and to the parent cell.

Monera One of the five biological kingdoms; it includes simple unicellular forms lacking nuclear membranes (e.g., bacteria).

Monohybrid cross A genetic cross between two organisms both heterozygous for a trait controlled by a single allele pair. The phenotypic ratio resulting is 3:1; the genotypic ratio is 1:2:1.

Monoploid chromosome number The number of chromosomes commonly found in the gametes of sexually reproducing species.

Monosaccharide A type of carbohydrate known also as a "simple sugar"; all monosaccharides have the molecular formula $C_6H_{22}O_6$.

Motor neuron A type of neuron that carries "command" impulses from the central nervous system to an effector organ.

Mucus A protein-rich mixture that bathes and moistens the respiratory surfaces.

Multicellular Having a body that consists of large groupings of specialized cells (e.g., human being).

Multiple alleles A pattern of inheritance in which the existence of more than two alleles is hypothesized, only two of which are present in the genotype of any one individual.

Muscle A type of tissue specialized to produce movement of body parts.

Mutagenic agent Any environmental condition that initiates or accelerates genetic mutation.

Mutation Any alteration of the genetic material, either a chromosome or a gene, in an organism.

Mutualism A type of symbiosis beneficial to both organisms in the relationship.

Nasal cavity A series of channels through which outside air is admitted to the body interior and is warmed and moistened before entering the lung.

Natural selection A concept, central to Darwin's theory of evolution, to the effect that the individuals best adapted to their environment tend to survive and to pass their favorable traits on to the next generation.

Negative feedback A type of endocrine regulation in which the effects of one gland may inhibit its own secretory activity, while stimulating the secretory activity of another gland.

Nephridium An organ found in certain organisms, including the earthworm, specialized for the removal of metabolic wastes.

Nephron The functional unit of the kidney. (See **glomerulus; Bowman's capsule**.)

Nerve A structure formed from the bundling of neurons carrying sensory or motor impulses.

Nerve impulse An electrochemical change in the surface of the nerve cell.

Nerve net A network of "nerve" cells in coelenterates such as the hydra.

Neuron A cell specialized for the transmission of nerve impulses.

Neurotransmitter A chemical substance secreted by a neuron that aids in the transmission of the nerve impulse to an adjacent neuron.

Niche The role that an organism plays in its environment.

Nitrifying bacteria In the nitrogen cycle, bacteria that absorb ammonia and convert it into nitrate salts.

Nitrogen cycle The process by which nitrogen is recycled and made available for use by other organisms.

Nitrogen-fixing bacteria A type of bacteria responsible for absorbing atmospheric nitrogen and converting it to nitrate salts in the soil.

Nitrogenous base A chemical unit composed of carbon, hydrogen, and nitrogen that is a component part of the nucleotide unit.

Nitrogenous waste Any of a number of nitrogen-rich compounds that result from the metabolism of proteins and amino acids in the cell. (See **ammonia; urea; uric acid**.)

Nondisjunction A type of chromosome mutation in which the members of one or more pairs of homologous chromosomes fail to separate during the disjunction phase of the first meiotic division.

Nonplacental mammal A species of mammal in which internal development is accomplished without the aid of a placental connection (marsupial mammals).

Nucleic acid An organic compound composed of repeating units of nucleotide.

Nucleolus A cell organelle located within the nucleus that is known to function in protein synthesis.

Nucleotide The repeating unit making up the nucleic acid polymer (e.g., DNA, RNA).

Nucleus A cell organelle that contains the cell's genetic information in the form of chromosomes.

Nutrition The life function by which living things obtain food and process it for their use.

Omnivore A heterotrophic organism that consumes both plant and animal matter as sources of nutrition.

One gene-one polypeptide A scientific hypothesis concerning the role of the individual gene in protein synthesis.

Oogenesis A type of meiotic cell division in which one ovum and three polar bodies are produced from each primary sex cell.

Open transport system A type of circulatory system in which the transport fluid is *not* always enclosed within blood vessels (e.g., grasshopper).

Oral cavity In human beings, the organ used for the ingestion of foods.

Oral groove The ingestive organ of the paramecium.

Organ transplant An application of the study of immunity in which an organ or tissue of a donor is transplanted into a compatible recipient.

Organelle A small, functional part of a cell specialized to perform a specific life function (e.g., nucleus, mitochondrion).

Organic compound A chemical compound that contains the elements carbon and hydrogen (e.g., carbohydrate, protein).

Organic evolution The mechanism thought to govern the changes in living species over geologic time.

Osmosis A form of passive transport by which water is absorbed or released by cells.

Ovary A female gonad that secretes the hormone estrogen, which regulates female secondary sex characteristics; the ovary also produces ova, which are used in reproduction.

Overcropping A negative aspect of human involvement with the environment in which soil is overused for the production of crops, leading to exhaustion of soil nutrients.

Overgrazing The exposure of soil to erosion due to the loss of stabilizing grasses when it is overused by domestic animals—a negative aspect of human involvement with the environment.

Overhunting A negative aspect of human involvement with the environment in which certain species have been greatly reduced or made extinct by uncontrolled hunting practices.

Oviduct A tube that serves as a channel for conducting mature ova from the ovary to the uterus; the site of fertilization and the earliest stages of embryonic development.

Ovulation The stage of the menstrual cycle in which the mature ovum is released from the follicle into the oviduct.

Ovule A structure located within the flower ovary that contains a monoploid egg nucleus and serves as the site of fertilization.

Ovum A type of gamete produced as a result of oogenesis in female animals; the egg, the female sex cell.

Oxygen-18 A radioactive isotope of oxygen that is used to trace the movement of this element in biochemical reaction sequences.

Oxygenated blood Blood that contains a high percentage of oxyhemoglobin.

Oxyhemoglobin Hemoglobin that is loosely bound to oxygen for purposes of oxygen transport.

Palisade layer A cell layer found in most leaves that contains high concentrations of chloroplasts.

Pancreas An accessory organ which produces enzymes that complete the hydrolysis of foods to soluble end-products; also the site of insulin and glucagon production.

Parasitism A type of symbiosis from which one organism in the relationship benefits, while the other (the "host") is harmed, but not ordinarily killed.

Parathormone A hormone of the parathyroid gland that regulates the metabolism of calcium in the body.

Parathyroid gland An endocrine gland whose secretion, parathormone, regulates the metabolism of calcium in the body.

Passive immunity A temporary immunity produced as a result of the injection of preformed antibodies.

Passive transport Any process by which materials are absorbed into the cell interior from an area of high concentration to an area of low concentration, without the expenditure of cell energy (e.g., osmosis, diffusion).

Penis A structure that permits internal fertilization through direct implantation of sperm into the female reproductive tract.

Peptide bond A type of chemical bond that links the nitrogen atom of one amino acid with the terminal carbon atom of a second amino acid in the formation of a dipeptide.

Peripheral nerves Nerves in the earthworm and grasshopper that branch from the ventral nerve cord to other parts of the body.

Peripheral nervous system A major subdivision of the nervous system that consists of all the nerves of all types branching through the body. (See **autonomic nervous system; somatic nervous system**.)

Peristalsis A wave of contraction of the smooth muscle lining; the digestive tract that causes ingested food to pass along the food tube.

Petal An accessory part of the flower that is thought to attract pollinating insects.

pH A chemical unit used to express the concentration of hydrogen ion (H^+), or the acidity, of a solution.

Phagocyte A type of white blood cell that engulfs and destroys bacteria.

Phagocytosis The process by which the ameba surrounds and ingests large food particles for intracellular digestion.

Pharynx The upper part of the digestive tube that temporarily stores food before digestion.

Phenotype The observable trait that results from the action of an allele pair.

Phenylketonuria (PKU) A genetically related human disorder in which the homozygous combination of a particular mutant gene prevents the normal metabolism of the amino acid phenylalanine.

Phloem A type of vascular tissue through which water and dissolved sugars are transported in plants from the leaf downward to the roots for storage.

Phosphate group A chemical group made up of phosphorus and oxygen that is a component part of the nucleotide unit.

Phosphoglyceraldehyde (PGAL) An intermediate product formed during photosynthesis that acts as the precursor of glucose formation.

Photochemical reactions A set of biochemical reactions in photosynthesis in which light is absorbed and water molecules are split. (See **photolysis**.)

Photolysis The portion of the photochemical reactions in which water molecules are split into hydrogen atoms and made available to the carbon fixation reactions.

Photosynthesis A type of autotrophic nutrition in which green plants use the energy of sunlight to convert carbon dioxide and water into glucose.

Phylum A level of biological classification that is a major subdivision of the kingdom level, containing fewer organisms with greater similarity (e.g., Chordata).

Pinocytosis A special type of absorption by which liquids and particles too large to diffuse through the cell membrane may be taken in by vacuoles formed at the cell surface.

Pioneer autotrophs The organisms supposed by the heterotroph hypothesis to have been the first to evolve the ability to carry on autotrophic nutrition.

Pioneer species In an ecological succession, the first organisms to inhabit a barren environment.

Pistil The female sex organ of the flower. (See **stigma; style; ovary**.)

Pituitary gland An endocrine gland that produces hormones regulating the secretions of other endocrine glands; the "master gland."

Placenta In placental mammals, a structure composed of both embryonic and maternal tissues that permits the diffusion of soluble substances to and from the fetus for nourishment and the elimination of fetal waste.

Placental mammal A mammal species in which embryonic development occurs internally with the aid of a placental connection to the female parent's body.

Plant One of the five biological kingdoms; it includes multicellular organisms whose cells are bounded by cell walls and which are capable of photosynthesis (e.g., maple tree).

Plasma The liquid fraction of blood, containing water and dissolved proteins.

Plasma membrane A cell organelle that encloses the cytoplasm and other cell organelles and regulates the passage of materials into and out of the cell.

Platelet A cell-like component of the blood that is important in clot formation.

Polar body One of three nonfunctional cells produced during oogenesis that contain monoploid nuclei and disintegrate soon after completion of the process.

Polio A disorder of the human regulatory system in which viral infection of the central nervous system may result in severe paralysis.

Pollen The male gamete of the flowering plant.

Pollen tube A structure produced by the germinating pollen grain that grows through the style to the ovary and carries the sperm nucleus to the ovule for fertilization.

Pollination The transfer of pollen grains from anther to stigma.

Pollution control The development of new procedures to reduce the incidence of air, water, and soil pollution—a positive aspect of human involvement with the environment.

Polyploidy A type of chromosome mutation in which an entire set of homologous chromosomes fail to separate during the disjunction phase of the first meiotic division.

Polysaccharide A type of carbohydrate composed of repeating units of monosaccharide that form a polymeric chain.

Polyunsaturated fat A type of fat in which many bonding sites are unavailable for the addition of hydrogen atoms.

Population All the members of a particular species in a given geographical location at a given time.

Population control The use of various practices to slow the rapid growth in the human population—a positive aspect of human interaction with the environment.

Population genetics A science that studies the genetic characteristics of a sexually reproducing species and the factors that affect its gene frequencies.

Postnatal development The growth and maturation of an individual from birth, through aging, to death.

Prenatal development The embryonic development that occurs within the uterus before birth. (See **gestation.**)

Primary consumer Any herbivorous organism that receives food energy from the producer level (e.g., mouse); the second stage of a food chain.

Primary sex cell The diploid cell that undergoes meiotic cell division to produce monoploid gametes.

Producer Any autotrophic organism capable of trapping light energy and converting it to the chemical bond energy of food (e.g., green plants); the organisms forming the basis of the food chain.

Progesterone A hormone produced by the corpus luteum and/or placenta that has the effect of maintaining the uterine lining and suppressing ovulation during gestation.

Protease Any protein-hydrolyzing enzyme.

Protein A complex organic compound composed of repeating units of amino acid.

Protista One of the five biological kingdoms; it includes simple unicellular forms whose nuclei are surrounded by nuclear membranes (e.g., ameba, paramecium).

Pseudopod A temporary, flowing extension of the cytoplasm of an ameba that is used in nutrition and locomotion.

Pulmonary artery One of two arteries that carry blood from the heart to the lungs for reoxygenation.

Pulmonary circulation Circulation of blood from the heart through the lungs and back to the heart.

Pulmonary vein One of four veins that carry oxygenated blood from the lungs to the heart.

Pulse Rhythmic contractions of the artery walls that help to push the blood fluid through the capillary networks of the body.

Punctuated equilibrium A theory of the time frame required for evolution which assumes that evolutionary change occurs in "bursts" with long periods of relative stability intervening.

Pyramid of biomass A construct used to illustrate the fact that the total biomass available in each stage of a food chain diminishes from producer level to consumer level.

Pyramid of energy A construct used to illustrate the fact that energy is lost at each trophic level in a food chain, being most abundant at the producer level.

Pyruvic acid An intermediate product in the aerobic or anaerobic respiration of glucose.

Receptor An organ specialized to receive a particular type of environmental stimulus.

Recessive allele (gene) An allele (gene) whose effect is masked by that of its dominant allele.

Recombinant DNA DNA molecules that have been moved from one cell to another in order to give the recipient cell a genetic characteristic of the donor cell.

Recombination The process by which the members of segregated allele pairs are randomly recombined in the zygote as a result of fertilization.

Rectum The portion of the digestive tract in which digestive wastes are stored until they can be released to the environment.

Red blood cell Small, nonnucleated cells in the blood that contain hemoglobin and carry oxygen to bodily tissues.

Reduction division See **meiosis**.

Reflex A simple, inborn, involuntary response to an environmental stimulus.

Reflex arc The complete path, involving a series of three neurons (sensory, interneuron, and motor), working together, in a reflex action.

Regeneration A type of asexual reproduction in which new organisms are produced from the severed parts of a single parent organism; the replacement of lost or damaged tissues.

Regulation The life process by which living things respond to changes within and around them, and by which all life processes are coordinated.

Replication An exact self-duplication of the chromosome during the early stages of cell division; the exact self-duplication of a molecule of DNA.

Reproduction The life process by which new cells arise from preexisting cells by cell division.

Reproductive isolation The inability of species varieties to interbreed and produce fertile offspring, because of variations in behavior or chromosome structure.

Respiration The life function by which living things convert the energy of organic foods into a form more easily used by the cell.

Response The reaction of an organism to an environmental stimulus.

Rhizoid A rootlike fiber produced by fungi that secrete hydrolytic enzymes and absorb digested nutrients.

Ribonucleic acid (RNA) A type of nucleic acid that operates in various ways to facilitate protein synthesis.

Ribose A five-carbon sugar found as a component part of the nucleotides of RNA molecules only.

Ribosomal RNA (r-RNA) The type of RNA that makes up the ribosome.

Ribosome A cell organelle that serves as the site of protein synthesis in the cell.

Root A plant organ specialized to absorb water and dissolved substances from the soil, as well as to anchor the plant to the soil.

Root hair A small projection of the growing root that serves to increase the surface area of the root for absorption.

Roughage A variety of undigestible carbohydrates that add bulk to the diet and facilitate the movement of foods through the intestine.

Runner A type of vegetative propagation in which an above-ground stem (runner) produces roots and leaves and establishes new organisms with identical characteristics.

Saliva A fluid secreted by salivary glands that contains hydrolytic enzymes specific to the digestion of starches.

Salivary gland The gland that secretes saliva, important in the chemical digestion of certain foods.

Salt A chemical composed of a metal and a nonmetal joined by means of an ionic bond (e.g., sodium chloride).

Saprophyte A heterotrophic organism that obtains its nutrition from the decomposing remains of dead plant and animal tissues (e.g., fungus, bacteria).

Saturated fat A type of fat molecule in which all available bonding sites on the hydrocarbon chains are taken up with hydrogen atoms.

Scrotum A pouch extending from the wall of the lower abdomen that houses the testes at a temperature optimum for sperm production.

Secondary consumer Any carnivorous animal that derives its food energy from the primary consumer level (e.g., a snake); the third level of a food chain.

Secondary sex characteristics The physical features, different in males and females, that appear with the onset of sexual maturity.

Seed A structure that develops from the fertilized ovule of the flower and germinates to produce a new plant.

Seed dispersal Any mechanism by which seeds are distributed in the environment so as to widen the range of a plant species. (See **fruit**.)

Segregation The random separation of the members of allele pairs that occurs during meiotic cell division.

Self-pollination A type of pollination in which the pollen of a flower pollinates another flower located on the same plant organism.

Sensory neuron A type of neuron specialized for receiving environmental stimuli, which are detected by receptor organs.

Sepal An accessory part of the flower that functions to protect the bud during development.

Sessile A term that relates to the "unmoving" state of certain organisms, including the hydra.

Seta One of several small, chitinous structures (setae) that aid the earthworm in its locomotor function.

Sex chromosomes A pair of homologous chromosomes carrying genes that determine the sex of an individual; these chromosomes are designated as X and Y.

Sex determination A pattern of inheritance in which the conditions of maleness and femaleness are determined by the inheritance of a pair of sex chromosomes (XX = female; XY = male).

Sex linkage A pattern of inheritance in which certain nonsex genes are located on the X sex chromosome, but have no corresponding alleles on the Y sex chromosome.

Sex-linked trait A genetic trait whose inheritance is controlled by the genetic pattern of sex linkage (e.g., color blindness).

Sexual reproduction A type of reproduction in which new organisms are formed as a result of the fusion of gametes from two parent organisms.

Shell An adaptation for embryonic development in many terrestrial, externally developing species that protects the developing embryo from drying and physical damage (e.g., birds).

Sickle cell anemia A genetically related human disorder in which the homozygous combination of a mutant gene leads to the production of abnormal hemoglobin and crescent-shaped red blood cells.

Skeletal muscle A type of muscle tissue associated with the voluntary movements of skeletal levers in locomotion.

Small intestine In human beings, the longest portion of the food tube, in which final digestion and absorption of soluble end-products occur.

Smooth muscle See **visceral muscle**.

Somatic nervous system A subdivision of the peripheral nervous system that is made up of nerves associated with voluntary actions.

Speciation The process by which new species are thought to arise from previously existing species.

Species A biological grouping of organisms so closely related that they are capable of interbreeding and producing fertile offspring (e.g., human being).

Species presentation The establishment of game lands and wildlife refuges that have permitted the recovery of certain endangered species—a positive aspect of human involvement with the environment.

Sperm A type of gamete produced as a result of spermatogenesis in male animals; the male reproductive cell.

Spermatogenesis A type of meiotic cell division in which four sperm cells are produced for each primary sex cell.

Spinal cord The part of the central nervous system responsible for reflex action, as well as impulse conduction between the peripheral nervous system and the brain.

Spindle apparatus A network of fibers that form during cell division and to which centromeres attach during the separation of chromosomes.

Spiracle One of several small pores in arthropods, including the grasshopper, that serve as points of entry of respiratory gases from the atmosphere to the tracheal tubes.

Spongy layer A cell layer found in most leaves that is loosely packed and contains many air spaces to aid in gas exchange.

Spore A specialized asexual reproductive cell produced by certain plants.

Sporulation A type of asexual reproduction in which spores released from special spore cases on the parent plant germinate and grow into new adult organisms of the species.

Staining A technique of cell study in which chemical stains are used to make cell parts more visible for microscopic study.

Stamen The male reproductive structure in a flower. (See **anther; filament**.)

Starch A type of polysaccharide produced and stored by plants.

Stem A plant organ specialized to support the leaves and flowers of a plant, as well as to conduct materials between the roots and the leaves.

Stigma The sticky upper portion of the pistil, which serves to receive pollen.

Stimulus Any change in the environment to which an organism responds.

Stomach A muscular organ that acts to liquefy food and that produces gastric protease for the hydrolysis of protein.

Stomate A small opening that penetrates the lower epidermis of a leaf and through which respiratory and photosynthetic gases diffuse.

Strata The layers of sedimentary rock that contain fossils, whose ages may be determined by studying the patterns of sedimentation.

Stroke A disorder of the human regulatory system in which brain function is impaired because of oxygen starvation of brain centers.

Stroma An area of the chloroplast within which the carbon-fixation reactions occur; stroma lie between pairs of grana.

Style The portion of the pistil that connects the stigma to the ovary.

Substrate A chemical that is metabolized by the action of a specific enzyme.

Succession A situation in which an established ecological community is gradually replaced by another until a climax community is established.

Survival of the fittest The concept, frequently associated with Darwin's theory of evolution, that in the intraspecies competition among naturally occurring species the organisms best adapted to the particular environment will survive.

Sweat glands In human beings, the glands responsible for the production of perspiration.

Symbiosis A term which refers to a variety of biotic relationships in which organisms of different species live together in close physical association.

Synapse The gap that separates the terminal branches of one neuron from the dendrites of an adjacent neuron.

Synapsis The intimate, highly specific pairing of homologous chromosomes that occurs in the first meiotic division, forming tetrads.

Synthesis The life function by which living things manufacture the complex compounds required to sustain life.

Systemic circulation The circulation of blood from the heart through the body tissues (except the lungs) and back to the heart.

Systole The higher pressure registered during blood pressure testing. (See **diastole**.)

Taiga A terrestrial biome characterized by long, severe winters and climax flora that includes coniferous trees.

Tay-Sachs A genetically related human disorder in which fatty deposits in the cells, particularly of the brain, inhibit proper functioning of the nervous system.

Technological oversight A term relating to human activities that adversely affect environmental quality due to failure to adequately assess the environmental impact of a technological development.

Teeth Structures located in the mouth that are specialized to aid in the mechanical digestion of foods.

Temperate deciduous forest A terrestrial biome characterized by moderate climatic conditions and climax flora that includes deciduous trees.

Template A pattern or design provided by the DNA molecule for the synthesis of protein molecules.

Tendon A type of connective tissue that attaches a skeletal muscle to a bone.

Tendonitis A disorder of the human locomotor system in which the junction between a tendon and a bone becomes irritated and inflamed.

Tentacle A grasping structure in certain organisms, including the hydra, that contains stinging cells and is used for capturing prey.

Terminal branch A cytoplasmic extension of the neuron that transmits a nerve impulse to adjacent neurons via the secretion of neurotransmitters.

Terrestrial biome A biome that comprises primarily land ecosystems, the characteristics of which are determined by the major climate zone of the earth.

Test cross A genetic cross accomplished for the purpose of determining the genotype of an organism expressing a dominant phenotype; the unknown is crossed with a homozygous recessive.

Testis A gonad in human males that secretes the hormone testosterone, which regulates male secondary sex characteristics; the testis also produces sperm cells for reproduction.

Testosterone A hormone secreted by the testis that regulates the production of male secondary sex characteristics.

Tetrad A grouping of four chromatids that results from synapsis.

Thymine A nitrogenous base found only in DNA.

Thyroid gland An endocrine gland that regulates the body's general rate of metabolism through secretion of the hormone thyroxin.

Thyroid-stimulating hormone (TSH) A pituitary hormone that regulates the secretions of the thyroid gland.

Thyroxin A thyroid hormone that regulates the body's general metabolic rate.

Tongue A structure that aids in the mechanical digestion of foods.

Trachea A cartilage-ringed tube that conducts air from the mouth to the bronchi.

Tracheal tube An adaptation in arthropods (e.g., grasshopper) which functions to conduct respiratory gases from the environment to the moist internal tissues.

Tracheophyta A phylum of the Plant Kingdom whose members (tracheophytes) contain vascular tissues and true roots, stems, and leaves (e.g., geranium, fern, bean, maple tree, corn).

Transfer RNA (t-RNA) A type of RNA that functions to transport specific amino acids from the cytoplasm to the ribosome for protein synthesis.

Translocation A type of chromosome mutation in which a section of a chromosome is transferred to a nonhomologous chromosome.

Transpiration The evaporation of water from leaf stomates.

Transpiration pull A force that aids the upward conduction of materials in the xylem by means of the evaporation of water (transpiration) from leaf surfaces.

Transport The life function by which substances are absorbed, circulated, and released by living things.

Triplet codon A group of three nitrogenous bases that provide information for the placement of amino acids in the synthesis of proteins.

Tropical forest A terrestrial biome characterized by a warm, moist climate and a climax flora that includes many species of broadleaf trees.

Tropism A plant growth response to an environmental stimulus.

Tuber A type of vegetative propagation in which an underground stem (tuber) produces new tubers, each of which is capable of producing new organisms with identical characteristics.

Tundra A terrestrial biome characterized by permanently frozen soil and climax flora that includes lichens and mosses.

Tympanum A receptor organ in arthropods (e.g., grasshopper) which is specialized to detect vibrational stimuli.

Ulcer A disorder of the human digestive tract in which a portion of its lining erodes and becomes irritated.

Ultracentrifuge A tool of biological study that uses very high speeds of centrifugation to separate cell parts for examination.

Umbilical cord In placental mammals, a structure containing blood vessels that connects the placenta to the embryo.

Unicellular Having a body that consists of a single cell (e.g., paramecium).

Uracil A nitrogenous base that is a component part of the nucleotides of RNA molecules only.

Urea A type of nitrogenous waste with moderate solubility and moderate toxicity.

Ureter In human beings, a tube that conducts urine from the kidney to the urinary bladder.

Urethra In human beings, a tube that conducts urine from the urinary bladder to the exterior of the body.

Uric acid A type of nitrogenous waste with low solubility and low toxicity.

Urinary bladder An organ responsible for the temporary storage of urine.

Urine A mixture of water, salts, and urea excreted from the kidney.

Use and disuse A term associated with the evolutionary theory of Lamarck, since proved incorrect.

Uterus In female placental mammals, the organ within which embryonic development occurs.

Vaccination An inoculation of dead or weakened disease organisms that stimulates the body's immune system to produce active immunity.

Vacuole A cell organelle that contains storage materials (e.g., starch, water) housed inside the cell.

Vagina In female placental mammals, the portion of the reproductive tract into which sperm are implanted during sexual intercourse and through which the baby passes during birth.

Variation A concept, central to Darwin's theory of evolution, that refers to the range of adaptation which can be observed in all species.

Vascular tissues Tubelike plant tissues specialized for the conduction of water and dissolved materials within the plant. (See **xylem; phloem.**)

Vegetative propagation A type of asexual reproduction in which new plant organisms are produced from the vegetative (nonfloral) parts of the parent plant.

Vein (human) A relatively thin-walled blood vessel that carries blood from capillary networks back toward the heart.

Vein (plant) An area of vascular tissues located in the leaf that aid the upward transport of water and minerals through the leaf and the transport of dissolved sugars to the stem and roots.

Vena cava One of two major arteries that return blood to the heart from the body tissues.

Ventral nerve cord The main pathway for nerve impulses between the brain and peripheral nerves of the grasshopper and earthworm.

Ventricle One of two thick-walled, muscular chambers of the heart that pump blood out to the lungs and body.

Villi Microscopic projections of the lining of the small intestine that absorb the soluble end-products of digestion. (See **lacteal**.)

Visceral muscle A type of muscle tissue associated with the involuntary movements of internal organs (e.g., peristalsis in the small intestine).

Vitamin a type of nutrient that acts as a coenzyme in various enzyme-controlled reactions.

Water cycle The mechanism by which water is made available to living things in the environment through the processes of precipitation, evaporation, runoff, and percolation.

Water pollution A type of technological oversight that involves the addition of some unwanted factor (e.g., sewage, heavy metals, heat, toxic chemicals) to our water resources.

Watson-Crick model A model of DNA structure devised by J. Watson and F. Crick that hypothesizes a "twisted ladder" arrangement for the DNA molecule.

White blood cell A type of blood cell that functions in disease control. (See **phagocyte; lymphocyte**.)

Xylem A type of vascular tissue through which water and dissolved minerals are transported upward through a plant from the root to the stems and leaves.

Yolk A food substance, rich in protein and lipid, found in the eggs of many animal species.

Yolk sac The membrane that surrounds the yolk food supply of the embryos of many animal species.

Zygote The single diploid cell that results from the fusion of gametes in sexual reproduction; a fertilized egg.

Regents Examinations, Answers, and Student Self-Appraisal Guides

Examination June 2018

Living Environment

PART A

Answer all questions in this part.   [30]

Directions (1–30): For *each* statement or question, record in the space provided the *number* of the word or expression that, of those given, best completes the statement or answers the question.

1 Producers are generally found at the beginning of a food chain. Which statement best explains why this is true?

(1) Producers are usually smaller in size than consumers.
(2) Producers do not rely on other organisms for food.
(3) There are always more consumers than producers in food chains.
(4) Consumers are always more complex organisms than producers.

1 _____

2 A lion cub resembles its parents because it inherits genes that produce

(1) DNA identical to all of the DNA found in both parents
(2) proteins identical to all of the proteins found in both parents
(3) ATP identical to some of the ATP found in each parent
(4) enzymes identical to some of the enzymes found in each parent

2 _____

3 If body temperature is too high, some blood vessels increase in size and sweat glands will excrete sweat, resulting in a lower body temperature. These changes are an example of

(1) a learned behavior   (3) an inherited disorder
(2) feedback mechanisms   (4) genetic mutations

3 _____

4 A farmer grows beans that he sells to local markets. Over a period of 40 years, the farmer has identified the plants that produced the most beans and only used those beans to produce new plants. This procedure is part of the process of

(1) selective breeding   (3) replication
(2) genetic engineering   (4) cloning   4 _____

5 Although we rely on coal, oil, and natural gas to produce energy, some environmental scientists have proposed that we use less fossil fuel. One reason to support this proposal is to

(1) enable us to preserve rain forests in tropical areas
(2) help us to reduce the production of carbon dioxide gas
(3) allow us to decrease the use of fertilizers on crops
(4) encourage us to end research on wind and water power sources   5 _____

6 The diagram below represents relationships in an ecosystem.

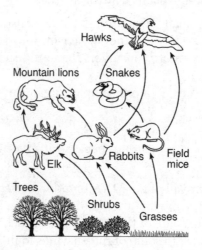

What is the primary source of energy in this environment?

(1) cellular respiration in the plants
(2) energy from minerals in the soil
(3) fossil fuels
(4) solar energy   6 _____

7 Research has shown that treadmill training increases the number of certain energy-releasing structures in the brain cells of rats.

The cellular structures referred to in this study are most likely

(1) mitochondria   (3) vacuoles
(2) nuclei   (4) ribosomes   7 _____

8 Which process must first take place in order for the proteins in foods to be used by body cells?

(1) digestion   (3) synthesis
(2) storage   (4) excretion   8 _____

9 Which statement is characteristic of reproduction in humans?

(1) The reproductive cells of males and females differ in chromosome number.
(2) Males and females produce gametes in the ovaries.
(3) Males and females produce the same number of gametes.
(4) The reproductive cycles of males and females are regulated by hormones.   9 _____

10 Which row in the chart below represents the most likely changes in the atmosphere due to widespread deforestation?

Row	Oxygen Concentration	Carbon Dioxide Concentration
(1)	increases ↑	increases ↑
(2)	increases ↑	decreases ↓
(3)	decreases ↓	increases ↑
(4)	decreases ↓	decreases ↓

10 _____

11 The chart below represents some of the events that occur during the cycling of nutrients in an ecosystem.

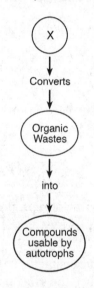

Which organisms would most appropriately complete the chart when written in the circle at *X*?

(1) producers   (3) carnivores
(2) herbivores   (4) decomposers   11 _____

12 The diagram below represents the formation of a cancerous growth.

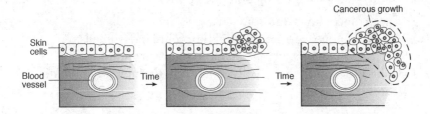

Which statement best explains the events represented in this diagram?

(1) A gene mutation caused the cells to become muscle cells.
(2) The growth resulted from the introduction of a vaccine.
(3) A gene mutation caused abnormal mitotic cell division.
(4) The growth resulted from uncontrolled meiotic cell division.   12 _____

13 A standard laboratory technique used to produce a new plant is represented in the diagram below.

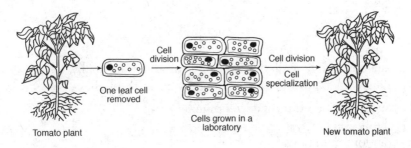

This technique is best identified as

(1) gene alteration   (3) replication
(2) selective breeding   (4) cloning   13 _____

14 An example of competition between members of two different species is

(1) mold growing on a dead tree that has fallen in the forest
(2) purple loosestrife plants growing in the same wet areas as cattail plants
(3) a coyote feeding on the remains of a deer that died of starvation
(4) two male turkeys displaying mating behaviors to attract a female turkey   14 _____

15 Which statement best explains why different body cells of the same individual look and function differently?

  (1) Each cell contains different genes.
  (2) Different genes are activated in different kinds of cells.
  (3) Cells are able to change to adapt to their surroundings.
  (4) Half of the genes in the cells came from the mother and half from the father.   15 _____

16 A diagram of the female reproductive system is shown below.

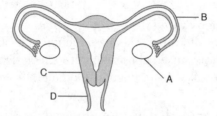

  Identify the structure within which the egg cell is normally fertilized.

  (1) *A*   (3) *C*
  (2) *B*   (4) *D*   16 _____

17 The ameba, a single-celled organism, reproduces asexually. Variations in an ameba would most commonly occur through

  (1) differentiation during development
  (2) the fusion of gametes
  (3) random mutations
  (4) recombination during fertilization   17 _____

18 The development of organs and tissues from a zygote includes

  (1) mitosis and differentiation
  (2) mitosis and gamete production
  (3) meiosis and gamete production
  (4) meiosis and fertilization   18 _____

19 In the 1920s, over 25 million acres of the American southern plains were stripped of prairie grasses to provide more land for farmers to grow wheat. The prairie grasses had served to hold the soil in place and prevent erosion. In the early 1930s, a series of severe dust storms eroded topsoil from more than 13 million acres of the southern plains and dumped it as tons of dust particles over many cities in the Northeast. Farmland was destroyed and people were sickened from "dust pneumonia." This occurrence illustrates that

(1) farmers should never clear land to grow crops as it always creates problems
(2) once an ecosystem has been altered, it cannot be restored to normal
(3) the farmers deliberately altered the equilibrium of the cities in the Northeast
(4) when humans alter ecosystems, serious consequences may result   19 _____

20 The Cornell University News Service reported, "The sugar maple is the most economically valuable tree in the eastern United States because of its high-priced lumber, syrup and tourist-attracting fall colors." The effects of acid rain now threaten the survival of these trees. This threat is the result of a human activity that has

(1) introduced a foreign species by accident
(2) stabilized a forest ecosystem through technology
(3) weakened an ecosystem through pollution
(4) weakened a species by direct harvesting   20 _____

21 The human male reproductive system is adapted for the production of

(1) sperm and the delivery of these cells for internal fertilization
(2) gametes that transport food to the egg
(3) zygotes and the development of these cells into a fetus
(4) hormones that stimulate placenta formation in the male   21 _____

22 The diagram below represents an important biological technique scientists rely on to produce replacement hormones.

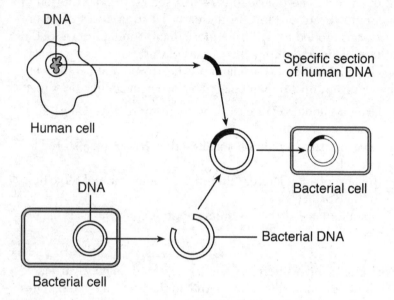

Which two processes are required for the technique to successfully produce hormones?

(1) replication of DNA in bacterial cells and cell division
(2) replication of DNA in bacterial cells and gamete formation
(3) meiosis and development
(4) mitosis and fertilization   22 _____

23 The diagram below summarizes some of the steps in the development of humans.

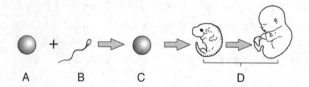

All the genetic information needed for the organism to develop is first present at

(1) A   (3) C
(2) B   (4) D   23 _____

24 Five different living organisms are represented below.

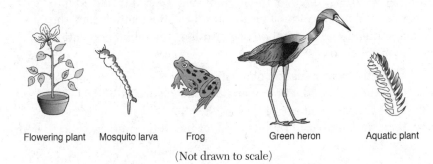

Flowering plant   Mosquito larva   Frog   Green heron   Aquatic plant

(Not drawn to scale)

Which statement about the organisms represented above is correct?

(1) All of the organisms are autotrophs.
(2) Only the flowering plant, green heron, and aquatic plant
     carry out photosynthesis.
(3) Only the frog and green heron can maintain homeostasis.
(4) All of the organisms pass on traits through reproduction.   24 _____

25 Mistletoe is an evergreen shrub that can produce most of its own
    food. Often, mistletoe can be found living on trees and taking
    water and nutrients away from the tissues of the trees.

Mistletoe

The relationship between mistletoe and trees is an example of

(1) consumer/herbivore   (3) scavenger/decomposer
(2) predator/prey   (4) parasite/host   25 _____

26 Rabbits are not native to Australia. They were imported by European settlers. In 1936, the myxoma virus was introduced into Australia as a means of biological control to infect and reduce the rabbit population. This method of controlling the rabbit population was an attempt to

(1) stop the overpopulation of a native species
(2) stop the overproduction of an introduced species
(3) limit the food sources of the rabbit
(4) limit the number of rabbits brought into the country   26 _____

27 The major role of carbohydrates in the human diet is to

(1) form the membranes that surround mitochondria
(2) act as a catalyst for cellular reactions
(3) supply energy for the body
(4) provide building blocks for amino acids   27 _____

28 Throughout New York State, some farmers have switched from growing a variety of vegetable crops to growing a single crop, such as corn. Other farmers are concerned that such a practice will make it more likely that an entire crop could be lost to disease or infestation by an insect pest. This is a valid concern because this practice

(1) reduces the biodiversity of their fields
(2) increases the number of decomposers in their fields
(3) decreases the need to import food
(4) increases the number of invasive species   28 _____

29 The breathing rate, heart rate, and blood hormone levels of an individual would directly provide information about that individual's

(1) cellular organization   (3) inheritance
(2) nutrition   (4) metabolic activity   29 _____

30 The diagram below represents an energy pyramid.

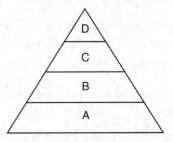

Which type of organism could occupy levels *B*, *C*, and *D* of this energy pyramid?

(1) consumer

(2) producer

(3) autotroph

(4) carnivore

30 ____

PART B–1

Answer all questions in this part. [13]

Directions (31–43): For *each* statement or question, record in the space provided the *number* of the word or expression that, of those given, best completes the statement or answers the question.

31 In the 1920s, two conflicting newspaper headlines called attention to a mysterious new illness.

> 1921–"Don't breathe the air! Mysterious disease affecting thousands is caused by breathing the air in swamps."
>
> 1922–"Don't drink the water! Mysterious disease affecting thousands is caused by drinking the water in swamps."

Another series of headlines appeared in the 1940s and 50s.

> 1945–"New technology finds tiny worms on swamp vegetation."
>
> 1950–"Tiny worms found in lungs of patients suffering from mysterious swamp disease."
>
> 1952–"Mysterious disease known to be caused by worms given name Swamp Lung Disease."

Headlines such as these best illustrate the concept that

(1) scientific explanations are tentative and subject to change
(2) some newspapers are not honest and report incorrect information on purpose
(3) worms can enter the body many different ways
(4) worms found in swamps should not be used for fishing

31 _____

Base your answer to question 32 on the information and diagram below and on your knowledge of biology.

In the early 1600s, a scientist planted a willow tree that weighed 5 pounds in 200 pounds of dry soil. He placed it outside and watered it for 5 years. At the end of that time, he observed that the tree had gained 164 pounds 3 ounces, while the soil had lost just 2 ounces.

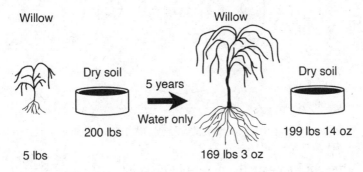

32 From this, he concluded that plants gain weight from the water they take in. His conclusion was based on.

(1) the input of scientists from many countries doing similar studies

(2) the application of advanced technologies to the study of a problem

(3) careful observation, measurements, and inferences from his data

(4) an extensive knowledge of the process of photosynthesis   32 _____

33 A student observed five living cells in the field of view of a micro-scope as represented below.

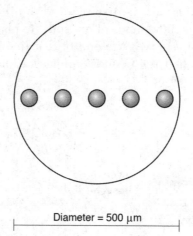

Diameter = 500 μm

What is the approximate diameter of one cell?

(1) 10 μm   (3) 250 μm
(2) 50 μm   (4) 500 μm   33 _____

34 Ecologists are concerned that the golden-winged warbler popula-tion is at a dangerously low level. One reason this could lead to extinction of this warbler is that

(1) after a species becomes extinct, it won't be able to carry out its role in the ecosystem
(2) there may not be enough diversity among the birds for the species to be able to survive an environmental change
(3) extinction always occurs when populations begin to decrease in number
(4) an increase in biodiversity within a population often causes the population to be classified as threatened or endangered   34 _____

35 One primary function of the cell membrane is

(1) regulating the flow of simple sugars into or out of the cell
(2) synthesizing substances by breaking down cell organelles
(3) storing carbohydrates, water, and starches for future use
(4) digesting carbohydrates, fats, and protein   35 _____

36 For several years now, there has been discussion of constructing a large oil pipeline across the United States. Which statement expresses a major concern many people are likely to have about the proposed pipeline?

 (1) The pipeline will bring a large number of jobs to the area where it is being constructed.
 (2) The oil pipeline will increase the amount of finite resources.
 (3) If this pipeline were to leak, the oil could contaminate soil, water, and wildlife.
 (4) The pipeline is a technological fix for ozone depletion.   36 _____

37 The rings in the diagram below represent the annual growth of a tree approximately 20 years old.

Tree trunks grow wider each year by continuous growth in a thin layer of cells just beneath the bark. Since one new layer is added each year, the number of rings in a tree can be used to tell its age. The thickness of the rings provides information about the environmental conditions in past years.

By observing the annual rings in the diagram, one can infer that

 (1) environmental conditions did not change over the last 20 years
 (2) trees grow faster on the side that faces the Sun
 (3) some years provide better conditions for growth than other years
 (4) tree rings are not reliable because trees must be cut down to see them   37 _____

Base your answers to questions 38 through 40 on the information below and on your knowledge of biology.

Harmless Skin Virus Fights Acne

...Acne is caused when hair follicles become blocked with an oily substance called sebum, which the body makes to stop the hair and skin from drying out.

Normally harmless bacteria, such as *Propioni-bacterium acnes*, that live on the skin can then contaminate and infect the plugged follicles.

Phages [a type of virus] appear to help counteract this.

When the scientists sequenced the DNA coding of the phages, they discovered that, as well as sharing most of their genetic material, the viruses all had some key features in common.

All carry a gene that makes a protein called endolysin—an enzyme thought to destroy bacteria by breaking down their cell walls.

And unlike antibiotics, which kill many types of bacteria, including "good" ones that live in our gut, phages are programmed to target only specific bacteria...

Source: BBC News
September 25, 2012

38 This treatment for acne, using phages, is effective because phages

(1) produce antibodies to clean out clogged pores and hair follicles
(2) eliminate bacteria by attacking specific cell structures
(3) carry genes and infect follicles
(4) attack every known type of bacteria   38 ____

39 The protein endolysin belongs to which group of chemical substances?

(1) hormones   (3) biological catalysts
(2) receptors   (4) molecular bases   39 ____

40 The typical response of the human body to an infection by bacteria is to

(1) stimulate the production of antigens
(2) decrease the number of enzymes in the blood
(3) ignore the organisms, unless they are pathogens
(4) produce white blood cells and antibodies   40 _____

41 Two biological processes that occur in certain organelles are represented in the diagrams below.

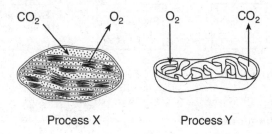

Process X   Process Y

Which statement is correct regarding the types of organisms able to carry out these processes?

(1) Process X occurs in heterotrophs, but not in autotrophs.
(2) Process Y occurs in consumers, but not in producers.
(3) Both processes X and Y occur in all living things.
(4) Both processes X and Y occur in green plants.   41 _____

Base your answers to questions 42 and 43 on the diagram below and on your knowledge of biology. The diagram shows how ATP is used by some cell structures to perform various functions.

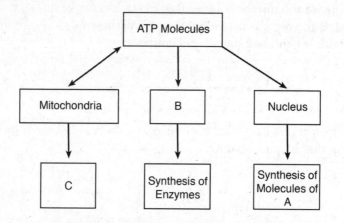

42 Which cell structure is represented by *B*?

    (1) vacuole   (3) cytoplasm

    (2) ribosome   (4) chloroplast   42 _____

43 The nucleus contains molecules of *A*, which

    (1) recycle waste products

    (2) remove water from the cell

    (3) store hereditary information

    (4) regulate the pH of cytoplasm   43 _____

PART B–2
Answer all questions in this part. [12]

Directions (44–55): For those questions that are multiple choice, record your answers in the space provided. For all other questions in this part, record your answers in accordance with the directions.

Base your answers to questions 44 through 47 on the information and data table below and on your knowledge of biology.

Moose-killing Winter Ticks

Moose habitat is determined by temperature. Moose prefer areas where the average summer temperature is around 15°C and does not exceed 27°C for too long. The reason for this temperature dependency: Moose cannot sweat.

Besides the cooling effect of water, which moose are almost always near, aquatic environments provide them with a good supply of food, and in the past, have protected them against biting insects. However, the North American moose population is facing a new threat: a parasite called the winter tick. These ticks lodge themselves in the animal's fur and hold on through the winter, sucking the animal's blood. Many infected moose end up dying of exhaustion and weakness as a result of the large number of ticks feeding on them.

Ticks are most active during dry days in the fall. Adult ticks that drop off moose in the spring and land on snow cover have a poorer survival rate. Climate change can be predicted to improve conditions for winter ticks due to longer and warmer falls, and earlier snowmelt in the spring.

Surveys of the moose population in northeastern Minnesota have recorded the change shown below in the moose population between 2005 and 2013.

Estimated Moose Population
In Northeastern Minnesota

Survey Year	Estimated Moose Population
2005	8160
2006	8840
2007	6860
2008	7890
2009	7840
2010	5700
2011	4900
2012	4230
2013	2760

Directions (44–45): Using the information in the data table, construct a line graph on the grid below, following the directions below.

44 Mark an appropriate scale, without any breaks in the data, on the axis labeled "Estimated Moose Population."   [1]

45 Plot the data for the estimated moose population on the grid. Connect the points and surround each point with a small circle.   [1]

Example:

**Estimated Moose Population
in Northeastern Minnesota**

46 Explain how climate change could result in an increased number of moose infested with winter ticks.   [1]

_____

_____

Note: The answer to question 47 should be recorded in the space provided.

47 Increased average yearly temperatures in regions presently inhabited by moose could result in a disruption in homeostasis in these
animals because

(1) a decrease in average temperatures will increase mutations
in their skin cells
(2) an increase in average temperatures will decrease the
amount of blood ticks can consume
(3) moose will not be able to maintain an appropriate body
temperature, since they do not sweat
(4) moose will sweat more and lose too much water from their
bodies

47 _____

Base your answers to questions 48 and 49 on the information and diagram
below and on your knowledge of biology. The diagram represents the evolutionary relationships among many organisms.

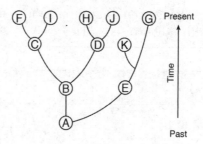

48 An environmental change severely affected the organism represented by species *K*. What was the result? Support your
answer. [1]

_____

_____

_____

Note: The answer to question 49 should be recorded in the space provided.

49 Three species with the most similar traits are most likely

(1) *F, I, G*   (3) *B, D, G*

(2) *D, H, J*   (4) *F, A, J*   49 _____

Base your answers to questions 50 and 51 on the diagram below and on your knowledge of biology. The diagram represents trophic levels in an ocean environment.

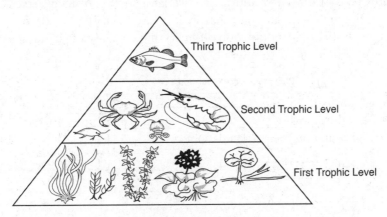

Note: The answer to question 50 should be recorded in the space provided.

50 The organisms found at the second trophic level of this pyramid would be

(1) producers   (3) carnivores

(2) decomposers   (4) herbivores   50 _____

51 State *one* reason why there is less energy available at each trophic
level going from the first to the third trophic level.   [1]

_____

_____

_____

52 Stable predator-prey relationships are necessary to maintain a
healthy ecosystem. The removal of a predator species from an area
caused the deer population to sharply increase from 1910 to 1925.
Changes in the deer population and carrying capacity of the area
are represented in the graph below.

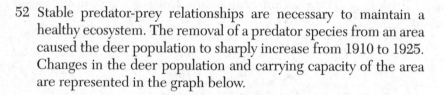

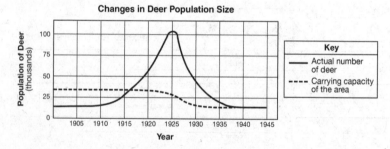

Based on the information provided, explain how the sharp population increase from 1910 to 1925 might have resulted in the
decrease in the carrying capacity after 1925.   [1]

_____

_____

Base your answers to questions 53–54 on the diagram below and on your knowledge of biology. The diagram indicates a change in an ecosystem.

53–54 Identify some of the key events associated with the change. In your answer, be sure to:

- identify *one* natural event that could cause the disruption indicated in the diagram   [1]
- state what would most likely happen to the new stable ecosystem in future years if no further disruptions occur   [1]

_____

_____

_____

_____

_____

_____

_____

55 Explain why a mutation that occurs in a body cell will *not* contribute to the evolution of a species.   [1]

_____

_____

PART C
Answer all questions in this part. [17]

Directions (56–72): Record your answers in the spaces provided.

Base your answers to questions 56 through 58 on the passage below and on your knowledge of biology.

Our [Nitrogen] Fertilized World

It is the engine of agriculture, the key to plenty in our crowded, hungry world....

...Enter modern chemistry. Giant factories capture inert nitrogen gas from the vast stores in our atmosphere and force it into a chemical union with the hydrogen in natural gas, creating the reactive compounds that plants crave. That nitrogen fertilizer—more than a hundred million tons applied worldwide every year—fuels bountiful harvests. Without it, human civilization in its current form could not exist. Our planet's soil simply could not grow enough food to provide all seven billion of us our accustomed diet. In fact, almost half of the nitrogen found in our bodies' muscle and organ tissue started out in a fertilizer factory.

Source: National Geographic, May 2013

56 Nitrogen fertilizers are used by plants to synthesize amino acids. State *one* reason why a supply of amino acids is important for the survival of complex organisms. [1]

_____

_____

57 Identify *one* possible effect on the human population if nitrogen fertilizers were not available. [1]

_____

_____

58 Explain how the building of factories to produce fertilizer is an example of a trade-off. [1]

_____

_____

_____

_____

Base your answers to questions 59–60 on the information and diagram below and on your knowledge of biology.

An experiment was carried out to determine the effect of exposure to ultraviolet (UV) light on the growth of bacteria. Equal quantities of bacterial cells were spread on Petri dishes that are used to grow colonies of bacteria. Half of each dish was shielded from the UV light with a UV screen. The other half was exposed to UV light for various amounts of time. After the UV treatment, the bacteria were grown in an incubator for 24 hours and the number of colonies was counted.

The diagram below represents the setup of the experiment.

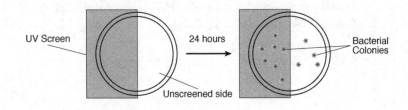

The table below contains the data collected at different exposure times by counting the number of bacterial colonies on both the screen-covered side and unscreened side.

Bacterial Growth

Exposure Time to UV Light (min)	Colonies on Screened Side	Colonies on Unscreened Side
0 (No exposure)	20	22
0.5	21	19
1.0	23	16
2.0	22	10
5.0	24	5
10.0	23	1

59–60 Analyze the experiment that produced the data in the table. In your answer, be sure to:

- state a hypothesis for the experiment   [1]
- state whether the results of the experiment support or fail to support your hypothesis. Support your answer   [1]

_____

_____

_____

_____

Base your answers to questions 61 and 62 on the information below and on your knowledge of biology.

Evolution leads to changes in how frequently certain traits appear in a population.

61 Explain the importance of the presence of variations within a population.   [1]

_____

_____

_____

62 Describe how the process of natural selection can result in an increase in frequencies of certain traits found in a population.   [1]

_____

_____

_____

_____

_____

Base your answers to questions 63 through 65 on the information below and on your knowledge of biology.

A typical human liver cell can have over 90,000 insulin receptors. Due to a genetic difference, some people have liver cells that contain only about 1000 insulin receptors.

63 Describe the importance of receptors in cellular communication.   [1]

_____

_____

64 Describe the importance of the shape of receptor molecules for carrying out their function. [1]

_____

_____

65 Identify *one* effect a reduced number of insulin receptors might have on an individual. [1]

_____

_____

_____

Base your answers to questions 66 through 68 on the passage below and on your knowledge of biology. Biologists have been studying the genes present in newborn twins.

Twins Don't Share Everything

...Chemicals called epigenetic markers can be attached to those [inherited] genes, like flags or balloons hanging off the sides of the DNA ladder. These don't just change the look of the genes. Like pieces of tape stuck over a light switch, these markers can force a gene to remain turned on or off. The type of marker scientists studied in the twins generally sticks the switch in the off position so that some proteins don't get made. And that means the proteins' jobs won't get done.

Every time a cell divides, new epigenetic markers may form. Foods, pollutants, and stress may all contribute to the development of new markers. So throughout our lives we tend to accumulate more and more. But a few are there from the day we're born.

...His [Jeffrey Craig's] team found that newborn twins have markers attached to different genes from the very start. It's true in identical twins, which come from the same fertilized egg. It's also true in fraternal twins, which come from different fertilized eggs. However, fraternal twins had more such differences than identical twins did.

Source: Science News for Students; July 31, 2012

66 Explain why the genetic material in an offspring produced by sexual reproduction contains genetic material that is *not* identical to the genetic material of either parent. [1]

_____

_____

67 State *one* reason why identical twins should have fewer genetic differences than fraternal twins. [1]

_____

_____

68 Identify *two* environmental factors that can lead to an increase in the number of epigenetic markers that modify gene expression. [1]

_____ and _____

_____

Base your answers to questions 69 and 70 on the information below and on your knowledge of biology.

With the emotional roller coaster that pregnancy brings, it can be daunting [challenging] for pregnant women to take on the additional pressure of eating the "perfect" pregnancy diet. The good news: there is no single perfect diet for pregnancy. The best way for expectant mothers to meet their nutritional needs is to focus on consuming an overall healthy diet, with a variety of vegetables, fruits, whole grains, lean meats or meat substitutes, and low-fat dairy or dairy substitutes....

Source: US News Health 11/9/2012

69 Describe how nutrients move from the mother to the fetus. [1]

_____

_____

70 State *one* other way, in addition to consuming a balanced diet, pregnant women can help ensure proper development of the fetus.   [1]

_____

_____

_____

Base your answer to question 71 on the information below and on your knowledge of biology.

...Bacteria often evolve clever ways of evading chemical assaults, but they will always struggle to resist the old-fashioned way of killing them; heating them up. It takes only a relatively mild warming to kill bugs [bacteria] without discomfort or harm to tissues. So imagine if little electric heaters could be implanted into wounds and powered wirelessly to fry bacteria during healing before dissolving harmlessly into body fluids once their job is done....

Source: BBC Future, May 24, 2013

71 State *one* way the use of these new "little electric heaters" might represent a long-term benefit over using antibiotics to treat bacterial infections.   [1]

_____

_____

_____

72 A child became ill with the measles. Measles is a disease that is highly contagious. The child's mother did not get sick, even though she and the child were close while the child was ill. State *one* reason why the mother did not get sick with the measles.   [1]

_____

PART D
Answer all questions in this part.   [13]

Directions (73–85): For those questions that are multiple choice, record your answer in the space provided. For all other questions in this part, record your answers in accordance with the directions.

Note: The answer to question 73 should be recorded in the space provided.

73 The diagram below represents the major parts of a growing onion plant. Nutrients are represented in the soil around the onion.

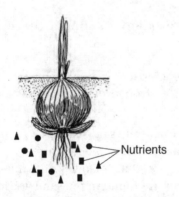

Which statement best describes how nutrients enter the root cells of the onion plant?

(1) Only nutrients needed by the plant enter root cells.
(2) The nutrients usually move from an area of high concentration in the soil to an area of low concentration in root cells.
(3) Nutrients always move into the plant cells by active transport.
(4) The nutrients always move from an area of low concentration in the soil to an area of high concentration in root cells.   73 ____

Note: The answer to question 74 should be recorded in the space provided.

74 Which concept is correctly matched with an example from *The Beaks of Finches* lab?
- (1) Variation—different "beaks" were available.
- (2) Adaptation—different types of foods were available.
- (3) Selecting Agent—an insecticide was used to kill insects on one island.
- (4) Environment—"beaks" with similar qualities were used to gather seeds.   74 _____

Note: The answer to question 75 should be recorded in the space provided.

75 When comparing characteristics of two organisms, which evidence would be considered the strongest for supporting a possible evolutionary relationship?
- (1) The two organisms are the same color.
- (2) The two organisms are the same height.
- (3) The two organisms produce many of the same proteins.
- (4) The two organisms are found in the same locations.   75 _____

Note: The answer to question 76 should be recorded in the space provided.

76 A and B below represent two different slide preparations of elodea leaves. Elodea is a plant found in streams and ponds in New York State.

The water used on slide A contained 1% salt and 99% water.
The salt solution used on slide B contained 6% salt and 94% water.
Elodea cells normally contain 1% salt.

A

Elodea leaf mounted in 1% salt solution

B

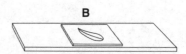

Elodea leaf mounted in 6% salt solution

Five minutes after the slides were prepared, a student using a compound light microscope to observe the cells in leaves *A* and *B* would most likely see that

(1) water had moved out of the cells of the leaf on slide *A*
(2) salt had moved into the cells of the leaf on slide *A*
(3) water had moved out of the cells of the leaf on slide *B*
(4) salt had moved out of the cells of the leaf on slide *B*   76 _____

77 The table below shows the food sources for two different species of Galapagos finches on an island.

Two Galapagos Finches and Their Sources of Food

Name	Foods
Vegetarian finch *Platyspiza crassirostris*	Buds, leaves, fruit of trees
Cactus finch *Geospiza scandens*	Cactus flowers and nectar

State *one* reason why these two species probably do *not* live in the same area of this island.   [1]

_____

_____

_____

Base your answers to questions 78 and 79 on the information below and on your knowledge of biology.

During a lab activity, a 14-year-old student took his resting pulse rate. He counted 20 beats in 20 seconds. He calculated his pulse rate for a minute and compared the result to the data shown in the table below.

Normal Pulse Rate Ranges

Age Group	Resting Heart Rate (beats per minute)
Children (ages 6-15)	70 – 100
Adults (ages 18 and over)	60 – 100

78 According to the data table, does the student's pulse rate fall within the normal range? Circle yes or no and support your answer. [1]

yes　　　　　　　　no

_____

_____

79 Using a biological explanation, state *one* reason why a person's heart rate increases during exercise. [1]

_____

_____

80 The chart below shows the molecular comparison between several species.

Molecular Comparison Chart

Botana curus	DNA	GTG	GAC	TGA	GGA	CTC
	mRNA	CAC	CUG	ACU	CCU	GAG
	Amino acid	His	Leu	Thr	Pro	Glu

Species X	DNA	GTG	GAC	AGA	GGA	CAC
	mRNA	CAC	CUG	UCU	CCU	GUG
	Amino acid	His	Leu	Ser	Pro	Val

Species Y	DNA	GTG	GAC	AGA	GGA	CAC
	mRNA	CAC	CUG	UCU	CCU	GUG
	Amino acid	His	Leu	Ser	Pro	Val

Species Z	DNA	GTA	GAC	TGA	GGA	CTC
	mRNA	CAU	CUG	ACU	CCU	GAG
	Amino acid	His	Leu	Thr	Pro	Glu

Identify which species is likely to be more closely related to *Botana curus*. Support your answer.   [1]

Species: _____

Support: _____

_____

Note: The answer to question 81 should be recorded in the space provided.

81 A factor that contributed to the evolution of finches on the Galapagos Islands was most likely the

(1) lack of variation in beak structure of the finches
(2) isolation of the finches on separate islands
(3) relatively constant atmospheric temperature
(4) total lack of competition for food   81 _____

Base your answers to questions 82 and 83 on the diagram below and on your knowledge of biology. The diagram represents a laboratory setup.

A starch solution in a test tube was separated from the water in a beaker by a dialysis membrane. One hour later, it was observed that the liquid had risen in the test tube.

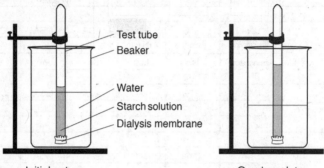

Initial set-up   One hour later

Note: The answer to question 82 should be recorded in the space provided.

82 The rise of the liquid in the test tube that was observed after one hour can be explained as a result of the

   (1) starch solution moving into the test tube and out of the beaker
   (2) water moving from the beaker into the test tube
   (3) large starch molecules blocking the dialysis membrane
   (4) dialysis membrane acting as a barrier to the water molecules   82 _____

83 If a starch indicator solution was initially added to the water in the beaker, describe *one* observation that would be made after one hour.   [1]

_____

_____

84 The diagram below represents an electrophoresis gel that was used to separate DNA fragments. Lanes 1, 2, and 3 contain DNA samples that were treated with the same restriction enzyme.

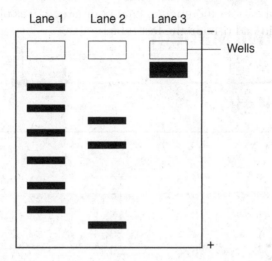

Explain why the DNA sample in lane 3 did *not* separate into fragments.   [1]

_____

_____

85 An experiment is performed to determine the effect of watching basketball games on pulse rates. Ten students agreed to wear devices that monitor pulse rates while watching a basketball game between competitive opponents. Their pulse rates were measured every minute for five minutes in the first quarter of the game. The data collected indicated that pulse rates did not change significantly during the monitored period. State *one* way that this experiment could be improved to obtain a valid conclusion.   [1]

_____

_____

Answers
June 2018
Living Environment

Answer Key

PART A

1. 2	6. 4	11. 4	16. 2	21. 1	26. 2
2. 4	7. 1	12. 3	17. 3	22. 1	27. 3
3. 2	8. 1	13. 4	18. 1	23. 3	28. 1
4. 1	9. 4	14. 2	19. 4	24. 4	29. 4
5. 2	10. 3	15. 2	20. 3	25. 4	30. 1

PART B–1

31. 1	34. 2	36. 3	38. 2	40. 4	42. 2
32. 3	35. 1	37. 3	39. 3	41. 4	43. 3
33. 2					

PART B–2

44. *See* Answers Explained.
45. *See* Answers Explained.
46. *See* Answers Explained.
47. 3
48. *See* Answers Explained.
49. 2

50. 4
51. *See* Answers Explained.
52. *See* Answers Explained.
53. *See* Answers Explained.
54. *See* Answers Explained.
55. *See* Answers Explained.

PART C *See* **Answers Explained.**

PART D

73. 2
74. 1
75. 3
76. 3
77. *See* Answers Explained.
78. *See* Answers Explained.
79. *See* Answers Explained.

80. *See* Answers Explained.
81. 2
82. 2
83. *See* Answers Explained.
84. *See* Answers Explained.
85. *See* Answers Explained.

Answers Explained

PART A

1. **2** *Producers do not rely on other organisms for food* is the statement that best explains why producers are generally found at the beginning of a food chain. Food chains are graphical representations of simple nutritional relationships commonly found in nature. Food chains illustrate such nutritional relationships in terms of the transfer of energy from producers (green plants), which manufacture their own food via photosynthesis, to sequential levels of consumer organisms. For this reason, producers are generally represented at the beginning of all food chains.

WRONG CHOICES EXPLAINED:

(1) *Producers are usually smaller in size than consumers* is *not* the statement that best explains why producers are generally found at the beginning of a food chain. Many producers are far larger in size (e.g., redwood trees, oak trees) than the consumers (e.g., elk, whitetail deer) that inhabit the same environment. Relative body size is not a criterion for placement in a food chain representation.

(3) *There are always more consumers than producers in food chains* is *not* the statement that best explains why producers are generally found at the beginning of a food chain. In fact in any stable environment, producers always far outnumber consumers by a factor of 10:1 or greater. The relative number of organisms is not a criterion for placement in a food chain representation.

(4) *Consumers are always more complex organisms than producers* is *not* the statement that best explains why producers are generally found at the beginning of a food chain. Many producers are far more complex than the consumers with which they interact (e.g., white oak tree (producer) vs. oak wilt fungus (consumer)). Relative body complexity is not a criterion for placement in a food chain representation.

2. **4** A lion cub resembles its parents because it inherits genes that produce *enzymes identical to some of the enzymes found in each parent.* According to the one gene–one enzyme hypothesis, each functional gene in the genome of an organism is responsible for the synthesis of a single unique enzyme. The role of each of these enzymes is the catalysis of a single biochemical reaction, resulting in a single genetic phenotype. Since a lion cub receives half of its genetic information from each of its parents, the enzymes produced in that cub should represent half of the enzymes produced by each parent.

WRONG CHOICES EXPLAINED:

(1), (2) It is *not* true that a lion cub resembles its parents because it inherits genes that produce *DNA identical to all of the DNA found in both parents* or *proteins identical to all of the proteins found in both parents*. Sexual reproduction ensures that offspring receive exactly half, not all, of the genes available in each parent's genome. By extension, this principle means that half of the proteins found in an offspring may be traced back to one of its parents and the other half of the proteins may be traced back to the other parent.

(3) It is *not* true that a lion cub resembles its parents because it inherits genes that produce *ATP identical to some of the ATP found in each parent*. ATP (adenosine triphosphate) is a biochemical found in most, if not all, living organisms. It has the function of transferring cellular energy from one biochemical reaction to another. The chemical form of ATP is identical in all living things.

3. **2** These changes are an example of *feedback mechanisms*. In this example, the human body senses its own temperature and initiates responses to modify its temperature. When the body's temperature is reduced to an appropriate level, the body senses this also and the responses are reversed. This stimulus-response (feedback) mechanism helps the body to stabilize its temperature within a relatively narrow range that promotes the survival of the individual.

WRONG CHOICES EXPLAINED:

(1) These changes are *not* an example of *a learned behavior*. Feedback mechanisms are reflex reactions that are beneficial inherited traits, not learned behaviors.

(3) These changes are *not* an example of *an inherited disorder*. Feedback mechanisms are reflex reactions that are beneficial inherited traits, not disorders.

(4) These changes are *not* an example of *genetic mutations*. Feedback mechanisms are reflex reactions that are beneficial inherited traits, not mutations.

4. **1** This procedure is part of the process of *selective breeding*. Selective breeding is a technique used by animal and plant breeders in which organisms with desirable traits are cross-bred with the hope of producing offspring that also display those traits. The resulting offspring would theoretically be capable of passing on these desirable traits to future generations.

WRONG CHOICES EXPLAINED:

(2) This procedure is *not* part of the process of *genetic engineering*. Genetic engineering is a laboratory technique in which a gene for a desired trait is snipped from the DNA of a donor cell and inserted into the genome of a recipient cell. This technique is not used by traditional farmers to select the best beans to plant to produce better bean crops.

(3) This procedure is *not* part of the process of *replication*. Replication is the exact self-duplication of DNA that occurs in the processes of mitotic and meiotic cell division. This technique is not used by traditional farmers to select the best beans to plant to produce better bean crops.

(4) This procedure is *not* part of the process of *cloning*. Cloning is a laboratory technique that involves the removal of a diploid nucleus from a donor organism and its insertion into a denucleated fertilized egg cell of the same or related species. This nucleus contains the DNA of the two parents of the donor. This technique is not used by traditional farmers to select the best beans to plant to produce better bean crops.

5. **2** One reason to support this proposal is to *help us to reduce the production of carbon dioxide gas*. Fossil fuels, such as coal, oil, and natural gas, are naturally occurring hydrocarbon substances that contain an abundance of stored energy. When burned, this stored energy is released. The burning of fossil fuels has been used for hundreds or thousands of years to power human industrial processes. It also produces polluting by-products, such as carbon dioxide, particulate matter, and sulfur and nitrogen oxides. All of these by-products pose a threat to the worldwide environment and to human health. Proposals aimed at reducing our use of fossil fuels help to reduce these threats.

WRONG CHOICES EXPLAINED:

(1) One reason to support this proposal is *not* to *enable us to preserve rain forests in tropical areas*. In fact, the burning of fossil fuels and its resulting pollution have been scientifically linked to global climate change. Pollution that occurs in the United States can easily diffuse into the atmosphere and cause unanticipated changes in the global climate. We can no longer afford to assume that pollution produced locally affects only the local environment. Instead, we should understand that it can and does adversely affect the global environment, including tropical rain forests.

[NOTE: This is a potential correct answer and may even be the best answer to the question as written.]

(3) One reason to support this proposal is *not* to *allow us to decrease the use of fertilizers on crops*. Crops grown on commercial farms require the judicious application of nutrient fertilizer in order to maximize production potential. The production of such fertilizers does not directly depend on the burning of fossil fuels. This proposal would probably not negatively affect the use of agricultural fertilizers.

(4) One reason to support this proposal is *not* to *encourage us to end research on wind and water power sources*. Wind and water power sources provide economically viable alternatives to the burning of fossil fuels. Research and development of alternative energy sources should be supported, not discouraged, by proposals such as this one. This proposal would probably not negatively affect the future of research on wind and water power sources.

6. **4** The primary source of energy in this environment is *solar energy*. Food webs are graphical representations of interactive nutritional relationships found commonly in nature. Food webs illustrate such relationships in terms of the transfer of energy from producers (green plants), which manufacture their own food via photosynthesis, to sequential levels of consumer organisms. In turn, the photosynthetic reactions depend on the energy of the sun (solar energy) to supply the energy needed to forge the chemical bonds in glucose. In this way, solar energy is known to be the primary source of energy in all natural environments.

WRONG CHOICES EXPLAINED:
(1) It is *not* true that the primary source of energy in this environment is *cellular respiration in the plants*. Cellular respiration in plants is a biochemical process by which the energy stored in the chemical bonds of glucose molecules is released in a controlled manner and is transferred to molecules of ATP for use in other biochemical processes. Although essential to the survival of individual plants, cellular respiration does not provide primary energy to green plants or to the environment in any circumstance.

(2) It is *not* true that the primary source of energy in this environment is *energy from minerals in the soil*. Soil minerals supply nitrogen, potassium, and phosphorus compounds to plants and are essential to their healthy growth. Soil minerals do not provide primary energy to green plants or to the environment in any circumstance.

(3) It is *not* true that the primary source of energy in this environment is *fossil fuels*. Fossil fuels, such as coal, oil, and natural gas, are naturally occurring hydrocarbon substances that contain an abundance of stored energy but not in a form that can be accessed by green plants. Fossil fuels do not provide primary energy to green plants or to the environment in any circumstance.

7. **1** The cellular structures referred to in this study are most likely *mitochondria*. Mitochondria are cell organelles that contain the enzymes that catalyze the respiratory reactions. These reactions are responsible for releasing the stored energy in glucose for use in other cellular reactions.

WRONG CHOICES EXPLAINED:
    (2) The cellular structures referred to in this study are *not* most likely *nuclei*. Nuclei are cell organelles that contain the genetic material in the form of chromosomes. Nuclei and their contents are not directly responsible for releasing the stored energy in glucose for use in other cellular reactions.
    (3) The cellular structures referred to in this study are *not* most likely *vacuoles*. Vacuoles are cell organelles that store nutrients or cellular wastes. Vacuoles are not directly responsible for releasing the stored energy in glucose for use in other cellular reactions.
    (4) The cellular structures referred to in this study are *not* most likely *ribosomes*. Ribosomes are cell organelles that serve as platforms for protein synthesis in cells. Ribosomes are not directly responsible for releasing the stored energy in glucose for use in other cellular reactions.

8. **1** *Digestion* is the process that must first take place in order for the proteins in foods to be used by body cells. The digestive reactions convert complex proteins to amino acids. These amino acids can be absorbed into the bloodstream, transported throughout the body, and delivered to cells for the process of protein synthesis.

WRONG CHOICES EXPLAINED:
    (2) *Storage* is *not* the process that must first take place in order for the proteins in foods to be used by body cells. Storage is a general term that relates to the temporary or permanent placement of materials in a defined space.
    (3) *Synthesis* is *not* the process that must first take place in order for the proteins in foods to be used by body cells. Synthesis is a term that describes the formation of complex molecules from simple subcomponents.
    (4) *Excretion* is *not* the process that must first take place in order for the proteins in foods to be used by body cells. Excretion is a term that describes the elimination of toxic metabolic waste from the body.

9. **4** *The reproductive cycles of males and females are regulated by hormones* is the statement that is characteristic of reproduction in humans. In human females, the hormones estrogen, progesterone, follicle-stimulating hormone (FSH), and luteinizing hormone (LH), among others, are integral to the formation, fertilization, and maintenance of egg cells. In human males, the hormone testosterone is integral to the formation of sperm cells.

WRONG CHOICES EXPLAINED:

(1) *The reproductive cells of males and females differ in chromosome number* is *not* the statement that is characteristic of reproduction in humans. The reproductive cells of both male and female humans carry the haploid (*n*) number of chromosomes. In humans, the haploid number of chromosomes is 23.

(2) *Males and females produce gametes in the ovaries* is *not* the statement that is characteristic of reproduction in humans. Female humans produce egg cells in the ovaries. Male humans produce sperm cells in the testes.

(3) *Males and females produce the same number of gametes* is *not* the statement that is characteristic of reproduction in humans. Males produce functional sperm cells at a rate approximately four times greater than the rate at which females produce functional egg cells.

10. **3** Row *3* in the chart represents the most likely changes in the atmosphere due to widespread deforestation. Widespread deforestation reduces the populations of producer organisms that are known to consume carbon dioxide gas and produce oxygen gas in the process of photosynthesis. The most likely result of such deforestation is a decrease in atmospheric oxygen content and an increase in atmospheric carbon dioxide content.

WRONG CHOICES EXPLAINED:

(1), (2), (4) Rows *1*, *2*, and *4* in the chart do *not* represent the most likely changes in the atmosphere due to widespread deforestation. Each of these rows contains information inconsistent with the correct answer shown above.

11. **4** *Decomposers* are the organisms that would most appropriately complete the chart when written in the circle at *X*. In every healthy natural environment, decomposers such as soil bacteria and fungi are active in reducing complex organic matter to simple components that can be absorbed by green plants as fertilizing nutrients.

WRONG CHOICES EXPLAINED:

(1) *Producers* are *not* the organisms that would most appropriately complete the chart when written in the circle at *X*. Producers (green plants) are autotrophic organisms that manufacture their own food by means of photosynthesis. Producers do not convert organic wastes into compounds usable by autotrophs.

(2) *Herbivores* are *not* the organisms that would most appropriately complete the chart when written in the circle at *X*. Herbivores (plant eaters) consume green plants growing in their environment, not dead and decomposing plant matter such as dead leaves and trees. Herbivores do not convert organic wastes into compounds usable by autotrophs.

(3) *Carnivores* are *not* the organisms that would most appropriately complete the chart when written in the circle at X. Carnivores (flesh eaters) consume prey animals in their environment, not dead and decomposing plant matter such as dead leaves and trees. Carnivores do not convert organic wastes into compounds usable by autotrophs.

12. **3** *A gene mutation caused abnormal mitotic cell division* is the statement that best explains the events represented in the diagram. Skin cells are known to reproduce themselves via mitotic cell division. In the presence of mutagenic agents, this cell division may become abnormal, resulting in the formation of a cancerous growth. Abnormal skin growths, such as moles and scabs, should be checked by a doctor to ensure that they are not cancerous growths.

WRONG CHOICES EXPLAINED:
(1) *A gene mutation caused the cells to become muscle cells* is *not* the statement that best explains the events represented in the diagram. Gene mutations are random events that may introduce small changes to the genome of an affected cell. Gene mutations are not capable of converting one cell type (skin) to a different cell type (muscle).
(2) *The growth resulted from the introduction of a vaccine* is *not* the statement that best explains the events represented in the diagram. Vaccines are substances created in laboratories and are targeted to provide immunity to specific microbial diseases. Vaccines are not known to promote cancer.
(4) *The growth resulted from uncontrolled meiotic cell division* is *not* the statement that best explains the events represented in the diagram. The diagram illustrates an occurrence in the skin. Meiotic cell division occurs only in the ovaries or the testes, not in the skin.

13. **4** This technique is best identified as *cloning*. Plant cloning is a laboratory technique that involves the removal of a viable diploid cell from a donor organism and its growth in laboratory culture to produce a genetically identical "offspring" organism.

WRONG CHOICES EXPLAINED:
(1) This technique is *not* best identified as *gene alteration*. Gene alterations (mutations) are random events in which genetic material is altered through the effects of mutagenic agents including radiation. Gene alterations are not represented in this diagram as a causative agent.
(2) This technique is *not* best identified as *selective breeding*. Selective breeding is a technique used by animal and plant breeders in which organisms with desirable traits are cross-bred with the hope of producing offspring that

also display those traits. Selective breeding is not represented in this diagram as a causative agent.

(3) This technique is *not* best identified as *replication*. Replication is the exact self-duplication of DNA that occurs in the processes of mitotic and meiotic cell division. Except for the fact that replication occurs each and every time a cell divides by mitotic cell division, replication is not represented in this diagram as a causative agent.

14. **2** An example of competition between members of two different species is *purple loosestrife plants growing in the same wet areas as cattail plants*. In this example, the invasive purple loosestrife plant is in direct interspecies (between different species) competition with the native cattail plant because the two species have very similar environmental requirements. In this competition, purple loosestrife often has the advantage as it has few natural enemies in its adopted North American environment.

WRONG CHOICES EXPLAINED:

(1) It is *not* true that an example of competition between members of two different species is *mold growing on a dead tree that has fallen in the forest*. This is an example of the actions of a decomposer.

(3) It is *not* true that an example of competition between members of two different species is *a coyote feeding on the remains of a deer that died of starvation*. This is an example of scavenging behavior.

(4) It is *not* true that an example of competition between members of two different species is *two male turkeys displaying mating behaviors to attract a female turkey*. This is an example of intraspecies (within a species) competition.

15. **2** *Different genes are activated in different kinds of cells* is the statement that best explains why different body cells of the same individual look and function differently. Differentiation is the process by which embryonic cells undergo specialization to become specific body tissues. Differentiation is accomplished when genes relating to a particular tissue type are switched on while those not related to that particular tissue type are switched off. This process allows each tissue type to function at maximum efficiency within the body.

WRONG CHOICES EXPLAINED:

(1) *Each cell contains different genes* is *not* the statement that best explains why different body cells of the same individual look and function differently. All somatic (body) cells of an individual organism contain exactly the same genetic complement.

(3) *Cells are able to change to adapt to their surroundings* is *not* the statement that best explains why different body cells of the same individual look and function differently. Although cells are able to monitor and respond to stimuli in their surroundings, they do not adapt their specialized functions as a result.

(4) *Half of the genes in the cells came from the mother and half from the father* is *not* the statement that best explains why different body cells of the same individual look and function differently. Although this is a true statement, it does not explain the process of differentiation as defined above.

16. **2** Structure *B* is the structure within which the egg cell is normally fertilized. Structure *B* is the oviduct (Fallopian tube). It transports a mature egg from the ovary (structure *A*) to the uterus (structure *C*) during the human female reproductive cycle. Sperm cells implanted in the vagina (structure *D*) travel upward through the female reproductive tract and normally encounter and fertilize the egg cell within the oviduct.

WRONG CHOICES EXPLAINED:
(1) Structure *A* is *not* the structure within which the egg cell is normally fertilized. Structure *A* is the ovary, within which the egg cell matures and from which it is released into the oviduct (structure *B*) following ovulation.

(3) Structure *C* is *not* the structure within which the egg cell is normally fertilized. Structure *C* is the uterus, within which the fertilized egg cell develops into a fetus prior to birth via the vagina (structure *D*).

(4) Structure *D* is *not* the structure within which the egg cell is normally fertilized. Structure *D* is the vagina (birth canal), within which the sperm cells are implanted prior to fertilization and through which the fetus passes during birth.

17. **3** Variations in an ameba would most commonly occur through *random mutations*. Because the ameba is a single-celled organism that reproduces via mitotic cell division, the only way its genetic information can be altered is through the action of mutagenic agents (e.g., radiation, chemicals). Any changes that occur within the genome of any individual ameba will be passed on to the offspring cells that result from future mitotic cell divisions.

WRONG CHOICES EXPLAINED:
(1) Variations in an ameba would *not* most commonly occur through *differentiation during development*. The concept of differentiation refers to the specialization of tissues in multicellular organisms. However, the ameba is a single-celled, not a multicellular, organism.

(2), (4) Variations in an ameba would *not* most commonly occur through *the fusion of gametes* or through *recombination during fertilization*. The concepts of gametic fusion and genetic recombination relate to sexually reproducing organisms. However, the ameba is an asexually reproducing, not a sexually reproducing, organism.

18. **1** The development of organs and tissues from a zygote includes *mitosis and differentiation*. A zygote (fertilized egg cell) reproduces asexually via mitotic cell division until a critical mass of embryonic cells is achieved. Once the critical cell mass has formed, these generic embryonic cells undergo the process of differentiation to convert themselves into the specialized tissue types needed to produce a viable fetus.

WRONG CHOICES EXPLAINED:
(2), (3), (4) The development of organs and tissues from a zygote does *not* include *mitosis and gamete production, meiosis and gamete production*, or *meiosis and fertilization*. Gamete (sex cell) production via meiotic cell division is a process that occurs only during maturation of the offspring. This maturation occurs significantly after (often years after) the basic development of organs and tissues via differentiation.

19. **4** This occurrence illustrates that *when humans alter ecosystems, serious consequences may result*. The human species has often altered, and continues to alter, the natural environment in multiple ways to serve its economic and survival needs. Before such an alteration occurs, the potential consequences to the natural environment should be taken into consideration. As a result of this consideration, humans should attempt to minimize any negative impact the alterations will have on the natural environment.

WRONG CHOICES EXPLAINED:
(1) This occurrence does *not* illustrate that *farmers should never clear land to grow crops as it always creates problems*. Farming is an example of an economic activity that significantly alters the natural environment. Farming practices can be adjusted to minimize their impact on the natural environment while still providing the food resources necessary for human survival.
(2) This occurrence does *not* illustrate that *once an ecosystem has been altered, it can not be restored to normal*. Any environment, no matter how extensively altered by humans, can recover and restore itself, if given time and the absence of human activity, via the process of ecological succession.
(3) This occurrence does *not* illustrate that *the farmers deliberately altered the equilibrium of the cities in the Northeast*. Midwestern farmers in the 1920s

had little understanding of the effects that their farming practices were having on the national and global climates. Today, after continuing extensive scientific research, we should have a much better understanding of the ways that human activities can negatively affect the natural environment.

20. **3** This threat is the result of a human activity that has *weakened an ecosystem through pollution*. The decades-long assault of acid rain on formerly unpolluted forest environments has altered soil chemistry to the point that some long-established native plant communities, such as the sugar maple, can no longer thrive. As a result, the New York State woodland ecosystem and the human economy it supports are in danger of collapsing.

WRONG CHOICES EXPLAINED:
(1) It is *not* true that this threat is the result of a human activity that has *introduced a foreign species by accident*. The sugar maple tree is a species native to New York State, not a foreign invasive species. Therefore, the sugar maple tree should be protected by any means necessary.
(2) It is *not* true that this threat is the result of a human activity that has *stabilized a forest ecosystem through technology*. Acid rain is a national and global environmental problem that needs to be corrected immediately to save our treasured natural environment and the human industries that depend on it. Industries that create acid rain are not stabilizing but, instead, are destabilizing technologies.
(4) It is *not* true that this threat is the result of a human activity that has *weakened a species by direct harvesting*. Direct harvesting is a human practice in which plants or animals are removed (harvested) from a natural environment for their economic value. Although sugar maple trees are often directly harvested for their wood and sap, this practice is not under question in this case.

21. **1** The human male reproductive system is adapted for the production of *sperm and the delivery of these cells for internal fertilization*. Sperm cells are gametes that contain the haploid (n) number of chromosomes. Sperm cells are produced in the male testes and are transported via the vasa deferentia and urethra to the female reproductive tract where fertilization of an egg cell may occur. These structures and processes enable the human species to survive from generation to generation.

WRONG CHOICES EXPLAINED:
(2) It is *not* true that the human male reproductive system is adapted for the production of *gametes that transport food to the egg*. An egg cell produced in the ovary in females contains sufficient food to power multiple cell divisions prior to the formation of the placenta during development. Male gametes do not transport food. This is a nonsense distracter.

(3) It is *not* true that the human male reproductive system is adapted for the production of *zygotes and the development of these cells into a fetus*. A zygote (fertilized egg) is formed following fertilization of an egg cell by a sperm cell in the oviduct of the female. Fetal development occurs in the uterus of the female. The oviduct and uterus are not part of the male reproductive system.

(4) It is *not* true that the human male reproductive system is adapted for the production of *hormones that stimulate placenta formation in the male*. The placenta forms in the uterus of the female following fertilization and embedding of the zygote in the uterine lining. The placenta is not part of the male reproductive system.

22. **1** *Replication of DNA in bacterial cells and cell division* are the two processes required for the technique to successfully produce hormones. The technique illustrated is genetic engineering. Genetic engineering is a laboratory technique in which a gene for a desired trait is snipped from the DNA of a donor cell and inserted into the genome of a recipient cell (in this case, a bacterial cell). The recipient bacterial cell is then cultured and allowed to reproduce rapidly by mitotic cell division such that a bacterial colony is established. All the cells within this colony will be capable of producing the desired hormone, which can be drawn off and purified for use in treating a human hormone deficiency.

WRONG CHOICES EXPLAINED:

(2), (3), (4) *Replication of DNA in bacterial cells and gamete formation, meiosis and development*, and *mitosis and fertilization* are *not* the two processes required for the technique to successfully produce hormones. Bacterial cells reproduce asexually by the process of mitosis followed by binary fission. Gametes (formed by meiotic cell division) and fertilization (fusion of male and female gametes) are sexual reproductive processes that are not involved in this technique.

23. **3** All the genetic information needed for the organism to develop is first present at step C. At this stage, the structure illustrated is the zygote, which results from the fusion of a haploid (n) egg cell (step A) with a haploid (n) sperm cell (step B). The zygote contains the diploid ($2n$) number of chromosomes and all the genetic information needed to form a new human offspring.

WRONG CHOICES EXPLAINED:

(1) It is *not* true that all the genetic information needed for the organism to develop is first present at step A. Step A represents the unfertilized egg cell, which contains only half of the genetic information needed to form a new human offspring.

(2) It is *not* true that all the genetic information needed for the organism to develop is first present at step *B*. Step B represents the sperm cell, which contains only half of the genetic information needed to form a new human offspring.

(4) It is *not* true that all the genetic information needed for the organism to develop is first present at step *D*. Step *D* represents the embryo/fetus, whose cells contain all of the genetic information needed to form a new human offspring. However, step D is not the first stage in which this full genetic complement is found.

24. **4** *All of the organisms pass on traits through reproduction* is the statement about the organisms represented that is correct. Every viable species, whether plant, animal, or single-celled organism, must reproduce in order to pass on the genetic traits that describe the species.

WRONG CHOICES EXPLAINED:
(1) *All of the organisms are autotrophs* is *not* the statement about the organisms represented that is correct. Autotrophs are organisms that manufacture their own food by means of photosynthesis. Of the organisms represented, only the flowering plant and the aquatic plant are autotrophs.

(2) *Only the flowering plant, green heron, and aquatic plant carry out photosynthesis* is *not* the statement about the organisms represented that is correct. The green heron is an animal and a heterotroph. So the green heron cannot carry out photosynthesis.

(3) *Only the frog and green heron can maintain homeostasis* is *not* the statement about the organisms represented that is correct. All organisms of all types in all environments must maintain homeostasis (steady state) in order to survive.

25. **4** The relationship between mistletoe and trees is an example of *parasite/host*. A parasite is defined as an organism that lives on or in another organism (host) and derives benefit from that organism while causing harm to it. The description given of the mistletoe plant and trees clearly meets this definition.

WRONG CHOICES EXPLAINED:
(1) The relationship between mistletoe and trees is *not* an example of *consumer/herbivore*. A consumer is an organism that derives its nutrition from the tissues of other plant or animal organisms. An herbivore is an animal consumer that derives its nutrition solely from plant matter. The description given of the mistletoe plant and trees does not meet this definition.

(2) The relationship between mistletoe and trees is *not* an example of *predator/prey*. A predator is an animal that hunts, kills, and consumes the bodies of

other animals. Prey organisms are animals that are hunted and consumed by predators. The description given of the mistletoe plant and trees does not meet this definition.

(3) The relationship between mistletoe and trees is *not* an example of *scavenger/decomposer*. A scavenger is a consumer animal that consumes the dead bodies of other animals for nutrition. A decomposer is a bacterium or fungus that breaks down dead plants and animals for their nutrition. The description given of the mistletoe plant and trees does not meet this definition.

26. **2** This method of controlling the rabbit population was an attempt to *stop the overproduction of an introduced species*. Although the exact effect of the myxoma virus is not stated in the paragraph, it is mentioned that the virus was introduced to infect the rabbits for the purpose of reducing the population of this introduced (invasive) species.

WRONG CHOICES EXPLAINED:

(1) It is *not* true that his method of controlling the rabbit population was an attempt to *stop the overproduction of a native species*. The paragraph states clearly that rabbits are not native to Australia.

(3) It is *not* true that his method of controlling the rabbit population was an attempt to *limit the food sources of the rabbit*. No information is presented in the paragraph that would support this inference.

(4) It is *not* true that his method of controlling the rabbit population was an attempt to *limit the number of rabbits brought into the country*. No information is presented in the paragraph that would support this inference.

27. **3** The major role of carbohydrates in the human diet is to *supply energy for the body*. The digestive process functions to reduce complex carbohydrates to simple sugars such as glucose. This glucose enters the body's cells where it is broken down chemically. This chemical breakdown releases energy that is transferred to molecules of ATP. ATP, in turn, releases this energy to fuel cellular reactions.

WRONG CHOICES EXPLAINED:

(1) It is *not* true that the major role of carbohydrates in the human diet is to *form the membranes that surround the mitochondria*. The membranes surrounding mitochondria are composed of proteins and lipids, not carbohydrates.

(2) It is *not* true that the major role of carbohydrates in the human diet is to *act as a catalyst for cellular reactions*. Cellular reactions are catalyzed by enzymes (specialized proteins), not by carbohydrates.

(4) It is *not* true that the major role of carbohydrates in the human diet is to *provide building blocks for amino acids*. Amino acids are the simple building blocks of proteins and are not composed of carbohydrates.

28. **1** This is a valid concern because this practice *reduces the biodiversity of their fields*. Biodiversity is a measure of the total number of different species that inhabit and interact in a given environment. When forests or fallow (resting) fields are cleared for the purpose of planting crops, biodiversity is significantly reduced. The planting of a single species in such areas (known as monocropping) maximizes the lack of biodiversity. When the environment changes, monocrop farm fields are more susceptible to drought, disease, and soil nutrient depletion than if the fields were planted in alternating rows of compatible crops (e.g., corn, beans, and squash).

WRONG CHOICES EXPLAINED:
    (2) It is *not* true that this is a valid concern because this practice *increases the number of decomposers in their fields*. Decomposers are most abundant in soils that contain an abundance of organic matter and that are kept shaded and moist. Monocrop fields may lack these conditions and so may be less likely to support decomposers than fields planted in multiple crop species.
    (3) It is *not* true that this is a valid concern because this practice *decreases the need to import food*. Planting only one crop may reduce the availability of vegetables and fruits within the state and so would increase the need to import these foods from other states or countries.
    (4) It is *not* true that this is a valid concern because this practice *increases the number of invasive species*. Invasive species enter the environment in multiple ways, including through purposeful introduction by humans.

29. **4** The breathing rate, heart rate, and blood hormone levels of an individual would directly provide information about that individual's *metabolic activity*. The term *metabolism* refers to the sum total of all the chemical and physical activities that occur in the body. When a doctor takes measurements such as those listed, he or she can make informed judgments about a patient's overall metabolic health.

WRONG CHOICES EXPLAINED:
    (1) It is *not* true that the breathing rate, heart rate, and blood hormone levels of an individual would directly provide information about that individual's *cellular organization*. Cellular organization is determined by the processes of growth and differentiation and cannot be determined through measures such as those listed.

(2) It is *not* true that the breathing rate, heart rate, and blood hormone levels of an individual would directly provide information about that individual's *nutrition*. A doctor could best determine a patient's nutrition by asking the patient to answer questions concerning his or her dietary habits, not through measures such as those listed.

(3) It is *not* true that the breathing rate, heart rate, and blood hormone levels of an individual would directly provide information about that individual's *inheritance*. Inheritance of genetic traits is determined at the moment of fertilization and might best be determined through DNA testing, not through measures such as those listed.

30. **1** *Consumer* is the type of organism that could occupy levels *B*, *C*, and *D* of this energy pyramid. In any traditional energy pyramid, the base of the pyramid is reserved for producer organisms whose photosynthetic activity supports the upper levels. The upper levels of the energy pyramid are reserved for the consumer organisms, including herbivores (level *B*), carnivores (level *C*), and top predators (level *D*).

WRONG CHOICES EXPLAINED:
(2), (3) *Producer* and *autotroph* are *not* the types of organisms that could occupy levels *B*, *C*, and *D* of this energy pyramid. Producers, also known as autotrophs, are listed in a traditional energy pyramid at level *A*.

(4) *Carnivore* is *not* the type of organism that could occupy levels *B*, *C*, and *D* of this energy pyramid. Carnivores, including top predators, are listed in a traditional energy pyramid at levels *C* and *D* but not *B*. Level *B* traditionally contains only herbivores.

PART B–1

31. **1** Headlines such as these best illustrate the concept that *scientific explanations are tentative and subject to change*. Humans' understanding of natural phenomena is constantly improving as more and better scientific information derived from professional research is published in scientific journals by competent scientists. The newspaper headlines indicate that this understanding improved significantly over an approximately 30-year period during which competent scientific research was carried out. Newspapers (and, in more modern times, the Internet) are not necessarily the best sources of scientific information. Rather, the best sources of scientific information are the scientific journals in which actual research results are published and vetted by other scientists.

WRONG CHOICES EXPLAINED:

(2), (3), (4) It is *not* true that headlines such as these best illustrate the concept that *some newspapers are not honest and report incorrect information on purpose, worms can enter the body many different ways,* or *worms found in swamps should not be used for fishing.* No information is presented in the question that supports any of these inferences.

32. **3** The scientist's conclusion was based on *careful observation, measurements, and inferences from his data.* Although his inferences were only partially correct, the scientist applied what he understood to be careful scientific methodologies (observation and measurement) to the problem. He followed through on a long-term experiment and drew inferences from his results. We know now, after 400 years of additional scientific experimentation, that the plant's weight gain was actually the result of biochemical activity that chemically combined water, carbon dioxide, and soil minerals to produce molecules of carbohydrates, proteins, and other essential organic components.

WRONG CHOICES EXPLAINED:

(1), (2), (4) It is *not* true that the scientist's conclusion was based on *the input of scientists from many countries doing similar studies, the application of advanced technologies to the study of a problem,* or *an extensive knowledge of the process of photosynthesis.* Such peer collaboration, advanced technologies, and biochemical knowledge were not available to scientists working in the early 1600s.

33. **2** The approximate diameter of one cell is *50 μm.* This value can be estimated by placing a piece of paper against one of the cells and marking its diameter along the paper edge, transferring that dimension to the 500 μm diameter line in the diagram, and estimating that approximately 10 of these cells can fit across the diameter of the field of view. Dividing 500 μm by 10 cell diameters yields a cell diameter of 50 μm.

WRONG CHOICES EXPLAINED:

(1), (3), (4) It is *not* true that the approximate diameter of one cell is *10 μm, 250 μm,* or *500 μm.* None of these values can be supported by application of the method described above or by any similar method.

34. **2** One reason this could lead to extinction of this warbler is that *there may not be enough diversity among the birds for the species to be able to survive an environmental change.* In general, a species with many members and with a large and diverse gene pool contains sufficient variation to allow a large proportion of

its members to survive any change in environmental conditions that might threaten its existence. By contrast, a species with few members and with a shrinking gene pool contains fewer variations and is less likely to survive such changes. This latter condition increases the likelihood of extinction.

WRONG CHOICES EXPLAINED:

(1) One reason this could lead to extinction of this warbler is *not* that *after a species becomes extinct, it won't be able to carry out its role in the ecosystem.* Although a species that has become extinct can no longer carry out its role in the environment, this phenomenon cannot lead to extinction of that species. This is a nonsense distracter.

(3) One reason this could lead to extinction of this warbler is *not* that *extinction always occurs when populations begin to reduce in number.* All natural populations experience periods of population decline, especially under changing environmental conditions. The majority of such population declines do not result in extinction. Instead, these declines are usually reversed as surviving members reproduce and pass on their favorable traits to ensure the continuing survival of future generations.

(4) One reason this could lead to extinction of this warbler is *not* that *an increase in biodiversity within a population often causes the population to be classified as threatened or endangered.* Biodiversity is defined as the number of different species that inhabit an environment. Using the term *biodiversity* to describe genetic variation within a species is not scientifically correct. This is a nonsense distracter.

35. **1** One primary function of the cell membrane is *regulating the flow of simple sugars into or out of the cell.* In addition to bounding the cell and its contents, the cell membrane implements several methods to control the movement of substances into and out of the cell. Simple sugars are normally brought into a cell by diffusion though the cell membrane.

WRONG CHOICES EXPLAINED:

(2) It is *not* true that one primary function of the cell membrane is *synthesizing substances by breaking down cell organelles.* This is not a function of the cell membrane or any other cell component. This is a nonsense distracter.

(3) It is *not* true that one primary function of the cell membrane is *storing carbohydrates, water, and starches for future use.* This is a function of a food vacuole in a cell, not the cell membrane.

(4) It is *not* true that one primary function of the cell membrane is *digesting carbohydrates, fats, and protein.* The process of digestion of complex organic molecules normally occurs in the cell's external environment or within the digestive tracts of complex animals but not in the cell membrane.

36. **3** *If this pipeline were to leak, the oil could contaminate soil, water, and wildlife* is the statement that expresses a major concern many people are likely to have about the proposed pipeline. Most people understand the complex and fragile relationships that exist in nature. They do not want ecosystems to be disrupted or destroyed indiscriminately by human commercial and industrial activities.

WRONG CHOICES EXPLAINED:
(1) *The pipeline will bring a large number of jobs to the area where it is being constructed* is *not* the statement that expresses a major concern many people are likely to have about the proposed pipeline. Oil pipelines provide a short-term advantage in the form of construction jobs. However, they also provide long-term risks to the environments they cross. Job creation is not usually the subject of environmental concerns about such projects.

(2) *The oil pipeline will increase the amount of finite resources* is *not* the statement that expresses a major concern many people are likely to have about the proposed pipeline. Oil and other fossil fuels are finite resources that are dwindling and will eventually become exhausted. Although this oil pipeline would increase access to these finite resources, it would not increase the overall amount of oil.

(4) *The pipeline is a technological fix for ozone depletion* is *not* the statement that expresses a major concern many people are likely to have about the proposed pipeline. Ozone is known to be depleted by the release of CFC (chlorofluorocarbons) and other aerosol propellants. No known relationship exists between oil pipelines and ozone depletion.

37. **3** By observing the annual rings in the diagram, one can infer that *some years provide better conditions for growth than other years*. An examination of the diagram reveals that some annual rings in this tree are wider or narrower than others. Narrow rings indicate relatively slow growth, while wide rings indicate relatively rapid growth. It can reasonably be inferred from this examination that this tree encountered environmental conditions that limited its growth in years 10–12. It may also be inferred that from year 13 to the present, the tree had relatively rapid growth, indicating more favorable environmental conditions in those years.

WRONG CHOICES EXPLAINED:
(1) It is *not* true that by observing the annual rings in the diagram, one can infer that *environmental conditions did not change over the last 20 years*. If one accepts the relationship between a tree's annual rings and its growth patterns, then one should infer that environmental conditions have varied as described above.

(2) It is *not* true that by observing the annual rings in the diagram, one can infer that *trees grow faster on the side that faces the Sun*. No information provided in the question would lead one to make this inference.

(4) It is *not* true that by observing the annual rings in the diagram, one can infer that *tree rings are not reliable because trees must be cut down to see them*. In fact, it is possible to examine a tree's annual ring pattern without cutting down the tree. This technique involves taking a core sample through the tree's bark to its center and then sealing the hole to prevent entry by fungi or insects.

38. **2** This treatment for acne, using phages, is effective because phages *eliminate bacteria by attacking specific cell structures*. According to the paragraph, phages produce a protein, endolysin, that breaks down the cell walls of acne bacteria.

WRONG CHOICES EXPLAINED:
(1), (3) It is *not* true that this treatment for acne, using phages, is effective because phages *produce antibodies to clean out clogged pores and hair follicles* or *carry genes and infect follicles*. No information is presented in the paragraph concerning the production of antibodies or the infection of follicles.

(4) It is *not* true that this treatment for acne, using phages, is effective because phages *attack every known type of bacteria*. The paragraph specifically states that phages attack only acne bacteria, not all types of bacteria.

39. **3** The protein endolysin belongs to the group of chemical substances known as *biological catalysts*. A catalyst is a substance that speeds up or slows down the rate of a chemical reaction. In this case, the biological catalyst endolysin targets and breaks apart the specific materials that make up the cell walls of acne bacteria.

WRONG CHOICES EXPLAINED:
(1) The protein endolysin does *not* belong to the group of chemical substances known as *hormones*. Hormones are specific proteins that help to regulate metabolic activities of plants and animals. Hormones are not known to break apart the cell walls of bacteria.

(2) The protein endolysin does *not* belong to the group of chemical substances known as *receptors*. Receptors are specific proteins embedded in the membranes of cells and that regulate cell-to-cell communication. Receptors are not known to break apart the cell walls of bacteria.

(4) The protein endolysin does *not* belong to the group of chemical substances known as *molecular bases*. Molecular bases, such as adenine, thymine,

cytosine, and guanine, are components of large nucleic acids, such as DNA and RNA. Molecular bases are not known to break apart the cell walls of bacteria.

40. **4** The typical response of the human body to an infection by bacteria is to *produce white blood cells and antibodies*. The immune response in humans detects the antigens of foreign invaders. It responds to these antigens by manufacturing antibodies that specifically target and neutralize the infectious agent. Likewise, the body programs white blood cells to engulf and destroy these invaders.

WRONG CHOICES EXPLAINED:

(1) The typical response of the human body to an infection by bacteria is *not* to *stimulate the production of antigens*. Antigens are proteins produced by the cells of all living things and are specific to the species that produces them. Antigens are not produced to protect the body from infection.

(2) The typical response of the human body to an infection by bacteria is *not* to *decrease the number of enzymes in the blood*. Enzymes are biological catalysts produced by the cells of all living things and that help to regulate the body's metabolism. The number of enzymes in the blood is unaffected by the immune response.

(3) The typical response of the human body to an infection by bacteria is *not* to *ignore the organisms, unless they are pathogens*. The human immune response does not discriminate between infectious and noninfectious agents in the exercise of its protective function.

41. **4** *Both processes X and Y occur in green plants* is the correct statement regarding the types of organisms able to carry out these processes. Process X is photosynthesis, in which green plants take in CO_2 (and water and solar energy) and release O_2 (and sugar). Process Y is respiration, in which both green plants and animals take in O_2 (and sugar) and release CO_2 (and water and ATP energy).

WRONG CHOICES EXPLAINED:

(1) *Process X occurs in heterotrophs, but not in autotrophs* is *not* the correct statement regarding the types of organisms able to carry out these processes. Process X (photosynthesis) does not occur in heterotrophs (animals) but does occur in autotrophs (green plants).

(2) *Process Y occurs in consumers, but not in producers* is *not* the correct statement regarding the types of organisms able to carry out these processes. Process Y (respiration) is carried out by both consumers (animals) and producers (green plants).

(3) *Both processes X and Y occur in all living things* is *not* the correct statement regarding the types of organisms able to carry out these processes. Process X (photosynthesis) is carried out by green plants (producers, autotrophs) but not by animals (consumers, heterotrophs).

42. **2** *Ribosome* is the structure represented by *B*. In the process of protein (enzyme) synthesis, the DNA code is read by molecules of mRNA and transferred to ribosomes for translation into specific protein molecules.

WRONG CHOICES EXPLAINED:

(1) *Vacuole* is *not* the structure represented by *B*. Vacuoles in cells serve to store materials that are too large to readily pass through the cell membrane.

(3) *Cytoplasm* is *not* the structure represented by *B*. Cytoplasm is the watery medium of the cell interior that supports the organelles and assists in the transport of materials within the cell.

(4) *Chloroplast* is *not* the structure represented by *B*. Chloroplasts in green plant cells function to convert solar energy to the chemical bonds of glucose molecules.

43. **3** The nucleus contains molecules of *A*, which *store hereditary information*. Molecule *A* is DNA, which carries the genetic code that provides information for the production of proteins in the cell.

WRONG CHOICES EXPLAINED:

(1) It is *not* true that the nucleus contains molecules of *A*, which *recycle waste products*. DNA is not known to recycle waste products.

(2) It is *not* true that the nucleus contains molecules of *A*, which *remove water from the cell*. This is a function of organelles known as contractile vacuoles, not DNA.

(4) It is *not* true that the nucleus contains molecules of *A*, which *regulate the pH of cytoplasm*. DNA is not known to regulate the pH of the cytoplasm.

PART B–2

44. One credit is allowed for correctly marking an appropriate scale, without any breaks in the data, on the axis labeled "Estimated Moose Population." [1]

45. One credit is allowed for correctly plotting the data for the estimated moose population on the grid, connecting the points, and surrounding each point with a small circle. [1]

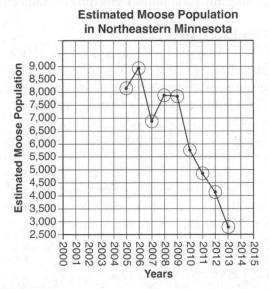

Estimated Moose Population in Northeastern Minnesota

46. One credit is allowed for correctly explaining how climate change could result in an increased number of moose infected with winter ticks. Acceptable responses include but are not limited to: [1]

- *The fall seasons are longer and warmer, increasing the chances of ticks surviving the winters and infesting moose.*
- *Climate change can be predicted to improve conditions for the survival of winter ticks through earlier snowmelt in the spring.*
- *Ticks that fall off moose onto soil instead of snow are more likely to survive and then reinfect moose in the spring.*
- *In the past, moose in northern climates were protected from winter ticks by long periods of cold weather that killed the ticks. However, this is no longer the case due to global climate change that has warmed average temperatures worldwide.*

47. **3** Increased average yearly temperatures in regions presently inhabited by moose could result in a disruption in homeostasis in these animals because *moose will not be able to maintain an appropriate body temperature, since they do not sweat.* The sweating mechanism present in humans enables us to regulate our body temperature as sweat evaporates from the skin surface and cools the blood. Since they lack this mechanism, moose can cool their bodies only by exposure to cool air or cool water. Failure to regulate body temperature within an acceptable range can significantly disrupt the body's homeostatic balance.

WRONG CHOICES EXPLAINED:

(1), (2), (4) It is *not* true that increased average yearly temperatures in regions presently inhabited by moose could result in a disruption in homeostasis in these animals because *a decrease in average temperatures will increase mutations in their skin cells, an increase in average temperatures will decrease the amount of blood ticks can consume,* or *moose will sweat more and lose too much water from their bodies.* No information is presented in the paragraph that would support any of these inferences.

48. One credit is allowed for correctly describing the probable result of an environmental change that severely affected the organism represented by species *K*. Acceptable responses include but are not limited to: [1]

- *Species K did not survive because they were unable to adapt to their new environment.*
- *Species K became extinct because they could not survive the changed environment.*
- *They became extinct. The diagram shown that they did not survive to the present.*

49. **2** Three species with the most similar traits are most likely *D, H, J*. The diagram indicates that species *D* is the most direct common ancestor of species *H* and *J*. These three species likely share a similar gene pool and, therefore, likely share similar traits.

WRONG CHOICES EXPLAINED:

(1), (3), (4) Three species with the most similar traits are *not* most likely *F, I, G; B, D, G;* or *F, A, J*. The diagram indicates that these species are widely separated in the evolutionary history of this genetic grouping. The only common ancestor linking these species is long-extinct species *A*. So it is unlikely that *F, A, J* are the three species with the most similar traits.

50. **4** The organisms found at the second trophic level of this pyramid would be *herbivores*. The diagram represents a food (energy) pyramid. Traditionally, organisms shown in the second trophic level of a food pyramid are herbivores, which derive their nutritional energy primarily from the producers found at the first trophic level.

WRONG CHOICES EXPLAINED:

(1) The organisms found at the second trophic level of this pyramid would *not* be *producers*. Producers (green plants) are traditionally represented as inhabiting the first trophic level in a food pyramid.

(2) The organisms found at the second trophic level of this pyramid would *not* be *decomposers*. Despite their crucial role in the ecological food web, decomposers are often not included in representations of a food pyramid, including this one.

(3) The organisms found at the second trophic level of this pyramid would *not* be *carnivores*. Carnivores are traditionally represented as inhabiting the third trophic level in a food pyramid.

51. One credit is allowed for correctly stating *one* reason why there is less energy available at each trophic level going from the first to the third trophic level. Acceptable responses include but are not limited to: [1]

- *Energy is lost at each level as heat that dissipates into the environment.*
- *Some energy is used by the organisms at each level and is not available to the organisms at the next level.*
- *An amount of energy is used at each level to support metabolic activities. This energy is lost and cannot be used by organisms at the next level of the food/energy pyramid.*

52. One credit is allowed for correctly explaining how the sharp population increase from 1910 to 1925 might have resulted in the decrease in the carrying capacity after 1925. Acceptable responses include but are not limited to: [1]

- *The overpopulation of deer in 1925 caused their habitat to be overgrazed. This did not allow the food sources in the area to regrow for the future deer population.*
- *The interference of humans in the predator-prey relationship was the cause of this problem. By removing natural predators of the deer, the deer population exploded. There was not enough vegetation to support the increased population of deer after 1925.*
- *In 1925, the deer ate too much food, reducing the amount of food available in future years.*

- *By removing the predator species that had kept the deer population in check, the deer population increased geometrically. This increase exceeded the environment's carrying capacity because the additional deer overgrazed the habitat from 1920 through 1925. From 1925 to 1937, the deer population dropped at a rapid rate, stabilizing after 1937.*
- *Overpopulation of deer in 1925 destroyed some of the resources deer needed. So fewer deer could live there.*

53–54.   Two credits are allowed for identifying some of the key events associated with the change. In your answer, be sure to:

- Identify *one* natural event that could cause the disruption indicated in the diagram.   [1]
- State what would most likely happen to the new stable ecosystem in future years if no further disruptions occur.   [1]

Acceptable responses include but are not limited to: [2]

- *A possible disruption in this ecosystem is forest fire. [1] The new ecosystem will undergo ecological succession until a new climax community compatible with the larger environment is achieved. [1]*
- *Volcanic eruption might have disrupted this ecosystem. [1] A long period of change will precede the formation of any new stable ecosystem. Once established, the new stable ecosystem will last until the next disruption. [1]*

55.   One credit is allowed for correctly explaining why a mutation that occurs in a body cell will not contribute to the evolution of a species. Acceptable responses include but are not limited to: [1]

- *Genetic information in body cells isn't passed on to offspring.*
- *The mutation isn't in sperm or egg cells.*
- *Mutations must be in gametes to be passed on to offspring.*

PART C

56.   One credit is allowed for correctly stating *one* reason why a supply of amino acids is important for the survival of complex organisms. Acceptable responses include but are not limited to: [1]

- *Amino acids are the building blocks of protein molecules.*
- *Amino acids are found in enzymes, which regulate chemical activity in complex organisms.*
- *Amino acids are the chemical components of the proteins found in our bodies' muscle and organ tissues.*

57. One credit is allowed for correctly identifying *one* possible effect on the human population if nitrogen fertilizers were not available. Acceptable responses include but are not limited to: [1]

- *People would starve.*
- *Farmers could not grow enough crops to feed the population.*
- *The human population would decrease.*
- *The food supply would decrease.*
- *Human civilization as we know it could not exist.*
- *Humans would have deficiencies in their diets.*
- *Humans would become malnourished.*

58. One credit is allowed for correctly explaining how the building of factories to produce fertilizer is an example of a trade-off. Acceptable responses include but are not limited to: [1]

- *The factories produce useful products to help increase the food supply but they also produce pollutants.*
- *The factories provide fertilizers for the production of more food but they also disrupt the habitats of some organisms.*
- *Fertilizers are useful for the growth of healthy plants but can also be harmful if they wash off into lakes/waterways.*
- *Fertilizers help plants grow but could harm the environment.*

59–60. Two credits are allowed for correctly analyzing the experiment that produced the data in the table. In your answer, be sure to:

- State a hypothesis for the experiment. [1]
- State whether the results of the experiment support or fail to support your hypothesis; support your answer. [1]

Acceptable responses include but are not limited to: [2]

- *Hypothesis: If bacteria are exposed to UV light, they will form fewer colonies than those not exposed. [1] All UV-screened areas of exposed Petri dishes formed an average of 22.2 colonies, whereas unscreened (UV-exposed) areas formed an average of only 12.2 colonies. Therefore, my hypothesis appears to be supported by the experimental data. [1]*
- *My hypothesis for this experiment is that longer exposure time to UV light will limit bacterial growth more than shorter exposure time. [1] My analysis of the data in the table is that the data support my hypothesis. The data show that 0.0 min. of UV exposure allowed 22 colonies to grow, while*

10.0 min. of UV exposure allowed only 1 colony to grow. Each increase in exposure time had more and more impact on the growth of the bacteria. [1]

- *I think that if bacteria are exposed to UV light, they will not form colonies (hypothesis). [1] This hypothesis isn't supported by the data, unfortunately. The data indicate that bacterial colonies grew in each UV-exposed Petri dish, even the one that received 10 minutes of exposure time! [1]*

61. One credit is allowed for correctly explaining the importance of the presence of variations within a population. Acceptable responses include but are not limited to: [1]

- *More variation increases the chances that some members of the population will survive in a changing environment.*
- *Differences among offspring increase the likelihood that some individuals will have adaptations to increase their chances of survival in a new environment.*
- *Variations provide raw material for natural selection when environments change.*
- *More variation increases the stability of the population.*
- *Without variations, the population may not survive in a changing environment.*

62. One credit is allowed for correctly describing how the process of natural selection can result in an increase in frequencies of certain traits found in a population. Acceptable responses include but are not limited to: [1]

- *More individuals with favorable traits usually survive, passing on those traits to the next generation.*
- *As a result of natural selection, the number of individuals with favorable traits increases.*
- *The fittest individuals will survive, causing an increase in the frequency of favorable traits.*
- *Individuals with unfavorable traits will succumb to environmental pressures and so will not be able to pass on those unfavorable traits to the next generation.*

63. One credit is allowed for correctly describing the importance of receptors in cellular communication. Acceptable responses include but are not limited to: [1]

- *Receptors receive messages sent by other cells.*
- *They enable cells to respond appropriately.*

- *They assist cells to recognize and admit hormones and neurotransmitters.*
- *In white blood cells, receptors assist in the recognition of antigens on the surfaces of foreign invaders.*

64. One credit is allowed for correctly describing the importance of the shape of receptor molecules for carrying out their function. Acceptable responses include but are not limited to: [1]

- *Receptor molecules have specific shapes that influence how they interact with other molecules.*
- *The shape of the receptor molecule determines with which molecules they can interact.*
- *The shape of the receptor molecule and messenger molecule must match.*
- *The shapes must match/fit together at the active site.*

65. One credit is allowed for correctly identifying *one* effect a reduced number of insulin receptors might have on an individual. Acceptable responses include but are not limited to: [1]

- *The individual might experience high levels of glucose in the blood/urine.*
- *The person might be/become diabetic as a result.*
- *If insulin receptors are reduced, less glucose will go to the cells.*
- *The blood glucose level will not be appropriately controlled.*

66. One credit is allowed for correctly explaining why the genetic material in an offspring produced by sexual reproduction contains genetic material that is *not* identical to the genetic material of either parent. Acceptable responses include but are not limited to: [1]

- *Offspring receive genetic information from each of the two parents.*
- *Each parent contributes half of the offspring's genetic material.*
- *Recombination of genes occurs at fertilization.*
- *The processes of meiosis, segregation, fertilization, and recombination ensure that each offspring receives a unique genetic profile.*
- *In rare cases, mutations such as deletion, addition, and crossing-over may occur to change the offspring's genetic complement.*

67. One credit is allowed for correctly stating *one* reason why identical twins should have fewer genetic differences than fraternal twins. Acceptable responses include but are not limited to: [1]

- *Fraternal twins develop from two separately fertilized egg cells.*
- *Fraternal twins are no more closely related than siblings born at different times because fraternal twins arise from separate fertilizations of two different egg cells by two different sperm cells.*
- *Identical twins are produced from a single zygote/fertilized egg, so all of their cells are identical.*
- *Identical twins come from a single fertilized egg, so they are genetically identical.*

68. One credit is allowed for correctly identifying *two* environmental factors that can lead to an increase in the number of epigenetic markers that modify gene expression. Acceptable responses include but are not limited to: [1]

- *Stress*
- *Pollutants*
- *Food*
- *Smoke*
- *Pesticides*
- *Alcohol*
- *Drugs*
- *Temperature*

69. One credit is allowed for correctly describing how nutrients move from the mother to the fetus. Acceptable responses include but are not limited to: [1]

- *Nutrients move across the placenta from mother to fetus.*
- *Most soluble substances pass readily across the placental membranes via diffusion.*
- *In the placenta, a series of maternal and fetal membranes separate the mother's blood from the fetus's blood but allow the diffusion of soluble materials between maternal and fetal bloodstreams.*

70. One credit is allowed for correctly stating *one* other way, in addition to consuming a balanced diet, pregnant women can help ensure proper development of the fetus. Acceptable responses include but are not limited to: [1]

- *Pregnant women should have regular doctor visits to ensure the health of mother and fetus.*

- *Pregnant women should continue to exercise appropriately during pregnancy.*
- *Pregnant women should avoid alcohol/drugs/tobacco products.*
- *Pregnant women should maintain an adequate waking/sleeping cycle.*
- *Pregnant women should avoid contact with infectious agents and their vectors, such as deer ticks carrying Lyme disease bacteria and mosquitos carrying Zika virus.*
- *Pregnant women should avoid unprotected sex and its associated risks, such as AIDS/gonorrhea/syphilis, that could affect the mother's health and the baby's development.*

71. One credit is allowed for correctly stating *one* way the use of these new "little electric heaters" might represent a long-term benefit over using antibiotics to treat bacterial infections. Acceptable responses include but are not limited to: [1]

- *The bacteria may not become resistant to the heaters.*
- *Antibiotics may have negative side effects.*
- *Heaters will work faster than the antibiotics.*

72. One credit is allowed for correctly stating *one* reason why the mother did not get sick with the measles. Acceptable responses include but are not limited to: [1]

- *The mother had antibodies against the measles.*
- *The mother had been vaccinated against the measles.*
- *The mother had measles when younger and still had a natural immunity to measles.*
- *The mother is immune to the measles.*
- *The mother had a mutation that made her resistant to the measles.*

PART D

73. **2** *The nutrients usually move from an area of high concentration in the soil to an area of low concentration in root cells* is the statement that best describes how nutrients enter the root cells of the onion plant. Dissolved nutrients, such as nitrates and phosphates, are absorbed from the soil into the root hair cells via the process of diffusion. Diffusion in the direction of the concentration gradient is accomplished without the expenditure of cell energy.

WRONG CHOICES EXPLAINED:

(1) *Only nutrients needed by the plant enter root cells* is *not* the statement that best describes how nutrients enter the root cells of the onion plant. Any dissolved substance of proper chemical composition may diffuse into the root cells of a plant.

(3), (4) *Nutrients always move into the plant cells by active transport* and *the nutrients always move from an area of low concentration in the soil to an area of high concentration in root cells* are *not* the statements that best describe how nutrients enter the root cells of the onion plant. Active transport is a process by which substances are transported into root cells from an area of low relative concentration to an area of high relative concentration. Active transport against the concentration gradient requires the expenditure of cell energy.

74. **1** *Variation—different "beaks" were available* is the concept correctly matched with an example from *The Beaks of Finches* lab. The various tools of different shapes and sizes used by students to gather seeds simulated the natural variation of beak shapes and sizes found among the Galapagos finches.

WRONG CHOICES EXPLAINED:

(2) *Adaptation—different types of food were available* is *not* the concept correctly matched with an example from *The Beaks of Finches* lab. The adaptations in this lab were the various tools of different shapes and sizes used by students to gather seeds, not the seeds themselves.

(3) *Selecting Agent—an insecticide was used to kill insects on one island* is *not* the concept correctly matched with an example from *The Beaks of Finches* lab. The selecting agent in this lab was the quantity of seeds available to be gathered by the various tools of different shapes and sizes.

(4) *Environment—"beaks" with similar qualities were used to gather seeds* is *not* the concept correctly matched with an example from *The Beaks of Finches* lab. In this lab, students used various tools of different shapes and sizes ("beaks"), not tools with similar qualities.

75. **3** *The two organisms produce many of the same proteins* is the evidence that would be considered the strongest for supporting a possible evolutionary relationship when comparing characteristics of two organisms. If two organisms share many proteins in common, this would indicate that they may share many genetic traits (many genes) in common and therefore may be related evolutionarily.

WRONG CHOICES EXPLAINED:

(1), (2) *The two organisms are the same color* or *the two organisms are the same height* are *not* the evidences that would be considered the strongest for supporting a possible evolutionary relationship when comparing characteristics of two organisms. These are superficial characteristics that may be shared among many organisms with widely different evolutionary histories.

(4) *The two organisms are found in the same locations* is *not* the evidence that would be considered the strongest for supporting a possible evolutionary relationship when comparing characteristics of two organisms. Every stable ecosystem is populated by hundreds or thousands of different species with widely varying evolutionary histories.

76. **3** Five minutes after the slides were prepared, a student using a compound light microscope to observe the cells in leaves *A* and *B* would most likely see that *water had moved out of the cells of the leaf on slide B*. Actually, the student would more likely see that the contents of the cells in leaf *B* had shrunk and pulled away from their cell walls. The student might then correctly infer from this observation and his/her knowledge of biology that water had moved out of those cells via osmosis to the surrounding 6% salt and 94% water on the slide.

WRONG CHOICES EXPLAINED:

(1), (2) Five minutes after the slides were prepared, a student using a compound light microscope to observe the cells in leaves *A* and *B* would *not* most likely see that *water had moved out of the cells of the leaf on slide A* or that *salt had moved into the cells of the leaf on slide A*. The leaf on slide *A* is mounted in a 1% salt and 99% water solution that is compatible with the Elodea's cell contents. The student would likely observe that the shape and size of these cells remained unchanged and correctly infer that their salt to water ratio was in equilibrium with the water on the slide.

(4) Five minutes after the slides were prepared, a student using a compound light microscope to observe the cells in leaves *A* and *B* would *not* most likely see that *salt had moved out of the cells of the leaf on slide B*. The student would likely make the observation noted in the correct answer above and use his/her knowledge of biology to correctly infer that water, not salt, had moved out of the cells of this leaf.

77. One credit is allowed for correctly stating *one* reason why these two species probably do *not* live in the same area of this island. Acceptable responses include but are not limited to: [1]

- *Trees and cacti do not usually grow in the same areas/environmental conditions.*
- *Their food may be available only in certain areas on the island and the foods are different.*
- *Their foods do not grow in the same place since these foods require different growing conditions.*
- *There is another resource one of the species needs that is not available where the other species lives.*

78. One credit is allowed for circling "no" and for correctly supporting the answer. Acceptable responses include but are not limited to: [1]

- *The student's pulse rate, at 60 bpm, is below the range for his age group.*
- *It falls in the range for people older than 14.*
- *His rate of 60 beats per minute puts him in the adult range.*

79. One credit is allowed for correctly stating *one* reason why a person's heart rate increases during exercise. Acceptable responses include but are not limited to: [1]

- *During exercise, there is an increase in carbon dioxide in the blood, which causes the heart to beat faster.*
- *There is a feedback mechanism that causes the heart to beat faster in response to an increase in physical activity.*
- *When the heart beats faster, it increases the flow of oxygen-rich blood to the muscles.*
- *The heart pumping faster removes wastes faster.*

80. One credit is allowed for correctly identifying which species is likely to be more closely related to *Botana curus* and for supporting the answer. Acceptable responses include but are not limited to: [1]

- *Species Z—This species has the same amino acid sequence as* Botana curus.
- *Species Z—*Botana curus *has more DNA bases in common with this species than it does with species X or species Y.*

81. **2** A factor that contributed to the evolution of finches on the Galapagos Islands was most likely the *isolation of the finches on separate islands.* Any species that experiences the geographic isolation of its population into several smaller

groups will undergo the phenomenon known as speciation. During speciation, genetic variations that exist among the parent species' populations will be subjected to natural selection in response to different environmental conditions. If separated long enough, these populations will continue to vary to the point that they can no longer interbreed. At this point, they may be considered different related species.

WRONG CHOICES EXPLAINED:

(1) It is *not* true that a factor that contributed to the evolution of finches on the Galapagos Islands was most likely the *lack of variation in beak structure of the finches*. In fact, the beak shape among the several Galapagos finch species exhibits considerable variation.

(3) It is *not* true that a factor that contributed to the evolution of finches on the Galapagos Islands was most likely the *relatively constant atmospheric temperature*. The more similar the temperatures of the environments into which the finch populations separated, the more slowly evolution would occur.

(4) It is *not* true that a factor that contributed to the evolution of finches on the Galapagos Islands was most likely the *total lack of competition for food*. In fact, the Galapagos Islands do not contain abundant food supplies. For this reason, competition has always been a factor in the evolution of the finches.

82. **2** The rise of the liquid in the test tube that was observed after one hour can be explained as a result of the *water moving from the beaker into the test tube*. Diffusion is a process in which a substance moves with the concentration gradient from an area of higher relative concentration to an area of lower relative concentration of that substance. This process does not require the expenditure of cellular energy. In the diagram, water molecules are more concentrated in the beaker and less concentrated in the tube. As a result, the net movement of water is from the breaker, through the membrane, and into the tube. As the water molecules increase in number inside the tube, the water-starch solution is displaced up the tube.

WRONG CHOICES EXPLAINED:

(1) It is *not* true that the rise of the liquid in the test tube that was observed after one hour can be explained as a result of the *starch solution moving into the test tube and out of the beaker*. Starch molecules are too large and complex to move through the extremely small pores in the dialysis membrane.

(3), (4) It is *not* true that the rise of the liquid in the test tube that was observed after one hour can be explained as a result of the *large starch molecules blocking the dialysis membrane* or the *dialysis membrane acting as a barrier to the water molecules.* If the dialysis membrane was being blocked or was actively blocking to prevent the movement of substances through it, the fluid level in the test tube would not change.

83. One credit is allowed for correctly describing *one* observation that would be made after one hour if a starch indicator solution was initially added to the water in the beaker. Acceptable responses include but are not limited to: [1]

- *The starch solution in the tube would turn blue-black.*
- *The water in the beaker would be amber/tan colored.*
- *The fluid in the test tube would be blue-black, indicating the presence of starch.*
- *A color change would occur in the test tube.*

84. One credit is allowed for correctly explaining why the DNA sample in lane 3 did not separate into fragments. Acceptable responses include but are not limited to: [1]

- *The restriction enzyme did not have a long enough time to cut the DNA sample.*
- *The DNA sample was not cut by this restriction enzyme.*
- *The restriction enzyme was denatured/did not work in the sample in lane 3.*

85. One credit is allowed for correctly stating *one* way that this experiment could be improved to obtain a valid conclusion. Acceptable responses include but are not limited to: [1]

- *The experimental group should have been larger.*
- *The data should have been collected for a longer time period.*
- *A control group should have been included in the study.*
- *Data should have been collected during the last quarter of the game.*
- *The experiment should be repeated by other competent scientists.*

Standards/Key Ideas	June 2018 Question Numbers	Number of Correct Responses
Standard 1		
Key Idea 1: The central purpose of scientific inquiry is to develop explanations of natural phenomena in a continuing and creative process.	31, 32	
Key Idea 2: Beyond the use of reasoning and consensus, scientific inquiry involves the testing of proposed explanations involving the use of conventional techniques and procedures and usually requiring considerable ingenuity.	33, 59	
Key Idea 3: The observations made while testing proposed explanations, when analyzed using conventional and invented methods, provide new insights into natural phenomena.	37, 44, 45, 60	
Laboratory Checklist	33, 44, 45, 59, 60	
Standard 4		
Key Idea 1: Living things are both similar to and different from each other and from nonliving things.	7, 8, 11, 14, 24, 29, 35, 42, 43, 63	
Key Idea 2: Organisms inherit genetic information in a variety of ways that result in continuity of structure and function between parents and offspring.	2, 4, 15, 17, 22, 66, 67, 68	
Key Idea 3: Individual organisms and species change over time.	34, 48, 49, 55, 61, 62	
Key Idea 4: The continuity of life is sustained through reproduction and development.	9, 13, 16, 18, 21, 23, 69, 70	
Key Idea 5: Organisms maintain a dynamic equilibrium that sustains life.	3, 12, 27, 38, 39, 40, 41, 47, 56, 64, 65, 71, 72	
Key Idea 6: Plants and animals depend on each other and their physical environment.	1, 6, 25, 30, 50, 51, 52, 53, 54	
Key Idea 7: Human decisions and activities have a profound impact on the physical and living environment.	5, 10, 19, 20, 26, 28, 36, 46, 57, 58	
Required Laboratories		
Lab 1: "Relationships and Biodiversity"	75, 80, 84	
Lab 2: "Making Connections"	78, 79, 85	
Lab 3: "The Beaks of Finches"	74, 77, 81	
Lab 5: "Diffusion Through a Membrane"	73, 76, 82, 83	

Examination August 2018
Living Environment

PART A

Answer all questions in this part. [30]

Directions (1–30): For *each* statement or question, record in the space provided the *number* of the word or expression that, of those given, best completes the statement or answers the question.

1 Which human activity most directly causes a significant increase in the amount of carbon dioxide in the atmosphere?

(1) growing corn for food
(2) not using products containing plastics
(3) driving cars long distances
(4) planting large numbers of trees

1 _____

2 An immune response is primarily due to the body's white blood cells recognizing

(1) a hormone imbalance
(2) abiotic organisms
(3) foreign antigens
(4) known antibiotics

2 _____

3 In an effort to reduce the number of deaths due to malaria, scientists have successfully introduced a gene into mosquitoes. The gene makes the mosquitoes unable to support the development of the parasite that causes malaria. The technique used to produce this new variety of mosquito is most likely

(1) chromatography
(2) genetic engineering
(3) electrophoresis of genes
(4) selective breeding 3 _____

4 The organic compounds that scientists use to cut, copy, and move segments of DNA are

(1) carbohydrates (3) hormones
(2) enzymes (4) starches 4 _____

Base your answers to questions 5 and 6 on the diagram below and on your knowledge of biology.

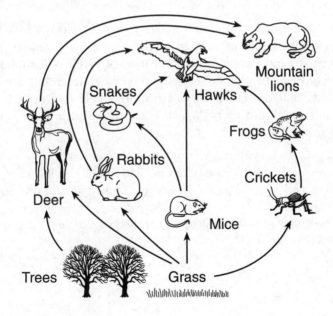

5 Which statement most accurately predicts the result of interfering with populations in the web?

 (1) Removing the cricket population will have little effect on the balance of the food web.

 (2) Removing all the mountain lions from the food web will benefit the ecosystem.

 (3) Removing the cricket and rabbit populations would cause the number of trees to decrease.

 (4) Removing deer from the food web will affect the rabbit and grass populations. 5 ____

6 A factor *not* shown in the diagram that provides energy for living organisms is

 (1) carbon dioxide   (3) the Sun

 (2) water   (4) oxygen 6 ____

7 Scientists have found a gene that makes a protein called PKG that controls certain behaviors in many types of ants. The soldier ant will help collect food when it has a low level of PKG. When it has a high level of PKG, the soldier ant will protect and defend its colony. Soldier ants that are given PKG are more likely to ignore food sources and attack intruders. Which conclusion can best be made from this information?

 (1) PKG protein is synthesized only by the soldier ants.

 (2) Genes control which type of amino acids a cell can make.

 (3) Eating too much protein makes some organisms very aggressive.

 (4) The behavior of soldier ants is controlled in part by the PKG protein. 7 ____

8 Genetic researchers have discovered a number of different gene mutations that have led to the development of cancer. These mutations affect how frequently a cell reproduces. Which process would be directly influenced by these mutations?

(1) differentiation of cells in an embryo
(2) meiotic cell division
(3) division of sperm and egg cells
(4) mitotic cell division   8 _____

9 Lobsters are crustaceans related to crayfish, crabs, and shrimp. Most lobsters are a reddish-brown color, but on rare occasions, they can be orange, blue, or even multicolored. These color differences can be caused by

(1) genetic variations
(2) different numbers of offspring
(3) overpopulation and excessive resources
(4) the instability of the ecosystem   9 _____

10 Which two factors could lead to the evolution of a species over time?

(1) overproduction of offspring and no variation
(2) changes in the genes of body cells and extinction
(3) struggle for survival and fossilization
(4) changes in the genes of sex cells and survival of the fittest   10 _____

11 As human red blood cells mature, they lose their nuclei. As a result of this loss, which process would be impossible for mature red blood cells to carry out?

(1) excretion   (3) reproduction
(2) respiration   (4) transport   11 _____

12 To clone a mammal, a cloned embryo is often put into an adult female of the same species to continue internal development. The structure in which the embryo will develop is the

(1) ovary   (3) uterus
(2) placenta   (4) egg   12 _____

13 Nuclear power plants, which produce electrical energy, use large quantities of water for cooling. Often, small fish, larvae, and fish eggs are sucked in along with the cooling water and destroyed. This example illustrates how

(1) industrialization can have positive and negative effects
(2) removal of these organisms has no effect on an ecosystem
(3) direct harvesting increases the natural fish population
(4) energy is generated without producing wastes   13 _____

14 Acid rain is a major problem in the Adirondack Mountains. Evidence that acid rain *negatively* affected the Adirondack ecosystem is that

(1) this rain has increased the amount of water in Adirondack lakes
(2) there has been a decrease in the variety of fish found in Adirondack lakes
(3) the amount of carbon dioxide in the air over the Adirondack Mountains has drastically decreased in recent years
(4) the number of heterotrophic organisms in Adirondack lakes has increased   14 _____

15 Which system in a multicellular organism functions most like the cytoplasm in a single-celled organism?

(1) immune   (3) nervous
(2) reproductive   (4) circulatory   15 _____

16 A common cycle in biology is represented below.

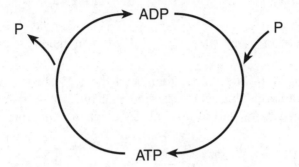

The ATP molecule above is commonly used to

(1) actively transport molecules in an organism
(2) diffuse water across a membrane
(3) move molecules from a high to a low concentration
(4) balance the nutrients in an ecosystem

16 ____

17 Fat molecules typically contain long chains of carbon atoms. Animals tend to store fats for use when food resources are scarce. This is an advantage to the animal because

(1) much energy can be gained by breaking the bonds between atoms in the fats
(2) fats give off carbon dioxide that can be used by the muscles
(3) amino acids from fat synthesis are more easily digested than carbohydrates
(4) energy can only be created by digesting fats

17 ____

18 While looking at the bottom surface of a leaf with a compound light microscope, a student notices pairs of cells with openings between them on the surface of the leaf. The main purpose of these openings and the cells that surround them is

(1) removing excess sugars
(2) synthesis of carbon dioxide
(3) regulating gas exchange
(4) purification of water

18 ____

19 An immune response to a usually harmless environmental substance is known as

(1) an antigen

(2) a vaccination

(3) an allergy

(4) a mutation

19 _____

20 Scientists at Penn State have sequenced the DNA of the extinct woolly mammoth. The data suggested that the woolly mammoth was more closely related to present-day elephants than previously believed.

Elephant

Woolly mammoth

Which statement could account for the similarities between the woolly mammoth and present-day elephants?

(1) Common gene mutations were caused by agents such as industrial chemicals and radiation.

(2) Present-day species developed from earlier, different species.

(3) Selective breeding results in offspring better able to survive.

(4) Both animals have identical genetic information.

20 _____

21 Single-celled organisms are able to maintain homeostasis, even though they lack higher levels of organization such as organs and organ systems, because

(1) single-celled organisms do not carry out the same life processes as multicellular organisms

(2) multicellular organisms do not rely on tissues or organs to carry out life processes

(3) cell structures work together to maintain homeostasis in single-celled organisms

(4) single-celled organisms are able to coordinate organ functions to maintain homeostasis

21 _____

22 Sharks are often followed by smaller fish that eat some of the scraps from the organisms eaten by the shark. These smaller fish are acting as

(1) decomposers   (3) producers

(2) scavengers   (4) herbivores   22 _____

23 The photographs below are side-by-side images of twins *A* and *B* with identical genetic information. Twin *A* is a nonsmoker, while twin *B* is a longtime smoker.

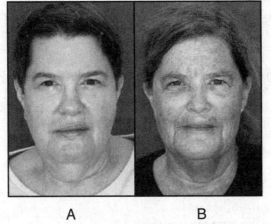

A   B

Source: http://holykaw.alltop.com

The best explanation for the differences in the appearance of the twins is that

(1) twin *B* is older than twin *A*

(2) they each inherited half of their DNA from each parent

(3) the expression of genes is influenced by the environment

(4) one twin resembles the mother and the other resembles the father   23 _____

24 Under the supervision of experts, certain areas in a nature preserve are regularly exposed to frequent, low-intensity fires. These controlled fires maintain specific populations of plants by directly

  (1) increasing the consumption of finite resources
  (2) decreasing the carbon dioxide level in the atmosphere
  (3) stopping the process of evolution
  (4) interfering with the process of ecological succession   24 _____

25 An invasive species, the spiny water flea, was recently found in a New York lake. These water fleas eat zooplankton, a food also consumed by native fishes. The fleas spread from lake to lake by attaching to fishing lines, anchor ropes, and boats. Which statement best describes the effect of the water flea on the lake?

  (1) It will not compete with animals in the local food chain.
  (2) It will feed on organisms that are important to other species.
  (3) The number of water fleas will decrease due to a lack of food.
  (4) There will be no effect on native species in the lake.   25 _____

26 A stable ecosystem can have high biodiversity because each species in that ecosystem

  (1) occupies a different niche
  (2) inhabits a different environment
  (3) is part of a different community
  (4) lives in a different biosphere   26 _____

27 Stability within an ecosystem is achieved partially by the presence of organisms that break down important molecules and make them available for other organisms to use. These organisms are

  (1) plants   (3) scavengers
  (2) herbivores   (4) decomposers   27 _____

28 The diagram below represents a feedback mechanism.

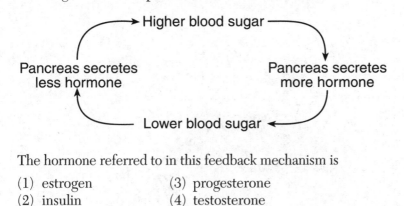

The hormone referred to in this feedback mechanism is

(1) estrogen   (3) progesterone

(2) insulin   (4) testosterone   28 _____

29 After a lake dried up during a severe drought, a section of undisturbed rock layers was exposed. The layers are represented below.

**Fossils in Undisturbed
Rock Layers**

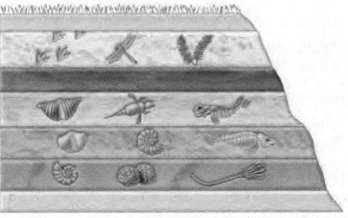

Source: https://www.superteachertools.net

This sequence of rock layers best illustrates the concept that

(1) the living and nonliving environment both change over time

(2) it is important to preserve the diversity of species and habitats

(3) new inheritable characteristics can result from the recombining of genes

(4) living organisms have the capacity to produce populations of unlimited size   29 _____

30 Some organisms in an ecosystem are represented in the pyramid below.

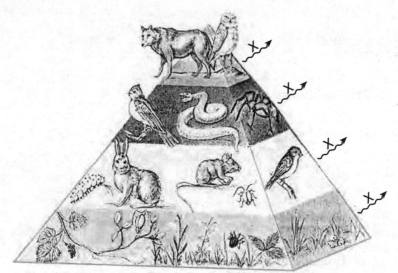

Source: Sylvia Mader, *Human Biology* (McGraw-Hill, 1998), p.476

In the pyramid, the arrows labeled *X* represent

(1) the loss of organisms due to predation
(2) a decrease in photosynthetic organisms
(3) the loss of energy in the form of heat
(4) a decrease in available oxygen

30 _____

PART B–1

Answer all questions in this part. [13]

Directions (31–43): For *each* statement or question, record in the space provided the *number* of the word or expression that, of those given, best completes the statement or answers the question.

Base your answers to questions 31 and 32 on the information below and on your knowledge of biology.

A student designed an experiment to determine if air temperature had an effect on the rate of photosynthesis in corn plants.

31 Which tool is correctly paired with a procedure that could be used during this experiment?

(1) an electronic balance to measure the volume of soil in which each corn plant is grown

(2) a graduated cylinder to measure 30 mL of water for each plant daily

(3) a metric ruler to determine the mass of each plant each week

(4) a Celsius thermometer to determine the pH of the soil   31 _____

32 The independent variable in this experiment is the

(1) air temperature at which the corn plants were grown

(2) amount of carbon dioxide used by the corn plants

(3) volume of oxygen produced by the corn plants

(4) number of corn plants used   32 _____

Base your answers to questions 33 and 34 on the information and graph below and on your knowledge of biology. The graph shows the number of animals in a population throughout the course of a year. The population migrated into the area at the beginning of 2011.

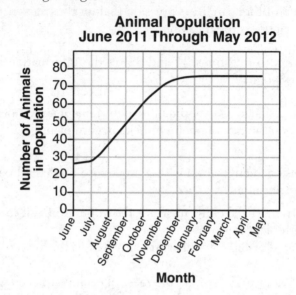

**Animal Population
June 2011 Through May 2012**

33 The graph can best be used to illustrate
   (1) a food chain   (3) natural selection
   (2) ecological succession   (4) carrying capacity   33 _____

34 The approximate number of animals that were found in June 2012
   was most likely
   (1) 16   (3) 76
   (2) 26   (4) 86   34 _____

35 To prepare for an experiment, ten different sources of food were sterilized and kept in a sterile container. Bacteria of the same species were placed on each of the ten different food sources and kept at 26°C for two days. During this time, bacteria grew in nine of the containers. Based on this observation, the scientist could conclude that

(1) all ten food sources used in the experiment are capable of supporting this species of bacteria

(2) the temperature varied greatly in nine of the containers during this experiment

(3) only the container that failed to grow any bacteria was prepared correctly

(4) this species of bacteria synthesizes enzymes needed to digest the food in nine of the ten containers   35 _____

36 The graph below shows three projections for future carbon dioxide (CO_2) levels.

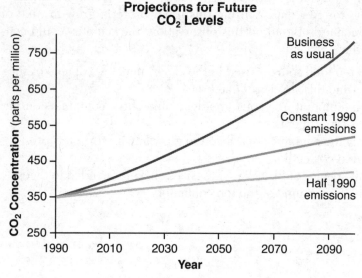

Projections for Future CO₂ Levels

Source: http://www.yourclimateyourlife.org.uk

- The "Business as usual" line shows CO_2 levels if emissions remain at current levels.
- The "Constant 1990 emissions" line shows CO_2 levels if emissions are cut to the same level that they were in 1990.
- The "Half 1990 emissions" line shows CO_2 levels if emissions are cut to half of the level that they were in 1990.

Which statement is supported by the graph?

(1) Climate change will result in the melting of polar ice caps.
(2) The increase in carbon dioxide levels will cause a decrease in global average temperature.
(3) Human activities have no effect on atmospheric carbon dioxide levels.
(4) Future generations can be affected by the choices of current generations.

36 _____

Base your answers to questions 37 through 39 on the information below and on your knowledge of biology.

Hydrogen peroxide (H_2O_2) is a toxic compound that is produced by plant and animal cells. These cells also produce the enzyme catalase, which converts H_2O_2 into water and oxygen gas, preventing the buildup of H_2O_2.

A student designed an experiment to test the effect of an acidic pH on the rate of the reaction of H_2O_2 with catalase. The data below summarize the outcome of the experiment.

pH Level	7 (neutral)	6	5	3
Reaction Rate (mL of oxygen/minute)	1.5	1.3	1.0	.55

37 Which graph most accurately represents the results obtained by this student?

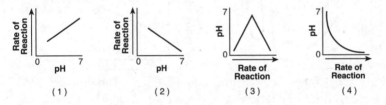

(1)   (2)   (3)   (4)   37 _____

38 Which conclusion is valid based upon the data collected by the student?

(1) The change in pH prevents catalase from breaking down water.
(2) Catalase has the greatest activity at a pH of 7.
(3) Oxygen production will increase if more water is added to the reaction.
(4) Catalase caused the greatest production of oxygen at a pH of 3.   38 _____

39 The best explanation for the change in catalase activity as the pH changed from 7 to 3 is that

(1) strong acid digests the catalase, causing the reaction rate to increase

(2) the student most likely cooled the H_2O_2 solution, causing the reaction rate to increase

(3) in acidic solutions, the shape of catalase changes, causing the reaction rate to decrease

(4) decreased oxygen production causes catalase to increase the rate of reaction

39 _____

40 Which graph best illustrates the body temperature in an individual maintaining dynamic equilibrium?

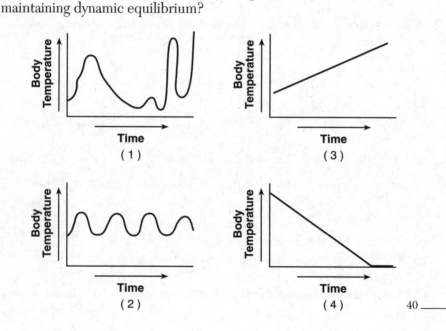

40 _____

Base your answers to questions 41 and 42 on the information and graph below and on your knowledge of biology.

Federal legislation establishes and updates energy-efficiency standards for consumer products, including refrigerators. The graph below shows the average annual energy consumption of similar types of refrigerators and the year they were manufactured.

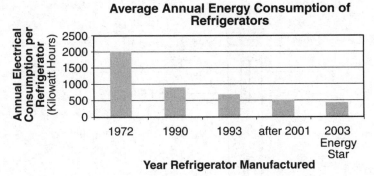

41 The 2003 Energy Star models of refrigerators use an average of about 450 kilowatt hours of electrical energy annually. Approximately how much energy is saved by these models annually when compared to the models produced in 1972?

(1) 500 kilowatt hours   (3) 1500 kilowatt hours
(2) 550 kilowatt hours   (4) 1550 kilowatt hours   41 _____

42 Which statement best represents an outcome of federal standards that require increasing the energy efficiency of appliances, such as refrigerators?

(1) More technological improvements in appliances can help conserve finite resources.
(2) Increased efficiency of appliances requires greater use of our energy resources.
(3) Newer appliances are manufactured from a greater number of finite resources.
(4) Manufacturing more efficient appliances will reduce the biodiversity of ecosystems.   42 _____

43 A student placed a test tube containing elodea (an aquatic plant) and pond water 10 cm from a light source. He observed that the plant gave off bubbles of a gas and counted how many bubbles were released in one minute. He moved the plant farther away from the light source to see if distance from the source made a difference. The data table below shows his results.

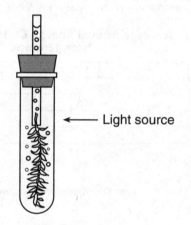

Light source

Gas Production

Distance from Light (cm)	Bubbles Produced per Minute
10	40
20	30
30	7
40	4

The gas that was being produced was most likely

(1) carbon dioxide as a product of the process of respiration
(2) carbon dioxide as a product of the process of photosynthesis
(3) oxygen as a product of the process of respiration
(4) oxygen as a product of the process of photosynthesis   43 _____

PART B–2

Answer all questions in this part. [12]

Directions (44–55): For those questions that are multiple choice, record your answer in the space provided. For all other questions in this part, record your answer in accordance with the directions.

Base your answers to questions 44 through 46 on the data table below and on your knowledge of biology. The data table shows the estimated number of species extinctions from 1960 to 2010.

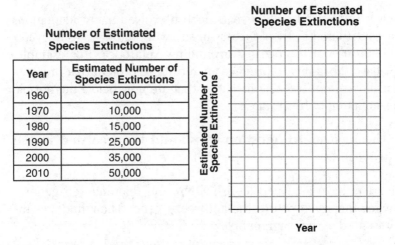

Number of Estimated Species Extinctions

Year	Estimated Number of Species Extinctions
1960	5000
1970	10,000
1980	15,000
1990	25,000
2000	35,000
2010	50,000

Number of Estimated Species Extinctions

Directions (44–45): Using the information in the data table, construct a line graph on the grid above, following the directions below.

44 Mark an appropriate scale, without any breaks in the data, on each labeled axis. [1]

45 Plot the data on the grid, connect the points, and surround each point with a small circle. [1]

Example:

46 State *one* possible cause for the increase in the number of species extinctions from 1960–2010.   [1]

_____

_____

Base your answers to questions 47 through 49 on the information and diagram below and on your knowledge of biology.

Icefish Evolution

Over the last 50 million years, icefish evolved many adaptations that contributed to their success in surviving the decreasing water temperatures of the ocean surrounding Antarctica. For example, they have the ability to produce an antifreeze protein that prevents their blood from freezing in waters that are now below the normal freezing point of fresh water.

Note: The answer to question 47 should be recorded in the space provided.

47 Scientists have analyzed icefish DNA and documented genetic changes that gave rise to the antifreeze gene. Their findings are represented in the diagram below.

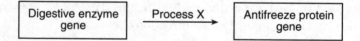

| Digestive enzyme gene | Process X → | Antifreeze protein gene |

Process *X* is referred to as

(1) mitosis   (3) differentiation

(2) mutation   (4) meiosis   47 _____

48 Explain how the process of natural selection can account for the increase in frequency of the antifreeze protein gene in the icefish population. [1]

_____

_____

Note: The answer to question 49 should be recorded in the space provided.

49 In addition to the appearance of the antifreeze gene, icefish have also been found to have DNA sequences similar to the DNA sequences in hemoglobin genes of other fish species. However, these DNA sequences are not complete and therefore not functional in icefish. This evidence makes it likely that

(1) icefish ancestors had hemoglobin
(2) icefish will soon produce offspring with hemoglobin
(3) hemoglobin is a molecule made by some fish that do not have genes for it
(4) soon all fish will stop producing hemoglobin   49 _____

Base your answers to questions 50 and 51 on the passage below and on your knowledge of biology.

Green sea slugs are animals that live in water and have developed the ability to produce their own chlorophyll. These creatures can also pass this ability to make chlorophyll to their offspring. Once the offspring have one meal of algae, they are able to make food using sunlight. This one meal provides the baby slugs with the chloroplasts needed to make use of the chlorophyll, and they are able to produce their own food in the future.

Note: The answer to question 50 should be recorded in the space provided.

50 The best explanation for why sea slugs are able to pass on this ability to make food to their offspring is that

    (1) the gene for making algae is in all their body cells
    (2) making food is beneficial, so the slugs needed to mutate
    (3) the environment causes the slugs to become green
    (4) the gene for chlorophyll production is part of their DNA   50 ____

51 Explain how green sea slugs can be considered both a producer and a consumer. [1]

_____

_____

52 State why fossil fuels are considered a finite resource. [1]

_____

_____

53 The diagram below represents an organelle.

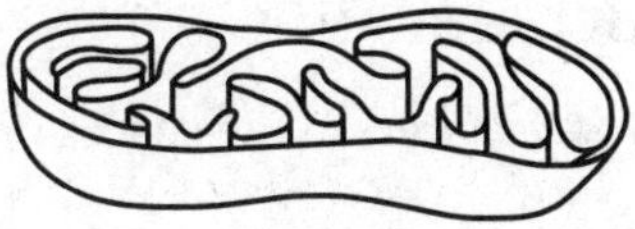

Identify the process that occurs in this organelle, and explain the importance of this process to the survival of organisms. [1]

Process: _____

Importance: _____

_____

Base your answers to questions 54 and 55 on the information below and on your knowledge of biology.

The testes of a human male produce gametes. The process that produces these gametes differs from the process that produces new skin cells in the same individual.

54 Identify the type of cell division involved in each process. [1]

Skin cells: _____

Gametes: _____

55 How does the genetic makeup of the skin cells differ from the genetic makeup of the gametes? [1]

_____

_____

PART C

Answer all questions in this part.   [17]

Directions (56–72): Record your answers in the spaces provided.

Base your answers to questions 56 through 58 on the information below and on your knowledge of biology.

> Placental mammals—as opposed to the kind that lay eggs, such as the platypus, or carry young in pouches, such as the kangaroo— are an extraordinarily diverse group of animals with more than 5000 species today. They [placental mammals] include examples that fly, swim, and run, and range in weight from a couple of grams to hundreds of tons. ...
>
> Source: "Earliest Placental Mammal Ancestor Pinpointed,"
> BBC News, February 7, 2013

56 Describe *one* function of the placenta during the internal development of an offspring.   [1]

_____

_____

57 Describe *one* advantage for an offspring to develop internally as
opposed to developing externally.   [1]

_____

_____

58 Identify *one* factor, besides genetics, that could influence the
development of human offspring.   [1]

_____

_____

Base your answers to questions 59 through 61 on the information below and on your knowledge of biology.

Birds Are Evolving Rapidly—Today

Many people think that evolutionary change occurs so slowly, we cannot observe it directly. Not so!

For example today in the U.S., house finches are evolving rapidly and visibly. In 1941, some captive house finches from California escaped near New York City. They spread rapidly and are now found across most of the United States and in southern Canada. Many of these areas have cold, snowy winters, during which many birds die. The finches have evolved, because those that survive differ from their parents. Size is one example.

Male house finches in recently established populations in Michigan and Montana are larger than the males that escaped. Large males outcompete small males for food, so are more likely to survive the winter. They also pair more successfully with females early in the spring.

Smaller females survive better than larger females as nestlings. Also, because they need less food to maintain their own bodies, they can breed earlier in spring. Females that breed earlier raise more young than those that start breeding later.

Rapid evolution of house finches reminds us that evolutionary changes are occurring visibly all around us.

Source: birdnote.org/show/birds-are-evolving-rapidly-today

59 Explain why it is an advantage for female finches to be small.   [1]

_____

_____

60 Explain why the population of large male house finches in Michigan and Montana continues to increase.   [1]

_____

_____

61 Predict what both the male and female finch populations might be like 10 years in the future, based on what is currently happening with the finch populations.   [1]

Male: _____

Female: _____

Base your answers to questions 62 through 64 on the information below and on your knowledge of biology.

New Threat to Endangered Sea Turtles

Endangered sea turtles in tropical areas are facing a new threat in the form of changing beach temperatures caused by climate change. The sex of sea turtle hatchlings is determined by the temperature inside the nest. Embryos develop into males when the temperature is approximately 28°C (82°F), whereas female embryos develop at approximately 31°C (88°F). If the temperature inside the nest is between these values, then both male and female turtles are produced.

If these endangered sea turtles are going to survive in the long term, we need to protect their nesting habitat, ensure that the potential nest sites have adequate amounts of shade-producing vegetation (such as palm trees) nearby, and ensure that they are not affected by tourist activities in the area.

62 Temperatures at beaches where these turtles nest are expected to slowly increase with global climate change. State *one* specific effect that a continuous rise in temperature could have on the sex ratio of the hatchling populations of these sea turtles.   [1]

_____

_____

63 Explain how having adequate amounts of shade-producing vegetation nearby, such as palm trees, can affect the nesting success of endangered sea turtles.   [1]

_____

_____

64 State *one* way that tourist activities in areas where turtles make their nests could have a *negative* impact on the nesting success of the turtles.   [1]

_____

_____

Base your answers to questions 65 through 68 on the information and photograph below and on your knowledge of biology. The photograph shows a grasshopper mouse howling after eating a scorpion.

Source: Michael and Patricia Fogden/Minden/NGS

Grasshopper Mice

In the Sonoran Desert in the southwestern United States, the grasshopper mouse is active at night, searching for crickets, rodents, tarantulas, and even scorpions. The mouse ignores the venom of the scorpion, kills it, and consumes its flesh. The ability of the mouse to ignore the pain normally associated with the venom of the scorpion is due to the presence of a mutated protein. This protein prevents the pain signal from reaching the brain.

These mice are born killers, capable of taking down prey that are much larger than themselves. They are also aggressive neighbors and take over nests by displacing other desert inhabitants rather than making their own. Under difficult environmental conditions, they may even eat members of their own species.

65 State the role of the population of grasshopper mice in the Sonoran Desert food web.   [1]

_____

_____

66 State *one* advantage grasshopper mice have over the other local populations when competing for resources.   [1]

_____

_____

67 Identify *one* advantage to grasshopper mice of being active during the night rather than daylight.   [1]

_____

_____

68 Explain how research on the mutated protein identified in the grasshopper mouse could benefit humans suffering from chronic pain.   [1]

_____

_____

Base your answers to questions 69 and 70 on the information below and on your knowledge of biology.

To attend public school in New York State, children need to be vaccinated against various diseases. The list below shows some required vaccinations.

Required Vaccinations

Polio

Tetanus

Pertussis

Measles

Mumps

Rubella

Diptheria

69 Explain how vaccinations protect against diseases.   [1]

_____

_____

70 The flu is a disease caused by a virus that can undergo frequent genetic changes. A different flu vaccination is needed each year. Explain why a single vaccination is *not* effective against all flu viruses.   [1]

_____

_____

Base your answers to questions 71–72 on the information below and on your knowledge of biology.

Survey Finds Invasive Snail in St. Lawrence River That Could Threaten Waterfowl

New research has found a larger presence of faucet snails in the Great Lakes than previously recognized, including the northern parts of Lake Ontario and the St. Lawrence River. The invasive species can carry three types of intestinal parasites that can injure and kill waterfowl such as ducks....

... When the waterfowl eat the snails, the parasites attack internal organs, causing lesions [sores] and hemorrhage [uncontrolled bleeding]. Birds affected by the snail will fly and dive erratically before their eventual death. The university said that the snails are about 12 to 15 millimeters in height at full size, brown to black with a distinctive whorl of concentric circles on the shell opening cover that looks like tree rings....

... Mr. Kosnicki [an ecologist] said the spread of snails, along with other invasive species, shows the need for increased awareness of possible contaminants coming from boats and in runoff from land. ...

Source: Watertown Daily Times, Monday, January 19, 2015,
by Gordon Block

71–72 Discuss how invasive species can harm an ecosystem. In your answer, be sure to:

- explain *one* negative effect that faucet snails have on the lake ecosystem [1]
- describe *one* human activity that can slow the spread of the faucet snail [1]

_____

_____

_____

_____

PART D

Answer all questions in this part. [13]

Directions (73–85): For those questions that are multiple choice, record your answer in the space provided. For all other questions in this part, record your answer in accordance with the directions.

Note: The answer to question 73 should be recorded in the space provided.

73 The diagram below represents one of many microscopic air sacs in a human lung. The alveolus (air sac) is the place where oxygen (O_2) and carbon dioxide (CO_2) move into or out of the blood, as represented in the diagram.

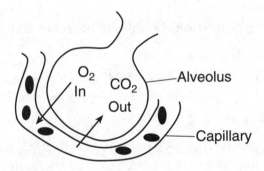

Which statement best explains why these gases are able to move in the directions shown in the diagram?

(1) The CO_2 moves out of the capillary and into the alveolus to make more room for the blood to carry O_2.

(2) The O_2 is needed by the cells, so it is actively transported into the blood. The CO_2, which is not needed, is actively transported out of the blood.

(3) The blood coming to the lungs is low in CO_2 and high in O_2, so the gases each diffuse from a lower to a higher concentration in this area.

(4) The blood coming to the lungs is high in CO_2 and low in O_2, so the gases each diffuse from a higher to a lower concentration in this area.

73 _____

Note: The answer to question 74 should be recorded in the space provided.

74 During the process of chromosome replication, a genetic error occurs. As a result, a sequence of events occurs as described below.

Event A: a protein with a new sequence of amino acids is produced

Event B: a DNA strand with an altered base sequence is formed

Event C: a new inheritable trait is expressed in an organism

Event D: an mRNA strand with a new sequence of bases is synthesized

The usual order in which these events would occur is

(1) $B - D - A - C$ (3) $D - A - B - C$

(2) $B - D - C - A$ (4) $D - C - B - A$ 74 _____

Base your answers to questions 75 through 77 on the information below and on your knowledge of biology. The evolutionary tree below represents possible relationships between several species of plants.

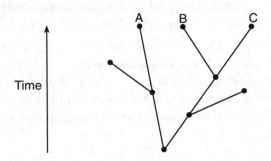

Note: The answer to question 75 should be recorded in the space provided.

75 According to the tree, species *B* and *C* are more closely related to each other than to species *A*. Which gel electrophoresis diagram would best support this statement?

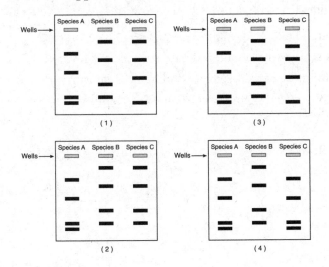

75 _____

Note: The answer to question 76 should be recorded in the space provided.

76 In addition to analyzing DNA, what other evidence could be used to best support the evolutionary relationship between species *B* and *C*?

   (1) Species *B* and *C* live in the same ecosystem.
   (2) Species *B* and *C* require the same amount of sunlight.
   (3) Species *B* and *C* possess many of the same enzymes.
   (4) Species *B* and *C* grow to the same maximum height
       as species *A*.   76 _____

77 On the diagram below, circle the dot that best represents the common ancestor of species *A*, *B*, and *C*. [1]

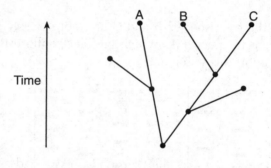

Base your answers to questions 78 and 79 on the information and chart below and on your knowledge of biology. The chart below shows variations in the beaks of finches in the Galapagos Islands.

Variations in Beaks of Galapagos Islands Finches

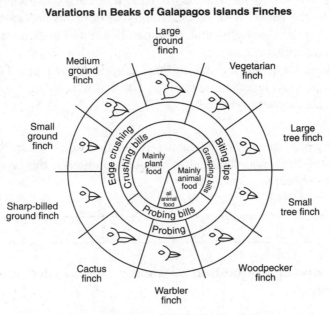

Source: *Galapagos: A Natural History Guide*

78 Explain *one* way that sharp-billed ground finches and small tree finches could possibly compete with each other if they lived on the same island. [1]

_____

_____

79 A small tree finch and a large tree finch inhabit the same island. Describe a situation that would allow both populations to live on the same island even though they both feed on animal food. [1]

_____

_____

Base your answers to questions 80 and 81 on the information below and on your knowledge of biology.

In his journey to the Galapagos Islands, Charles Darwin was amazed by the variation in the characteristics of plants and animals he encountered. In any habitat, food can be limited and the types of foods available may vary.

One year, there was no rain on these islands. Many plants failed to bloom and produced no new seeds. This left mostly large, tough seeds for the finches to eat.

80 Describe *one* change in beak characteristics that would most likely occur in the finch population after many generations if this change in seed size became permanent. [1]

_____

_____

Note: The answer to question 81 should be recorded in the space provided.

81 The different tools (such as spoons, chopsticks, or pliers) used during *The Beaks of Finches* laboratory activity represented variations in

(1) feeding adaptations   (3) finch migration

(2) seed size   (4) island ecosystems   81 _____

Base your answers to questions 82 through 84 on the information below and on your knowledge of biology.

Respiratory Rates

A student completed a lab activity that demonstrated the connection between pulse rate, heart rate, and flow of blood during exercise. She left her class wondering if there is a connection between breathing rate and exercise. With the help of her track coach, she conducted an investigation to try and find an answer to her question.

In the investigation, the respiratory rates of thirty athletes were measured before any exercise. To measure their rates, they counted the number of times they took a breath in one minute. Then, they ran one lap around the track, and their respiratory rates were determined again. Finally, they ran two laps around the track and checked their respiratory rates one last time. All of their data were recorded, and the averages were calculated. The data table shows the information obtained in this investigation.

The Effect of Exercise on the
Respiratory Rates of Thirty Athletes

Type of Activity	Average Respiratory Rates (breaths per minute)
At rest	13
After one lap	30
After two laps	38

Note: The answer to question 82 should be recorded in the space provided.

82 During this activity, what was the purpose of finding the respiratory rates of the students at rest?

(1) to determine if the students were healthy
(2) to practice using the timer
(3) to use as a comparison
(4) to use as the variable

82 _____

83 State *one* likely hypothesis that the student was testing in this investigation.   [1]

_____

_____

84 When determining the respiratory rates of the students at rest, the student found that the rates ranged from 10 to 20 breaths per minute. State *one* reason why students would have a range of respiratory rates.   [1]

_____

_____

85 The diagram below represents a container of water divided by a dialysis membrane into two areas, *X* and *Y*.

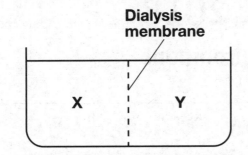

Starch solution was added to the water on side *X*. One hour later, amber-colored starch indicator solution was added to both sides. Identify the colors that could be observed at *X* and at *Y* after the addition of the starch indicator. [1]

Final color of *X*: _____

Final color of *Y*: _____

Answers
August 2018
Living Environment

Answer Key

PART A

1. 3	6. 3	11. 3	16. 1	21. 3	26. 1
2. 3	7. 4	12. 3	17. 1	22. 2	27. 4
3. 2	8. 4	13. 1	18. 3	23. 3	28. 2
4. 2	9. 1	14. 2	19. 3	24. 4	29. 1
5. 4	10. 4	15. 4	20. 2	25. 2	30. 3

PART B–1

31. 2	34. 3	36. 4	38. 2	40. 2	42. 1
32. 1	35. 4	37. 1	39. 3	41. 4	43. 4
33. 4					

PART B–2

44. *See* Answers Explained.
45. *See* Answers Explained.
46. *See* Answers Explained.
47. 2
48. *See* Answers Explained.
49. 1

50. 4
51. *See* Answers Explained.
52. *See* Answers Explained.
53. *See* Answers Explained.
54. *See* Answers Explained.
55. *See* Answers Explained.

PART C. *See Answers Explained.*

PART D

73. 4
74. 1
75. 2
76. 3
77. *See* Answers Explained.
78. *See* Answers Explained.
79. *See* Answers Explained.

80. *See* Answers Explained.
81. 1
82. 3
83. *See* Answers Explained.
84. *See* Answers Explained.
85. *See* Answers Explained.

PART A

1. **3** *Driving cars long distances* is the human activity that most directly causes a significant increase in the amount of carbon dioxide in the atmosphere. Cars with traditional internal-combustion engines burn carbon-based fuels such as gasoline and emit exhaust containing the byproducts of that combustion. One of the main byproducts in car exhaust is carbon dioxide (CO_2), which enters the atmosphere in large quantities and increases atmospheric CO_2 content significantly.

WRONG CHOICES EXPLAINED:
(1), (4) *Growing corn for food* and *planting large numbers of trees* are *not* the human activities that most directly cause a significant increase in the amount of carbon dioxide in the atmosphere. Any green plant crop consumes more carbon dioxide via photosynthesis than it produces via cellular respiration. For this reason, green plants are known to help decrease, not increase, atmospheric carbon dioxide concentrations.

(2) *Not using products containing plastics* is *not* the human activity that most directly causes a significant increase in the amount of carbon dioxide in the atmosphere. Plastics are highly modified carbon-based polymers that are usually chemically inert. Using or not using products that contain plastic has little direct influence on atmospheric carbon dioxide concentrations.

2. **3** An immune response is primarily due to the body's white blood cells recognizing *foreign antigens*. Antigens are proteins produced by the cells of all living things and are specific to the species that produces them. The immune response in humans detects the antigens of foreign invaders and responds to these antigens by manufacturing antibodies that specifically target and neutralize the infectious agent.

WRONG CHOICES EXPLAINED:
(1) It is *not* true that an immune response is primarily due to the body's white blood cells recognizing *a hormone imbalance*. Hormones are specific proteins that help to regulate metabolic activities of plants and animals. The immune response does not recognize or target the body's own hormones or imbalances in their concentrations.

(2) It is *not* true that an immune response is primarily due to the body's white blood cells recognizing *abiotic organisms*. The term abiotic is used to refer to nonliving factors (such as temperature, light, and minerals) in the environment that help to limit the growth of populations. By definition, organisms are biotic (living), not abiotic, environmental factors in the environment. The term "abiotic organisms" is a nonsense distracter.

(4) It is *not* true that an immune response is primarily due to the body's white blood cells recognizing *known antibiotics*. Antibiotics are complex biochemicals, produced naturally by molds and fungi, that inhibit the growth and reproduction of bacteria. The immune response does not normally recognize or target antibiotics in the blood.

3. **2** The technique used to produce this new variety of mosquito is most likely *genetic engineering*. Genetic engineering is a laboratory technique in which a gene for a desired trait is snipped from the DNA of a donor cell and inserted into the genome of a recipient cell. This technique may be used effectively by scientists to introduce the desired gene into the mosquito population.

WRONG CHOICES EXPLAINED:
(1) The technique used to produce this new variety of mosquito is *not* most likely *chromatography*. Chromatography is a laboratory technique that is used to help researchers study mixtures of organic chemicals such as plant pigments. This is not the technique described in the question.

(3) The technique used to produce this new variety of mosquito is *not* most likely *electrophoresis of genes*. Electrophoresis is a laboratory technique in which DNA fragments from an organism are placed in and drawn through an electrified gel to produce a unique banding pattern of the DNA for that organism. This is not the technique described in the question.

(4) The technique used to produce this new variety of mosquito is *not* most likely *selective breeding*. Selective breeding is a technique used by animal and plant breeders in which organisms with desirable traits are cross-bred with the hope of producing offspring that also display those traits. This is not the technique described in the question.

4. **2** The organic compounds that scientists use to cut, copy, and move segments of DNA are *enzymes*. Enzymes are proteins that function as biological catalysts in the cells of all living things and that help regulate the body's metabolism. Scientists use a class of these catalysts known as restriction enzymes to perform these functions in the laboratory.

WRONG CHOICES EXPLAINED:
(1), (4) It is *not* true that the organic compounds that scientists use to cut, copy, and move segments of DNA are *carbohydrates* or *starches*. Carbohydrates are a class of biochemicals that differ chemically from proteins. Starches are a specific type of carbohydrate. Carbohydrates (and starches) do not function as biological catalysts.

(3) It is *not* true that the organic compounds that scientists use to cut, copy, and move segments of DNA are *hormones*. Hormones are a class of proteins that function to regulate the expression of genes in living cells. Hormones do not function as biological catalysts.

5. **4** *Removing deer from the food web will affect the rabbit and grass populations* is the statement that most accurately predicts the result of interfering with populations in the web. Food webs are graphical representations of interactive nutritional relationships found commonly in nature. If the entire deer population were removed from this food web, it is likely that the grasses in the environment would temporarily thrive and provide additional food for rabbits, which would temporarily support an increase in the rabbit population.

WRONG CHOICES EXPLAINED:
(1), (2) *Removing the cricket population will have little effect on the balance of the food web* and *Removing all the mountain lions from the food web will benefit the ecosystem* are *not* the statements that most accurately predict the result of interfering with populations in the web. Any interference with a balanced ecosystem will have the effect of disrupting the balance of that ecosystem. The ecosystem will adjust to such disruptions until ecological balance is restored.

(3) *Removing the cricket and rabbit populations would cause the number of trees to decrease* is *not* the statement that most accurately predicts the result of interfering with populations in the web. The food web as illustrated shows no direct nutritional links among crickets, rabbits, and trees.

6. **3** A factor *not* shown in the diagram that provides energy for living organisms is *the Sun*. Food webs illustrate nutritional relationships in terms of the transfer of energy from producers (green plants), which manufacture their own food via photosynthesis, to sequential levels of consumer organisms. In turn, the photosynthetic reactions depend on the energy of the Sun (solar energy) to supply the energy needed to forge the chemical bonds in glucose. In this way, the Sun is known to be the primary source of energy in nearly all Earth's ecosystems.

WRONG CHOICES EXPLAINED:
(1), (2), (4) It is *not* true that a factor *not* shown in the diagram that provides energy for living organisms is *carbon dioxide*, *water*, or *oxygen*. Although it is true that these materials are not shown in the diagram and that they are essential components of the abiotic environment supporting life, none of them provide energy for living organisms.

7. **4** *The behavior of soldier ants is controlled in part by the PKG protein* is the conclusion that can best be made from this information. The paragraph describes a small number of behaviors (food collecting and colony defense) observed in soldier ants that appear to be affected by varying levels of the specific protein PKG. These observations may lead scientists to reasonably conclude that the behavior of soldier ants is controlled to some degree by the concentration of PKG in their environment.

WRONG CHOICES EXPLAINED:

(1), (2), (3) *PKG protein is synthesized only by soldier ants*, *Genes control which type of amino acids a cell can make*, and *Eating too much protein makes some organisms very aggressive* are *not* the conclusions that can best be made from this information. No information is provided in the paragraph that would support any of these conclusions. Each of these statements may represent a separate hypothesis that could be tested though further scientific research.

8. **4** *Mitotic cell division* is the process that would be directly influenced by these mutations. Mutations are random events in which genetic material (DNA) is altered. In order for a cell to become cancerous, mutations must occur in the genes that control mitotic cell division. Once these genes are compromised in a cell, the mutated cell's division becomes rapid and abnormal and results in a mass of cancerous cells that crowd out and kill normal tissues.

WRONG CHOICES EXPLAINED:

(1) *Differentiation of cells in an embryo* is *not* the process that would be directly influenced by these mutations. Differentiation is the process by which embryonic cells undergo specialization to become specific body tissues. Although mutations may interfere with embryonic differentiation, this is a rare occurrence.

(2) *Meiotic cell division* is *not* the process that would be directly influenced by these mutations. Meiotic cell division is a process by which gametes (sperm and egg cells) are produced in sexually reproducing organisms. Although mutations may interfere with meiotic cell division, this is a rare occurrence.

(3) *Division of sperm and egg cells* is *not* the process that would be directly influenced by these mutations. Sperm and egg cells (gametes) are produced as end products of meiotic cell division. These cells do not divide further unless fertilization occurs. This is a nonsense distracter.

9. **1** These color differences can be caused by *genetic variations*. Each living cell of a lobster (or of any organism) contains genetic information that helps to determine the physical and biochemical traits of that organism. These genetically determined traits include the color of a lobster's exoskeleton. Environmental conditions and diet may also play a part in modifying such traits.

WRONG CHOICES EXPLAINED:

(2), (3) It is *not* true that these color differences can be caused by *different numbers of offspring* or *overpopulation and excessive resources*. No competent scientific evidence exists that would support these claims. These are nonsense distracters.

(4) It is *not* true that these color differences can be caused by *the instability of the ecosystem*. The ocean environment in which lobsters live is among the most stable of Earth's ecosystems. It is possible that slight variations in the conditions of certain ocean microenvironments may have a part to play in determining the specific color of lobster exoskeletons living in those environments, but this is not a major factor.

10. **4** *Changes in the genes of sex cells and survival of the fittest* are the two factors that could lead to the evolution of a species over time. Genetic changes (mutations) that occur in sex cells (sperm and egg cells) that participate in fertilization and the formation of viable offspring can lead to variations in a species population. If a variation is favorable and promotes an increase in frequency of the gene producing it, they (both variation and gene) will become common in the population. In the face of environmental change, such favorable variations may provide an advantage that promotes the survival of the segment of the population expressing them (survival of the fittest). Over time, the population may continue to change by this mechanism until it becomes reproductively distinct from other populations of the same species. At that point, it can be said that evolution has occurred, and a new species has formed.

WRONG CHOICES EXPLAINED:

(1) *Overproduction of offspring and no variation* are *not* the two factors that could lead to the evolution of a species over time. Overproduction is a term that refers to the tendency of any species to produce more offspring than can possibly survive in a given environment. Variation is a term that refers to the genetic modifications that can be observed in any naturally occurring species. A species population that does not vary genetically cannot respond to environmental change and cannot evolve. It should be noted that no known species displays "no variation."

(2) *Changes in the genes of body cells and extinction* are *not* the two factors that could lead to the evolution of a species over time. Extinction is a term that refers to the complete and permanent elimination of a critical mass of breeding members of a species from the Earth. Body (somatic) cells that develop changes in genes (mutations) cannot pass these changes on to offspring because body cells do not participate in fertilization. If such changes cannot be passed on to offspring in a species population, that population cannot evolve.

(3) *Struggle for survival and fossilization* are *not* the two factors that could lead to the evolution of a species over time. Struggle for survival is a term that refers to the difficulty all living things experience in coping with environmental change. Fossilization is a term that refers to the preservation of body parts or impressions of body parts of dead organisms through mineralization, drying, freezing, or other means. Although fossilization provides scientists with evidence of species change over time, it has no direct effect on the evolution of a species population.

11. **3** *Reproduction* is the process that would be impossible for mature red blood cells to carry out. In order for a living cell to reproduce, it must be able to carry out the process of mitotic cell division. Mitotic cell division can only be accomplished in a living cell containing a nucleus containing the diploid ($2n$) number of chromosomes.

WRONG CHOICES EXPLAINED:
(1) *Excretion* is *not* the process that would be impossible for mature red blood cells to carry out. Excretion is an essential process by which waste materials are removed from living cells via diffusion through the cell membrane.

(2) *Respiration* is *not* the process that would be impossible for mature red blood cells to carry out. Respiration is an essential process by which chemical bond energy is converted to ATP energy in living cells via biochemical reactions in the mitochondria.

(4) *Transport* is *not* the process that would be impossible for mature red blood cells to carry out. Transport is an essential process by which materials are moved throughout living cells via diffusion through the cytoplasm.

12. **3** The structure in which the embryo will develop is the *uterus*. The uterus is a female mammalian organ within which the embryo implants itself after fertilization and that provides a stable environment for fetal development. A cell mass formed from cleavage of a cloned egg cell would also be placed in the uterus where it would complete its development.

WRONG CHOICES EXPLAINED:

(1) It is *not* true that the structure in which the embryo will develop is the *ovary*. The ovary is a female mammalian organ in which egg cells mature prior to fertilization. Embryos do not develop within this structure.

(2) It is *not* true that the structure in which the embryo will develop is the *placenta*. The placenta is a combined female/embryonic mammalian structure that provides developing embryos with essential materials and waste removal. Embryos do not develop within this structure.

(4) It is *not* true that the structure in which the embryo will develop is the *egg*. The egg is a female mammalian cell that combines with a sperm cell in the process of fertilization. Embryos do not develop within this structure.

13. **1** This example illustrates how *industrialization can have positive and negative effects.* The electrical energy produced by nuclear power plants assists human activities and is a positive consequence of this industrial process. The destruction of native plants and animals in the area surrounding a nuclear power plant, plus the warming of the waters surrounding them, represent a negative effect of this process. This is an example of a tradeoff in human decision-making.

WRONG CHOICES EXPLAINED:

(2) This example does *not* illustrate how *removal of these organisms has no effect on an ecosystem.* The removal of even small numbers of organisms from their natural environment reduces biodiversity and represents a negative consequence for that ecosystem.

(3) This example does *not* illustrate how *direct harvesting increases the natural fish population.* Direct harvesting is a human practice in which plants or animals are removed (harvested) from a natural environment for their economic value. The process as described in the question does not represent direct harvesting activity.

(4) This example does *not* illustrate how *energy is generated without producing wastes.* Although the question does not mention the production of wastes by nuclear power plants, these plants produce both waste heat and waste radioactivity, both of which can have negative consequences to the environment and to human health.

14. **2** Evidence that acid rain *negatively* affected the Adirondack ecosystem is that *there has been a decrease in the variety of fish in Adirondack lakes*. Multiyear surveys conducted by Department of Environmental Conservation (DEC) scientists indicate that Adirondack lakes have become increasingly acidic over the past 50–100 years. These surveys also show that native fish species preferring neutral/alkaline pH have declined in number or have been completely eliminated from many of these lakes over the same period. Only fish species that can tolerate acidic waters have been able to survive in these lakes, while some Adirondack lakes have become "dead" due to the extreme acidity.

WRONG CHOICES EXPLAINED:
(1) It is *not* true that evidence that acid rain *negatively* affected the Adirondack ecosystem is that *this rain has increased the amount of water in Adirondack lakes*. The amount of water in Adirondack lakes is a function of the annual snow pack and rainfall in this mountainous region but does not provide evidence of the effects of acid rain as such.

(3) It is *not* true that evidence that acid rain *negatively* affected the Adirondack ecosystem is that *the amount of carbon dioxide in the air over the Adirondack Mountains has drastically decreased in recent years*. Surveys conducted by the National Oceanic and Atmospheric Administration (NOAA) and others have provided evidence of increased, not decreased, carbon dioxide concentration in the atmosphere. Carbon dioxide dissolves in airborne water droplets to form carbonic acid, a known source of acid rain.

(4) It is *not* true that evidence that acid rain *negatively* affected the Adirondack ecosystem is that *the number of heterotrophic organisms in Adirondack lakes has increased*. Heterotrophic (other-feeding) organisms in and around Adirondack lakes include a wide variety of species, including but not limited to zooplankton, worms, mollusks, crustaceans, fish, amphibians, reptiles, birds, and mammals. DEC surveys have shown that the numbers and varieties of these species has decreased, not increased, as lake acidity has increased.

15. **4** The *circulatory* system in a multicellular organism functions most like the cytoplasm in a single-celled organism. In both cases, these circulatory structures function to move needed materials from the environment to the interior of the organisms, where they are used as the raw materials for cell processes. These circulatory structures also function to remove metabolic wastes from the cell interior to the environment for elimination.

WRONG CHOICES EXPLAINED:

(1) The *immune* system in a multicellular organism does *not* function most like the cytoplasm in a single-celled organism. In multicellular organisms, the immune system protects the body from intrusion by foreign pathogens. In single-celled organisms, a similar function is provided by the cell membrane.

(2) The *reproductive* system in a multicellular organism does *not* function most like the cytoplasm in a single-celled organism. In multicellular organisms, the reproductive system enables the body to produce offspring. In single-celled organisms, a similar function is provided by the cell nucleus.

(3) The *nervous* system in a multicellular organism does *not* function most like the cytoplasm in a single-celled organism. In multicellular organisms, the nervous system provides instructions for the coordinated functioning of the body. In single-celled organisms, a similar function is provided by the cell nucleus.

16. **1** The ATP molecule is commonly used to *actively transport molecules in an organism*. ATP is a compound that functions in cells to carry energy from the respiratory reactions in the mitochondria to the parts of the cell that require energy to perform their functions. Active transport, in which molecules are moved across the cell membrane against the concentration gradient, requires the input of ATP energy in order to function.

WRONG CHOICES EXPLAINED:

(2), (3) The ATP molecule is *not* commonly used to *diffuse water across a membrane* or to *move molecules from a high to a low concentration*. Diffusion is, by definition, the movement of molecules with the concentration gradient from a region of high relative concentration of that molecule to a region of low relative concentration of that molecule. Diffusion occurs in cells without the input of ATP energy.

(4) The ATP molecule is *not* commonly used to *balance the nutrients in an ecosystem*. Nutrients in an ecosystem may include anything from simple soil mineral nutrients absorbed by plants to complex organic food nutrients consumed by animals. The balance and distribution of these nutrients is subject to environmental, not cellular, forces. This is a nonsense distracter.

17. **1** This is an advantage to the animal because *much energy can be gained by breaking the bonds between atoms in the fats*. Fats represent a class of organic compounds that contain large numbers of carbon-carbon (C–C) bonds. Each of these C–C bonds contains a relatively large amount of energy that, when released, can be transferred to molecules of ATP. In turn, ATP transfers this energy to activate essential biochemical reactions within the cell.

WRONG CHOICES EXPLAINED:

(2) It is *not* true that this is an advantage to the animal because *fats give off carbon dioxide that can be used by the muscles*. Fats do not give off carbon dioxide. Carbon dioxide cannot be used by the muscles.

(3) It is *not* true that this is an advantage to the animal because *amino acids from fat synthesis are more easily digested than carbohydrates*. Amino acids do not result from fat synthesis. Amino acids are not digested, but rather are the products of protein digestion.

(4) It is *not* true that this is an advantage to the animal because *energy can only be created by digesting fats*. Energy is not created, only transferred, in the process of digestion. Many foods other than fats contain energy that can be transferred during digestion.

18. **3** The main purpose of these openings and the cells that surround them is *regulating gas exchange*. The openings, known as stomates, are each surrounded by a pair of guard cells that regulate the size of the opening. When the stomates are open during the day, carbon dioxide needed for the photosynthetic reactions diffuses into the leaf. Oxygen gas, a byproduct of photosynthesis, diffuses out of the leaf into the atmosphere at the same time. At night and during periods of dry weather, the stomates close to slow the evaporation of water from the leaf interior.

WRONG CHOICES EXPLAINED:

(1) It is *not* true that the main purpose of these openings and the cells that surround them is *removing excess sugars*. Plants convert sugar manufactured in the leaves into starch for storage in the roots; excess sugar is not removed by stomates.

(2) It is *not* true that the main purpose of these openings and the cells that surround them is *synthesis of carbon dioxide*. Carbon dioxide is an inorganic molecule that is a raw material for the photosynthetic reactions. Carbon dioxide is not synthesized in plants except as a byproduct of the respiratory reactions. Carbon dioxide is not synthesized by stomates.

(4) It is *not* true that the main purpose of these openings and the cells that surround them is *purification of water*. Water used by plants as a raw material for the photosynthetic reactions in the leaves is absorbed from the soil by the roots. Water is not purified by stomates.

19. **3** An immune response to a usually harmless environmental substance is known as *an allergy*. Common environmental substances such as mold spores, dust mites, plant pollen, nut flesh, and animal dander carry foreign proteins known as antigens. When these substances are breathed in or ingested, the body's immune system recognizes their antigens and responds to them by attempting to prevent them from entering the blood or interior tissues. This reaction often produces symptoms such as tearing, sneezing, congestion, headaches, and swelling of mucous membranes. The symptoms caused by the immune response are known collectively as an allergic reaction.

WRONG CHOICES EXPLAINED:
(1) An immune response to a usually harmless environmental substance is *not* known as *an antigen*. Antigens are proteins produced by the cells of all living things and are specific to the species that produces them.
(2) An immune response to a usually harmless environmental substance is *not* known as *a vaccination*. Vaccinations contain dead or weakened pathogens that are purposely injected into the body as a means of introducing the pathogen's antigens into the bloodstream.
(4) An immune response to a usually harmless environmental substance is *not* known as *a mutation*. Mutations are random events in which genetic material (DNA) is altered.

20. **2** *Present-day species developed from earlier, different species* is the statement that could account for the similarities between the woolly mammoth and present-day elephants. The physical and biochemical similarities between these two species indicate that they share a common ancestry. It is likely that both of these species arose from a more ancient ancestral species that was similar to both in terms of its DNA sequence but different from both in terms of its specific phenotypic characteristics.

WRONG CHOICES EXPLAINED:

(1) *Common gene mutations were caused by agents such as industrial chemicals and radiation* is *not* the statement that could account for the similarities between the woolly mammoth and present-day elephants. Mutations are random events in which genetic material (DNA) is altered. While mutagenic agents such as industrial chemicals may be responsible for mutations in present-day elephants, it is not possible that these same chemicals would have caused the same mutations in mammoths that lived 10,000 years ago.

(3) *Selective breeding results in offspring better able to survive* is *not* the statement that could account for the similarities between the woolly mammoth and present-day elephants. Selective breeding is a technique used by animal and plant breeders in which organisms with desirable traits are cross-bred with the hope of producing offspring that also display those traits. It is extremely unlikely that humans knew of or attempted to use this technique on mammoths 10,000 years ago.

(4) *Both animals have identical genetic information* is *not* the statement that could account for the similarities between the woolly mammoth and present-day elephants. If these two species had identical genetic information, they would be by definition the same species.

21. **3** Single-celled organisms are able to maintain homeostasis even though they lack higher levels of organization such as organs and organ systems because *cell structures work together to maintain homeostasis in single-celled organisms.* Known collectively as organelles, these cell structures are specialized for carrying out the essential life processes needed to maintain homeostatic balance in single-celled organisms.

WRONG CHOICES EXPLAINED:

(1) It is *not* true that single-celled organisms are able to maintain homeostasis even though they lack higher levels of organization such as organs and organ systems because *single-celled organisms do not carry out the same life processes as multicellular organisms.* All living organisms carry out the same essential life processes.

(2) It is *not* true that single-celled organisms are able to maintain homeostasis even though they lack higher levels of organization such as organs and organ systems because *multicellular organisms do not rely on tissues or organs to carry out life processes.* All multicellular organisms rely on complex organs and organ systems in order to carry out essential life processes.

(4) It is *not* true that single-celled organisms are able to maintain homeostasis even though they lack higher levels of organization such as organs and organ systems because *single-celled organisms are able to coordinate organ functions to maintain homeostasis*. The question clearly states that single-celled organisms "lack higher levels of organization such as organs and organ systems."

22. **2** These smaller fish are acting as *scavengers*. A scavenger is a consumer animal that consumes the dead bodies of other animals for its nutrition. The description of the behavior of the smaller fish in the question is consistent with this definition of scavenging behavior.

WRONG CHOICES EXPLAINED:
(1) These smaller fish are *not* acting as *decomposers*. A decomposer is a bacterium or fungus that breaks down the bodies of dead plants and animals for its nutrition. The description of the behavior of the smaller fish in the question is *not* consistent with this definition of a decomposer.

(3) These smaller fish are *not* acting as *producers*. A producer is a plant or other species that is capable of manufacturing its own food. The description of the behavior of the smaller fish in the question is *not* consistent with this definition of a producer.

(4) These smaller fish are *not* acting as *herbivores*. An herbivore is an animal consumer that derives its nutrition solely from plant matter. The description of the behavior of the smaller fish in the question is *not* consistent with this definition of an herbivore.

23. **3** The best explanation for the differences in the appearance of the twins is that *the expression of genes is influenced by the environment*. By definition, identical twins share exactly the same genetic information. This means that their zygotes (fertilized eggs) should contain exactly the same instructions to guide their parallel differentiation, growth, and maturation. However, numerous studies by competent scientists have provided data that indicate that environmental conditions can alter the way genetic traits are expressed in the adult. Since the twins' environments differed in at least one respect (smoking versus nonsmoking), it is possible that the presence or absence of smoking has played a role in altering their appearance as adults as shown in the question.

WRONG CHOICES EXPLAINED:
(1) The best explanation for the differences in the appearance of the twins is *not* that *twin B is older than twin A*. As a matter of medical fact, twins are born at very nearly the same time, the difference usually being measured in minutes. However, this fact does not explain the variation in their appearance as adults.

(2) The best explanation for the differences in the appearance of the twins is *not* that *they each inherited half of their DNA from each parent.* It is true that the twins both inherited half of their DNA from each parent. However, this fact does not explain the variation in their appearance as adults.

(4) The best explanation for the differences in the appearance of the twins is *not* that *one twin resembles the mother and the other resembles the father.* By definition, identical twins share exactly the same genetic information, which determines at the point of fertilization whether they resemble one or the other parent. However, this fact does not explain the variation in their appearance as adults.

24. **4** These controlled fires maintain specific populations of plants by directly *interfering with the process of ecological succession.* The controlled fires help eliminate or retard the growth of shrubs and trees that would overtake the grassy field environment if not burned back in this way. Herbs and grasses quickly recover from these burns, allowing the grassy field ecosystem to thrive.

WRONG CHOICES EXPLAINED:

(1) It is *not* true that these controlled fires maintain specific populations of plants by directly *increasing the consumption of finite resources.* Finite resources such as minerals and fossil fuels are not impacted by such controlled burns.

(2) It is *not* true that these controlled fires maintain specific populations of plants by directly *decreasing the carbon dioxide level in the atmosphere.* Such burns temporarily increase, not decrease, the carbon dioxide content of the atmosphere.

(3) It is *not* true that these controlled fires maintain specific populations of plants by directly *stopping the process of evolution.* The process of evolution is not affected directly by such activities.

25. **2** *It will feed on organisms that are important to other species* is the statement that best describes the effect of the water flea on the lake. The presence of this invasive species will disrupt the natural ecological balance among native species established over millennia. Its insertion into, and spread throughout, this environment will likely lead to the elimination of one or more native species from New York State's aquatic ecosystems.

WRONG CHOICES EXPLAINED:

(1), (4) *It will not compete with animals in the local food chain* and *There will be no effect on native species in the lake* are *not* the statements that best describe the effect of the water flea on the lake. The question states, "These water fleas eat zooplankton, a food also consumed by native fishes." This statement indicates that the water fleas do directly compete with/affect animals/native species in the local food chain/lake.

(3) *The number of water fleas will decrease due to a lack of food* is *not* the statement that best describes the effect of the water flea on the lake. There is no information provided in the question that would support this inference.

26. **1** A stable ecosystem can have high biodiversity because each species in that ecosystem *occupies a different niche*. An ecological niche is the environmental role played by the species that fill that niche. These roles can include things needed by, and things provided by, these species. In a balanced ecosystem, only one species occupies a particular environmental niche. This helps to minimize interspecies competition within the ecosystem, a situation that assists in the establishment and maintenance of environmental stability.

WRONG CHOICES EXPLAINED:

(2) It is *not* true that a stable ecosystem can have high biodiversity because each species in that ecosystem *inhabits a different environment*. The terms "ecosystem" and "environment" are often used interchangeably. All species within a given ecosystem inhabit the same environment.

(3) It is *not* true that a stable ecosystem can have high biodiversity because each species in that ecosystem *is part of a different community*. By definition, ecological communities are included within ecosystems. All species in the community of species in an environment are included within the ecosystem.

(4) It is *not* true that a stable ecosystem can have high biodiversity because each species in that ecosystem *lives in a different biosphere*. By definition, the term "biosphere" is inclusive of all ecosystems found anywhere on Earth. So, all species in all ecosystems live within the biosphere.

27. **4** These organisms are *decomposers*. A decomposer is a bacterium or fungus that solely breaks down the bodies of dead plants and animals for its nutrition. Byproducts of this process that are not used by the decomposers are left available to other organisms for their use.

WRONG CHOICES EXPLAINED:

(1) It is *not* true that these organisms are *plants*. A green plant (producer) is one that is capable of manufacturing its own food. While plants are essential in the establishment and maintenance of ecosystem stability, they do not solely break down important molecules and make them available for other organisms to use.

(2) It is *not* true that these organisms are *herbivores*. An herbivore is an animal consumer that derives its nutrition solely from plant matter. While herbivores are essential in the establishment and maintenance of ecosystem stability, they do not solely break down important molecules and make them available for other organisms to use.

(3) It is *not* true that these organisms are *scavengers*. A scavenger is a consumer animal that consumes the dead bodies of other animals for its nutrition. While scavengers are essential in the establishment and maintenance of ecosystem stability, they do not solely break down important molecules and make them available for other organisms to use.

28. **2** The hormone referred to in this feedback mechanism is *insulin*. Insulin is synthesized within the pancreas and released into the bloodstream as needed to reduce levels of sugar in the blood. When the correct blood sugar concentration is established, the pancreas secretes less insulin so that the blood sugar level is held constant. The action of this feedback mechanism helps the body maintain homeostatic balance.

WRONG CHOICES EXPLAINED:

(1), (3), (4) It is *not* true that the hormone referred to in this feedback mechanism is *estrogen, progesterone,* or *testosterone*. These hormones are synthesized within the human reproductive organs for the purpose of regulating the production of secondary sex characteristics, the manufacture and release of gametes, and/or the maintenance of the developing embryo. They have no direct role in the regulation of blood sugar concentration.

29. **1** This sequence of rock layers best illustrates the concept that *the living and nonliving environment both change over time*. In an undisturbed section of sedimentary rock, the deepest layers (and the fossils found within them) are presumed to be the oldest. A bottom-to-top review of the section of rock illustrates that the bands of rock appear to be different in terms of their textures and colors. This leads to a reasonable inference that the nonliving environment (mineral content and sedimentation rate) changed over time. An examination of the fossils embedded in these rock layers indicates that the living environment (types of

plants and animals) changed over time, as well. This leads to a reasonable inference that these organisms underwent evolutionary change to produce progressively more complex forms. In some cases, fossilized organisms appear suddenly, then disappear without a leaving a complete fossil record.

WRONG CHOICES EXPLAINED:
(2), (3), (4) This sequence of rock layers does *not* best illustrate the concepts that *it is important to preserve the diversity of species and habitats*, that *new inheritable characteristics can result from the recombining of genes*, or that *living organisms have the capacity to produce populations of unlimited size*. Each of these statements states a true biological fact, but none of them is illustrated by the sequence of rock layers illustrated in the question.

30. **3** In the pyramid, the arrows labeled X represent *the loss of energy in the form of heat*. In any traditional energy pyramid, the base of the pyramid (producer level) is presumed to contain the greatest quantity of energy. As producers are consumed by herbivores, some of this energy ($\sim$10%) is retained in the chemical bonds of the herbivore tissues. The remainder ($\sim$90%) is dissipated as heat to the surrounding environment. A similar energy transfer/dissipation occurs at each trophic level transition. The arrows labeled X represent such losses.

WRONG CHOICES EXPLAINED:
(1) In the pyramid, the arrows labeled X do *not* represent *the loss of organisms due to predation*. Organisms are lost due to predation at many places in a food web/pyramid. Such predation constitutes a transfer of materials and energy from one trophic level to another. The arrows labeled X do not represent such losses.

(2) In the pyramid, the arrows labeled X do *not* represent *a decrease in photosynthetic organisms*. A change in the number of photosynthetic organisms (producers) may occur for many reasons in a natural environment, including fire, flood, drought, and other natural occurrences. The arrows labeled X do not represent such losses.

(4) In the pyramid, the arrows labeled X do *not* represent *a decrease in available oxygen*. The oxygen content of a balanced ecosystem remains stable within a fairly narrow range due to the processes of photosynthesis and respiration. The arrows labeled X do not represent such changes.

PART B–1

31. **2** *A graduated cylinder to measure 30 mL of water for each plant daily* is the tool that is correctly paired with a procedure that could be used during this experiment. A graduated cylinder is used in a laboratory to measure the volume of a liquid in units of volume (in this case milliliters/mL).

WRONG CHOICES EXPLAINED:

(1) *An electronic balance to measure the volume of soil in which each corn plant is grown* is *not* the tool that is correctly paired with a procedure that could be used during this experiment. An electronic balance is designed to measure the mass (e.g., grams/g), not the volume, of materials.

(3) *A metric ruler to determine the mass of each plant each week* is *not* the tool that is correctly paired with a procedure that could be used during this experiment. A metric ruler is designed to measure the linear dimension (e.g., centimeters/cm), not the mass, of materials.

(4) *A Celsius thermometer to determine the pH of the soil* is *not* the tool that is correctly paired with a procedure that could be used during this experiment. A Celsius thermometer is designed to measure the temperature (e.g., degrees Celsius/°C), not the pH (acidity), of materials.

32. **1** The independent variable in this experiment is the *air temperature at which the corn plants were grown*. The independent variable in a scientific experiment is the variable that is controlled by the investigator. In this case the investigator would vary the temperature conditions (independent variable) under which the corn plants were grown, and then measure and record the differences in growth rate (dependent variable).

WRONG CHOICES EXPLAINED:

(2), (3) It is *not* true that the independent variable in this experiment is the *amount of carbon dioxide used by the corn plants* or the *volume of oxygen produced by the corn plants*. These values are examples of the sort of data that should be collected by the researcher in this experiment.

(4) It is *not* true that the independent variable in this experiment is the *number of corn plants used*. The number of corn plants used is an element of the experimental design of this study. This value should be a constant in the experiment, not a variable of any kind.

33. **4** The graph can best be used to illustrate *carrying capacity*. The carrying capacity of an environment is the maximum number of individuals of a species that can be supported given limited food and other resources. When the carrying capacity is exceeded, there will be a growing scarcity of the limited resource until individuals of the species begin to die from lack of that resource. These deaths will tend to reduce the species population and help bring it back into balance, although that may require several generations to accomplish. Eventually, a stable population is formed that is balanced by births and deaths of the species.

WRONG CHOICES EXPLAINED:
(1) The graph *cannot* best be used to illustrate *a food chain*. Food chains illustrate nutritional relationships among species in terms of the transfer of energy from producers (green plants), which manufacture their own food via photosynthesis, to sequential levels of consumer organisms. A graph would not be an appropriate way to illustrate a food chain.

(2) The graph *cannot* best be used to illustrate *ecological succession*. Ecological succession refers to the series of changes that occur to the dominant plant community of an area that has been significantly affected by an environmental change. A graph would not be an appropriate way to illustrate ecological succession.

(3) The graph *cannot* best be used to illustrate *natural selection*. Natural selection, sometimes referred to as survival of the fittest, postulates that certain members of a species are better adapted to their environment than others and that these organisms are more likely to survive and pass their favorable adaptations on to future generations. A graph would not be an appropriate way to illustrate natural selection.

34. **3** The approximate number of animals that were found in June 2012 was most likely 76. The graph illustrates the population levels of this species in this environment over time. The population level of the species population reached a stable maximum of about 76 individuals in February, remaining stable after that through May. Assuming that environmental conditions do not change drastically, it is likely that the population will remain at around 76 for the month of June.

WRONG CHOICES EXPLAINED:
(1), (2), (4) The approximate number of animals that were found in June 2012 was *not* most likely *16*, *26*, or *86*. None of these values can be supported by application of the method described above or any similar method.

35. **4** Based on this observation, the scientist could conclude that *this species of bacteria synthesizes enzymes needed to digest the food in nine of the ten containers*. Based on their DNA information, bacteria manufacture and secrete digestive enzymes that are capable of breaking down specific organic compounds (foods). The observation that the bacteria could not grow in one of the ten containers is an indication that this species of bacteria lacks the DNA information that would allow them to produce the digestive enzyme that could break down that specific food type.

WRONG CHOICES EXPLAINED:

(1) It is *not* true that, based on this observation, the scientist could conclude that *all ten food sources used in the experiment are capable of supporting this species of bacteria*. Based on the description of the experiment, one of the ten containers was not capable of supporting the species of bacteria.

(2) It is *not* true that, based on this observation, the scientist could conclude that *the temperature varied greatly in nine of the containers during this experiment*. Based on the description of the experiment, all containers were incubated at 26°C.

(3) It is *not* true that, based on this observation, the scientist could conclude that *only the container that failed to grow any bacteria was prepared correctly*. Based on the description of the experiment, it is unlikely that such an error would have been made by a competent scientist.

36. **4** *Future generations can be affected by the choices of current generations* is the statement that is supported by the graph. The graphed data project the likely outcomes by the year 2090 of three different courses of action relative to atmospheric carbon dioxide concentration. The information presented poses choices that can be made by governments today that will help to slow the increase of carbon dioxide content of the atmosphere in the future. The cost of doing nothing will very likely lead to catastrophic consequences by the year 2100.

WRONG CHOICES EXPLAINED:

(1) *Climate change will result in the melting of polar ice caps* is *not* the statement that is supported by the graph. Although this phenomenon is caused in part by atmospheric carbon dioxide concentrations, no data are presented in the graph that support this concept.

(2) *The increase in carbon dioxide levels will cause a decrease in global average temperature* is *not* the statement that is supported by the graph. Increasing carbon dioxide levels are known to be partly responsible for increasing, not decreasing, global average temperature. No data are presented in the graph that support this concept.

(3) *Human activities have no effect on atmospheric carbon dioxide levels* is *not* the statement that is supported by the graph. Although competent scientists have concluded that human activities do have a significant effect on atmospheric carbon dioxide concentrations, no data are presented in the graph that support this concept.

37. **1** Graph *1* most accurately represents the results obtained by this student. The data in the table demonstrate that as pH increases (acidity decreases) from 3 to 7, the reaction rate of catalase on H_2O_2 increases proportionally. When graphed, these data most closely resemble the graph labeled *1*.

WRONG CHOICES EXPLAINED:
(2), (3), (4) Graphs 2, 3, and 4 do *not* most accurately represent the results obtained by this student. None of these graphs accurately represents the data presented in the table. See the correct answer above.

38. **2** *Catalase has the greatest activity at a pH of 7* is the conclusion that is valid based upon the data collected by the student. According to the data presented in the table, the highest reaction rate (1.5 mL O_2/min) was observed in the experimental setup at pH 7.

WRONG CHOICES EXPLAINED:
(1) *The change in pH prevents catalase from breaking down water* is *not* the conclusion that is valid based upon the data collected by the student. Water is a byproduct, not a reactant, in this process. Catalase catalyzes a reaction in which hydrogen peroxide (H_2O_2) is broken down into water and oxygen gas.
(3) *Oxygen production will increase if more water is added to the reaction* is *not* the conclusion that is valid based upon the data collected by the student. Water is a byproduct, not a reactant, in this process. Adding water to the reaction setups will have the effect of reducing the concentration of catalase and hydrogen peroxide (H_2O_2), which will decrease not increase, the reaction rate.
(4) *Catalase has the greatest production of oxygen at a pH of 3* is *not* the conclusion that is valid based upon the data collected by the student. The data clearly indicate that this reaction occurs most rapidly at a pH of 7.

39. **3** The best explanation for the change in catalase activity as the pH changed from 7 to 3 is that *in acidic solutions, the shape of catalase changes, causing the reaction rate to decrease*. All enzymes (in this case catalase) have an area of their molecules (active site) that links to the substrate molecule (in this case H_2O_2). If the active site is altered (denatured) by conditions such as acidity, heat, or chemicals, this alteration may make the active site less effective or stop its catalytic action altogether.

WRONG CHOICES EXPLAINED:

(1) The best explanation for the change in catalase activity as the pH changed from 7 to 3 is *not* that *strong acid digests the catalase, causing the reaction rate to increase*. Strong acid may denature, not digest, an enzyme such as catalase. Denaturation would have the effect of decreasing, not increasing, the catalytic action of that enzyme.

(2) The best explanation for the change in catalase activity as the pH changed from 7 to 3 is *not* that *the student most likely cooled the H_2O_2 solution, causing the reaction rate to increase*. Cooling the solution in which this reaction occurs would most likely have the effect of decreasing, not increasing, the rate of reaction.

(4) The best explanation for the change in catalase activity as the pH changed from 7 to 3 is *not* that *decreased oxygen production causes catalase to increase the rate of reaction*. Oxygen (O_2) is a byproduct, not a reactant, in this process. A decrease in oxygen concentration would have no direct effect on the rate of this reaction.

40. **2** Graph 2 best illustrates the body temperature in an individual maintaining dynamic equilibrium. The body maintains many of its life processes through feedback mechanisms that sense conditions in the blood and make adjustments based on this input. When the mechanism detects decreasing blood temperature, it stimulates the metabolic processes to release additional heat via the respiratory reactions, which raises blood/body temperature. When the mechanism detects increasing body temperature, it signals the metabolic process to slow and stimulates the body's skin pores to open, skin capillaries to dilate, and sweat glands to secrete perspiration, which lowers body temperature. In this way, the feedback mechanism causes blood/body temperature to vary around a "normal" body temperature of 98.6°F (37°C). Graph 1 best illustrates the results of this mechanism.

WRONG CHOICES EXPLAINED:

(1), (3), (4) Graphs 1, 3, and 4 do *not* best illustrate the body temperature in an individual maintaining dynamic equilibrium. None of these graphs accurately illustrates the manner in which the body/blood normally vary in terms of temperature.

41. **4** Approximately *1550 kilowatt hours* of energy are saved by these models annually when compared to the models produced in 1972. This value is determined by subtracting the approximately 450 kwh measured in 2003 models from the approximately 2000 kwh measured in 1972 models. This difference is closest to 1550 kwh.

WRONG CHOICES EXPLAINED:

(1), (2), (3) It is *not* true that approximately *500 kilowatt hours, 550 kilowatt hours,* or *1500 kilowatt hours* of energy are saved by these models annually when compared to the models produced in 1972. None of these values accurately represent the difference between 1972 models and 2003 models. See the correct answer above.

42. **1** *More technological improvements in appliances can help conserve finite resources* is the statement that best represents an outcome of federal standards that require increasing the energy efficiency of appliances such as refrigerators. Finite resources such as coal, oil, and natural gas are used to power electrical generation plants. Increased efficiency of electrical appliances means that they require less energy/unit to operate, which helps conserve those finite resources for the future.

WRONG CHOICES EXPLAINED:

(2) *Increased efficiency of appliances requires greater use of our energy resources* is *not* the statement that best represents an outcome of federal standards that require increasing the energy efficiency of appliances such as refrigerators. Increased efficiency of appliances will lead to lesser, not greater, energy/unit used.

(3) *Newer appliances are manufactured from a greater number of finite resources* is *not* the statement that best represents an outcome of federal standards that require increasing the energy efficiency of appliances such as refrigerators. It is true that each new appliance requires the use of finite resources to manufacture it. However, this fact is not a direct outcome of federal energy standards.

(4) *Manufacturing more efficient appliances will reduce the biodiversity of ecosystems* is *not* the statement that best represents an outcome of federal standards that require increasing the energy efficiency of appliances such as refrigerators. The manufacture of new appliances will not reduce biodiversity of ecosystems unless that manufacture requires the loss or degradation of natural habitats in those ecosystems.

43. **4** The gas that was being produced was most likely *oxygen as a product of the process of photosynthesis.* The biochemical process of photosynthesis carried on by green plants recombines the atoms that make up water and carbon dioxide to form molecules of glucose and oxygen gas. Oxygen gas is released into the atmosphere as a byproduct of photosynthesis.

WRONG CHOICES EXPLAINED:

(1) The gas that was being produced was *not* most likely *carbon dioxide as a product of the process of respiration*. Under experimental conditions as illustrated, carbon dioxide produced by respiration in the plant is immediately taken up as a raw material for use in photosynthesis. For this reason, carbon dioxide gas is not released into the atmosphere under such conditions.

(2) The gas that was being produced was *not* most likely *carbon dioxide as a product of the process of photosynthesis*. Carbon dioxide is a raw material, not a byproduct, of photosynthesis.

(3) The gas that was being produced was *not* most likely *oxygen as a product of the process of respiration*. Oxygen is a raw material, not a byproduct, of respiration.

PART B–2

44. One credit is allowed for correctly marking an appropriate scale, without any breaks in the data, on each labeled axis. [1]

45. One credit is allowed for correctly plotting the data on the grid, connecting the points, and surrounding each point with a small circle. [1]

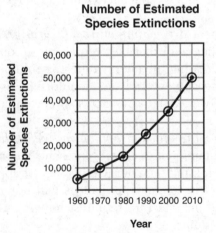

Number of Estimated Species Extinctions

46. One credit is allowed for correctly stating *one* possible cause for the increase in the number of species extinctions from 1960 to 2010. Acceptable responses include but are not limited to: [1]

- *There could have been a sudden increase in environmental changes during this period.*
- *It is possible that these species have inherited traits not favorable for a changing environment.*

- *Some of these species may lack sufficient genetic diversity to respond to environmental change.*
- *During that period, alterations of ecosystems due to human impact/population growth may have occurred.*
- *Food supplies these species depended on may have been disrupted.*
- *Pollution from human activities may have made survival of these species difficult/impossible.*
- *The natural habitats these species depended on for their survival could have been disrupted/destroyed.*

47. **2** Process X is referred to as *mutation*. Mutations are random events in which genetic material (DNA) is altered through the action of mutagenic agents. It can be inferred from the diagram and information provided that such an alteration/mutation of the digestive enzyme gene occurred in the past that gave rise to the antifreeze protein gene in this species. This alteration produced a genetic adaptation that has promoted the survival of this species in this extreme environment.

WRONG CHOICES EXPLAINED:

(1) Process X is *not* referred to as *mitosis*. Mitosis is a process in which homologous chromosome pairs of a parent nucleus are replicated exactly and separated into two genetically identical daughter nuclei. Mitosis is not the process by which genes are altered.

(3) Process X is *not* referred to as *differentiation*. Differentiation is the process by which embryonic cells undergo specialization to become specific body tissues. Differentiation is not the process by which genes are altered.

(4) Process X is *not* referred to as *meiosis*. Meiosis is a process in which homologous chromosome pairs and the genes they carry are separated into haploid gametes during gametogenesis. Meiosis is not the process by which genes are altered.

48. One credit is allowed for correctly explaining how the process of natural selection can account for the increase in frequency of the antifreeze protein gene in the icefish population. Acceptable responses include but are not limited to: [1]

- As the ocean cooled, fish with the antifreeze protein were more likely to survive and pass on this trait.
- Icefish that did not have the gene to keep their blood from freezing would die.

- The theory of natural selection states that variation in a species can provide opportunities for selection of favorable adaptations (antifreeze protein) over less favorable adaptations (digestive enzyme) within a species struggling to survive under extreme environmental conditions/change.

49. **1** This evidence makes it likely that *icefish ancestors had hemoglobin.* There is insufficient information provided in this question to arrive at any reasonable inference regarding the past or future genetic capabilities of the icefish species. The incomplete DNA sequences referenced may be precursors of future hemoglobin sequences, artifacts of former hemoglobin sequences, or copied/random sequences that mimic hemoglobin sequences. Of the choices given this is the most reasonable, but still speculative, assumption.

WRONG CHOICES EXPLAINED:
(2), (3), (4) This evidence does *not* make it likely that *icefish will soon produce offspring with hemoglobin,* that *hemoglobin is a molecule made by some fish that do not have genes for it,* or that *soon all fish will stop producing hemoglobin.* No information is provided in the question that would support any of these inferences.

50. **4** The best explanation for why sea slugs are able to pass on this ability to make food to their offspring is that *the gene for chlorophyll production is part of their DNA.* There is insufficient information provided in this question to arrive at any reasonable inference regarding the genetic capabilities of the green sea slug. In addition, the paragraph provides inaccurate scientific information concerning the green sea slug's synthesis of chlorophyll versus its ability to absorb and maintain DNA and chloroplasts ingested from algae. In fact, green sea slugs can perform photosynthesis only after ingesting algae in their environment and incorporating the algal chloroplasts and chlorophyll into their body cells. For this reason, the correct answer above is correct only insofar as it can be supported by the incomplete and inaccurate scientific information provided in the paragraph.

WRONG CHOICES EXPLAINED:
(1) The best explanation for why sea slugs are able to pass on this ability to make food to their offspring is *not* that *the gene for making algae is in all their body cells.* Algae are a large and diverse class of producer species that inhabit the Earth's aquatic and marine ecosystems. There is no information provided in the paragraph that would support the concept that green sea slugs make algae in their body cells.

(2) The best explanation for why sea slugs are able to pass on this ability to make food to their offspring is *not* that *making food is beneficial, so the slugs need to mutate*. Mutations are random events in which genetic material (DNA) is altered by the action of mutagenic agents. Mutations do not occur as a function of need in any known species.

(3) The best explanation for why sea slugs are able to pass on this ability to make food to their offspring is *not* that *the environment causes the slugs to become green*. Sea slugs inhabit marine environments that support their survival. Green sea slugs turn green only after ingesting algae in their environment and incorporating the algal chloroplasts and chlorophyll into their body cells. Sunlight is an environmental factor needed to maintain the chlorophyll's photosynthetic action and its green pigmentation. Note that this distracter may be considered a correct answer to the question as written, although it is not the best answer.

51. One credit is allowed for correctly explaining how green sea slugs can be considered both a producer and a consumer. Acceptable responses include but are not limited to: [1]

- The sea slug is considered a producer when it makes its own food using the chlorophyll and a consumer when it eats the algae.
- It is a producer when it makes food and a consumer when it eats algae.
- They can both eat other organisms and make their own food,
- They carry on photosynthesis and eat algae for food.

52. One credit is allowed for correctly stating why fossil fuels are considered a finite resource. Acceptable responses include but are not limited to: [1]

- Fossil fuels are not renewable.
- Fossil fuels will run out one day.
- Fossil fuels take millions of years to form.
- Coal, oil, and natural gas are fuels that were formed under unique circumstances that existed on Earth millions of years ago. These fuels are not being formed on Earth anymore and will eventually be exhausted by human uses.

53. One credit is allowed for correctly identifying the process that occurs in this organelle and for explaining the importance of this process to the survival of organisms. Acceptable responses include but are not limited to: [1]

- Process:
  - *respiration*
  - *cellular respiration*
  - *aerobic respiration*
  - *synthesis of ATP*

- Importance:
  - *produces ATP for the cell to carry out life functions*
  - *releases energy for use by the organism*
  - *ATP provides energy for life processes*
  - *converts the bond energy in glucose to the bond energy in ATP*
  - *makes energy available to activate cellular reactions*

54. One credit is allowed for correctly identifying the type of cell division involved in each process. Acceptable responses include but are not limited to: [1]

- Skin cells:
  - *mitosis*
  - *mitotic cell division*

- Gametes:
  - *meiosis*
  - *meiotic cell division*
  - *reduction division*
  - *gametogenesis*

55. One credit is allowed for correctly explaining how the genetic makeup of the skin cells differs from the genetic makeup of the gametes. Acceptable responses include but are not limited to: [1]

- Skin cells contain the full number of chromosomes for the individual, while gametes contain half the number of chromosomes.
- Skin cells of an individual are normally genetically identical, but gametes have variation in their genetic makeup.
- Human skin cells have the diploid/2n number of chromosomes/46 chromosomes; human gametes have the haploid/monoploid/n number of chromosomes/23 chromosomes.

- Skin cells contain twice as much genetic material as gametes.
- Skin cells contain two sets of homologous chromosomes, while gametes contain only one set of homologous chromosomes.

PART C

56. One credit is allowed for correctly describing *one* function of the placenta during the internal development of an offspring. Acceptable responses include but are not limited to: [1]

- Oxygen and nourishment for the developing offspring diffuse across the placenta.
- The placenta allows the exchange of materials, including wastes, between the mother's blood and her developing fetus.
- The placenta protects the fetus from some infections.
- The placental tissues help to ensure that the mother's blood and the fetus's blood do not comingle.
- The placenta provides a barrier that helps prevent an antigen-antibody reaction that could harm both mother and fetus.

57. One credit is allowed for correctly describing *one* advantage for an offspring to develop internally as opposed to developing externally. Acceptable responses include but are not limited to: [1]

- The developing offspring has a greater chance of survival because it is inside the mother.
- The offspring is more protected and therefore is more likely to survive.
- Internal environmental conditions are more constant/controlled than with external development.
- Offspring that develop externally are exposed to harsh/life-threatening environmental conditions, whereas offspring that develop internally are not.
- The moist/warm/protected conditions within the uterus provide a stable/controlled environment for the developing offspring.

58. One credit is allowed for correctly identifying *one* factor, besides genetics, that could influence the development of human offspring. Acceptable responses include but are not limited to: [1]

- hormones produced by the mother
- the environment within the uterus
- mother's health/nutrition/exposure to pollutants
- mother's use of drugs/alcohol/tobacco
- disease in the mother

59. One credit is allowed for correctly explaining why it is an advantage for female finches to be small. Acceptable responses include but are not limited to: [1]

- Small female finches need less food to survive.
- Small finches can mate earlier in the spring/produce more offspring.
- Small females survive better as nestlings.
- The smaller females survive, reproduce, and pass on this trait to offspring.

60. One credit is allowed for correctly explaining why the population of large male house finches in Michigan and Montana continues to increase. Acceptable responses include but are not limited to: [1]

- Large males are more likely to mate successfully with females, producing more large males.
- Large males are more likely to survive the winter.
- The large males survive, reproduce, and pass on this trait.
- They outcompete the smaller males for food.

61. One credit is allowed for correctly predicting what both the males and females in the finch population might be like 10 years in the future based on what is currently happening with the finch population. Acceptable responses include but are not limited to: [1]
[NOTE: The biological definition of *population* is "all the members of a particular species that inhabit a particular environment at a particular time." For this reason, it is biologically inaccurate to refer to a "male house finch population" or a "female house finch population."]

- Males:
  - *Based on the passage, I predict that almost all the male house finches will be large.*
  - *more large ♂ individuals than small ♂ individuals compared to now*
  - *a greater proportion of large and a lesser proportion of small males than today*
  - *The smaller males will gradually decrease as a percentage of the house finch population.*
- Females:
  - *Based on the passage, I predict that almost all the female house finches will be small.*
  - *more small ♀ individuals than large ♀ individuals compared to now*

○ *a greater proportion of small and a lesser proportion of large females than today*

○ *The smaller females will gradually increase as a percentage of the house finch population.*

62. One credit is allowed for correctly stating *one* specific effect that a continuous rise in temperature could have on the sex ratio of the hatchlings of the sea turtle population. Acceptable responses include but are not limited to: [1]

- As temperatures rise, more sea turtles will develop as females than as males.
- Females will begin to outnumber males in the sea turtle population.
- If temperatures rise to/above 31°C/88°F, the hatchlings will be almost all female.
- The ratio of male to female hatchlings will change in favor of the females.

63. One credit is allowed for correctly explaining how having adequate amounts of shade-producing vegetation nearby, such as palm trees, can affect the nesting success of endangered sea turtles. Acceptable responses include but are not limited to: [1]

- It is important to keep the nests from getting too hot and perhaps killing all the eggs before they hatch.
- If there are no palm trees to shade and cool the nests, more females than males will hatch.
- Without trees to shade them and moderate temperatures, fewer of the sea turtles will survive to hatch.

64. One credit is allowed for correctly stating *one* way that tourist activities in areas where turtles make their nests could have a *negative* impact on the nesting success of the turtles. Acceptable responses include but are not limited to: [1]

- Adult turtle females attempting to lay eggs could be disturbed by tourists, causing the turtles to abandon their nests prematurely.
- Tourists using the beach could unknowingly trample on or damage the nests.
- Tourists could deliberately damage the nests by exposing/destroying/collecting the sea turtle eggs.
- If beaches are disturbed by hotel staff making beaches nice for tourists, the nests could be disrupted.
- Hatchling sea turtles struggling down the beach could be killed/collected by tourists unaware of the hatchlings' need to reach the ocean waves for their survival.

65. One credit is allowed for correctly stating the role of the population of grasshopper mice in the Sonoran Desert food web. Acceptable responses include but are not limited to: [1]

- The mice are carnivores/consumers in their desert environment.
- They consume animals such as crickets, rodents, tarantulas, and scorpions, which helps to keep their numbers in check.
- They serve as predators in the food web.
- They inhabit nests of other animals and displace other animals from their living spaces.

66. One credit is allowed for correctly stating *one* advantage grasshopper mice have over the other local populations when competing for resources. Acceptable responses include but are not limited to: [1]

- They have a defense mechanism against scorpion venom that enables them to use the scorpion for food.
- They are very aggressive and take nests of other organisms, which enables them to conserve energy.
- They are capable of taking down animals much larger than themselves.
- They are able to gather food energy from several different sources unavailable to other species of mice.

67. One credit is allowed for correctly identifying *one* advantage to grasshopper mice of being active during the night rather than daylight. Acceptable responses include but are not limited to: [1]

- Their hunting is not affected by the extreme heat during the day.
- They are better able to sneak up on prey in the dark.
- They are not as likely to become dehydrated by the hot daytime conditions.
- Their prey is active at night.
- They are less visible at night to predators such as owls and foxes.

68. One credit is allowed for correctly explaining how research on the mutated protein identified in the grasshopper mouse could benefit humans suffering from chronic pain. Acceptable responses include but are not limited to: [1]

- If scientists can isolate the mutant protein, they could give it to patients and block the sensation of pain.
- If humans learn how the mutated protein works, scientists might have a better idea how to treat human pain.

- Mice are often used in research for human diseases because mice and humans are both mammals.
- If there is a chemical that works for blocking pain in the mouse, it might work in people.

69. One credit is allowed for correctly explaining how vaccinations protect against diseases. Acceptable responses include but are not limited to: [1]

- Vaccinations stimulate the immune system to produce antibodies against the disease.
- Vaccinations activate the human immune system.
- Vaccinations cause you to make antibodies against a specific disease.

70. One credit is allowed for correctly explaining why a single vaccination is *not* effective against all flu viruses. Acceptable responses include but are not limited to: [1]

- Because the virus is different in each flu strain, different antibodies would be needed.
- The old vaccine might not work because the flu virus changes/mutates each year.
- Vaccinations work against specific viruses. As the virus changes, the vaccination must change to match it.
- The flu virus mutates frequently, so the old vaccine is no longer effective.

71–72. Two credits are allowed for correctly discussing how invasive species can harm the ecosystem. In your answer, be sure to:

- explain *one* negative effect that faucet snails have on the lake ecosystem [1]
- describe *one* human activity that can slow the spread of the faucet snail [1]

Acceptable responses include but are not limited to: [2]

One negative effect that the faucet snail has on the ecosystem is that it can infect and kill waterfowl such as ducks and geese. Humans can help slow the spread of the snail by cleaning/sanitizing/decontaminating the bottoms of their boats after each use and before putting their boats in at another water body. [2]

Parasites found in faucet snails can attack ducks that eat the snails, causing the ducks to sicken and die. A human activity that can slow the spread of the snail is to present educational programs to boaters about the hazards posed by the snail and point out the importance of decontaminating their boats before moving them from one body of water to another. [2]

PART D

73. **4** *The blood coming to the lungs is high in CO_2 and low in O_2, so the gases each diffuse from a higher to a lower concentration in this area* is the statement that best explains why these gases are able to move in the directions shown in the diagram. The statement as written represents a reasonable definition of the process of diffusion. In any system, gaseous or dissolved materials tend to move from a region of higher concentration to a region of lower concentration (diffuse) until the material is evenly distributed throughout the system. This process enables CO_2 molecules to diffuse from the blood fluid into the alveolus at the same time that O_2 molecules diffuse from the alveolus into the blood fluid.

WRONG CHOICES EXPLAINED:

(1) *The CO_2 moves out of the capillary and into the alveolus to make room for the blood to carry O_2* is *not* the statement that best explains why these gases are able to move in the directions shown in the diagram. Dissolved gases of different molecular types may be held in a fluid system independent of the concentrations of each other. Molecules of O_2 that are absorbed into the blood fluid are quickly bound to and transported by hemoglobin in red blood cells, while CO_2 molecules are dissolved in and transported by the blood fluid.

(2) *The O_2 is needed by the cells, so it is actively transported into the blood. The CO_2, which is not needed, is actively transported out of the blood* is *not* the statement that best explains why these gases are able to move in the directions shown in the diagram. Active transport is a process by which substances are transported across cell membranes from an area of low relative concentration to an area of high relative concentration, requiring the expenditure of cell energy. Active transport is not represented in the diagram.

(3) *The blood coming to the lungs is low in CO_2 and high in O_2, so the gases each diffuse from a lower to a higher concentration in this area* is *not* the statement that best explains why these gases are able to move in the directions shown in the diagram. The statement as written represents a reasonable definition of active transport. Active transport against the concentration gradient requires the expenditure of cell energy and is not represented in the diagram.

74. **1** The usual order in which these events would occur is $B - D - A - C$. The sequence of actions in protein synthesis in a living cell starts with the formation (replication) of DNA in the nucleus, which carries the genetic information (in this case altered) to encode a protein (Event B). The second event that occurs is the synthesis of a strand of mRNA that is a complementary impression of the altered DNA (Event D). Next, the mRNA moves out of the nucleus to the ribosome,

where it attracts tRNA molecules with complementary codons and joins their amino acids in the specific (altered) order prescribed by the mRNA strand, in order to synthesize a specific (altered) protein molecule (Event A). Finally, this altered protein is released into the cytoplasm where it may attempt to catalyze a new biochemical process that may produce a new genetic phenotype in that cell (Event C). [Note: This genetic change would be inheritable *only* if the genetic change were to occur in a gamete that subsequently participated in a successful fertilization leading to the development of a fertile adult organism. Otherwise, the alteration and its phenotype would be lost when the cell and its offspring died.]

WRONG CHOICES EXPLAINED:
(2), (3), (4) It is *not* true that the usual order in which these events would occur is $B - D - C - A$, $D - A - B - C$, or $D - B - C - A$. None of these event orders is consistent with the sequence by which this process normally occurs. See the correct answer above.

75. **2** Gel electrophoresis diagram 2 best supports this statement. The diagram illustrates that the banding pattern of DNA fragments for species A varies considerably compared to those of species B and species C, while banding patterns for species B and species C are identical to each other. These patterns indicate that species B and species C are more closely related to each other than either is to species A.

WRONG CHOICES EXPLAINED:
(1), (3) Gel electrophoresis diagrams 1 and 3 do *not* best support this statement. In these diagrams, the banding patterns for species B and species C are neither identical nor nearly identical. These patterns would indicate that species B and species C are no more closely related to each other than either is to species A.
(4) Gel electrophoresis diagram 4 does *not* best support this statement. In this diagram, the banding patterns for species A and species C are identical. This pattern would indicate that species A and species C are more closely related to each other than either is to species B.

76. **3** In addition to analyzing DNA, evidence that *species B and C possess many of the same enzymes* could be used to best support the evolutionary relationship between species B and C. If two organisms share many enzymes in common, this would indicate that they share many genetic traits (many genes) in common and therefore are related evolutionarily.

WRONG CHOICES EXPLAINED:

(1) In addition to analyzing DNA, evidence that *species B and C live in the same ecosystem* could *not* be used to best support the evolutionary relationship between species *B* and *C*. Many diverse species may inhabit a particular ecosystem but still not be closely related evolutionarily.

(2) In addition to analyzing DNA, evidence that *species B and C require the same amount of sunlight* could *not* be used to best support the evolutionary relationship between species *B* and *C*. Many different species of plants may require similar amounts of sunlight but still not be closely related evolutionarily.

(4) In addition to analyzing DNA, evidence that *species B and C grow to the same maximum height as species A* could *not* be used to best support the evolutionary relationship between species *B* and *C*. Many divergent species may grow to the same maximum height but still not be closely related evolutionarily.

77. One credit is allowed for correctly circling the dot that best represents the common ancestor of species *A*, *B*, and *C*. Acceptable responses include: [1]

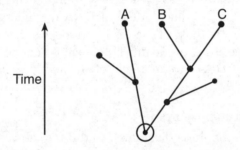

78. One credit is allowed for correctly explaining *one* way that sharp-billed ground finches and small tree finches could possibly compete with each other if they lived on the same island. Acceptable responses include but are not limited to: [1]

- *Because their diets are listed as "mainly" plant/animal food, they might compete if all/most plant foods, or all/most animal foods, available to these species were removed from their shared island environment.*
- *If food resources were scarce, one or both species might abandon their genetically programmed feeding spaces (ground versus tree) in order to enable such competition to occur.*
- *They might compete for abiotic resources (water/temperature/oxygen/ mineral) resources if they were limited in some way in their shared island environment.*

[NOTE: There is insufficient information provided in the chart to allow students to identify other biotic factors (e.g., nesting, mating, territorial, symbiotic) for which these species might compete.]

79. One credit is allowed for correctly describing a situation that would allow both the small tree finch population and the large tree finch population to live on the same island even though they both feed on animal food. Acceptable responses include but are not limited to: [1]

- *One species might feed in the evening and the other species might feed during the day.*
- *One species may feed higher in the trees and the other species lower in the trees.*
- *They might feed on different types/sizes of animal food.*
- *They might feed on animal food located in different areas of the island.*
- *There may be enough food available, so they do not need to compete.*

80. One credit is allowed for correctly describing *one* change in beak characteristics that would most likely occur in the finch population after many generations if this change in seed size became permanent. Acceptable responses include but are not limited to:　[1]

- *Finches with thicker, stronger beaks would survive the change and would pass their characteristics to future generations.*
- *Birds with larger, thicker beaks would become more common in the various finch species populations.*
- *Some finch species populations lacking variation in beak size and shape might be eliminated from the island altogether due to starvation.*
- *There would be fewer individual finches of any species displaying smaller beaks.*

81. **1** The different tools (such as spoons, chopsticks, or pliers) used during *The Beaks of Finches* laboratory activity represented variations in *feeding adaptations*. These tools were used in this activity to mimic the beak variations that could be observed among the various species of finches that inhabit the Galapagos Islands.

WRONG CHOICES EXPLAINED:
(2) The different tools (such as spoons, chopsticks, or pliers) used during *The Beaks of Finches* laboratory activity did *not* represent variations in *seed size*. This factor was represented by the objects of various sizes that students were challenged to pick up and move with the tools mentioned.

(3), (4) The different tools (such as spoons, chopsticks, or pliers) used during *The Beaks of Finches* laboratory activity did *not* represent variations in *finch migration* or *island ecosystems*. These factors were not represented in any way in this laboratory activity.

82. **3** During this activity, the purpose of finding the respiratory rates of the students at rest was *to use as a comparison* with the results of the students' respiratory rates during exercise. These data for the students at rest served as a control in this investigation. In any scientific investigation, the control group is set up in exactly the same manner as the experimental group except for the independent variable that is manipulated by the researcher. In this investigation, the independent variable was the type and duration of exercise. The dependent variable was the students' measured and averaged respiratory rates. The control data provided a baseline for comparison of resting versus active respiratory rates.

WRONG CHOICES EXPLAINED:
(1) It is *not* true that, during this activity, the purpose of finding the respiratory rates of the students at rest was *to determine if the students were healthy*. This might have been determined by having each student undergo a physical evaluation by a health professional.

(2) It is *not* true that, during this activity, the purpose of finding the respiratory rates of the students at rest was *to practice using the timer*. This activity should have been practiced independently of the actual investigation using random students as subjects.

(4) It is *not* true that, during this activity, the purpose of finding the respiratory rates of the students at rest was *to use as the variable*. The independent variable was the type and duration of activity/exercise in this investigation. The dependent variable was the students' measured and averaged respiratory rates. The data of the students' resting respiratory rates served as a control, not a variable, in this investigation.

83. One credit is allowed for correctly stating *one* likely hypothesis that the student was testing in this investigation. Acceptable responses include but are not limited to: [1]

- *If the athletes exercise, then their respiratory rates will increase.*
- *If the athletes exercise, then their breathing rates will decrease.*
- *Breathing rates are affected by exercise.*
- *Exercise will make people breathe faster than when they are at rest.*

84. One credit is allowed for correctly stating *one* reason why students would have a range of respiratory rates. Acceptable responses include but are not limited to: [1]

- *The students are different and would have different respiratory rates.*
- *Some students may have a disease or medical condition that causes them to have a faster or slower breathing rate than normal.*
- *Some may be in better physical condition than others.*
- *Some may have been active just before the measurement.*
- *Some students may not have been given an opportunity to rest fully after arriving at the track and before their breathing rates were measured.*

85. One credit is allowed for correctly identifying the colors that could be observed at X and at Y after the addition of the starch indicator. Acceptable responses include: [1]

- Final color of X:
  - *black*
  - *blue-black*
- Final color of Y:
  - *amber*
  - *amber colored*
  - *yellowish*

Standards/Key Ideas	August 2018 Question Numbers	Number of Correct Responses
Standard 1		
Key Idea 1: The central purpose of scientific inquiry is to develop explanations of natural phenomena in a continuing and creative process.	37, 38	
Key Idea 2: Beyond the use of reasoning and consensus, scientific inquiry involves the testing of proposed explanations involving the use of conventional techniques and procedures and usually requiring considerable ingenuity.		
Key Idea 3: The observations made while testing proposed explanations, when analyzed using conventional and invented methods, provide new insights into natural phenomena.	34, 41, 44, 45, 46	
Laboratory Checklist	31, 32	
Standard 4		
Key Idea 1: Living things are both similar to and different from each other and from nonliving things.	1, 5, 15, 16, 21, 51, 53, 65, 66, 67, 68	
Key Idea 2: Organisms inherit genetic information in a variety of ways that result in continuity of structure and function between parents and offspring.	3, 4, 7, 8, 9, 23, 47, 50, 55	
Key Idea 3: Individual organisms and species change over time.	10, 20, 48, 49, 59, 60, 61	
Key Idea 4: The continuity of life is sustained through reproduction and development.	11, 12, 54, 56, 57, 58	
Key Idea 5: Organisms maintain a dynamic equilibrium that sustains life.	2, 6, 17, 18, 19, 28, 35, 39, 40, 43, 69, 70	
Key Idea 6: Plants and animals depend on each other and their physical environment.	22, 24, 26, 27, 29, 30, 33, 62, 71	
Key Idea 7: Human decisions and activities have a profound impact on the physical and living environment.	13, 14, 25, 36, 42, 52, 63, 64, 72	
Required Laboratories		
Lab 1: "Relationships and Biodiversity"	74, 75, 76, 77	
Lab 2: "Making Connections"	82, 83, 84	
Lab 3: "The Beaks of Finches"	78, 79, 80, 81	
Lab 5: "Diffusion Through a Membrane"	73, 85	

Examination June 2019

Living Environment

PART A

Answer all questions in this part.   [30]

Directions (1–30): For *each* statement or question, record in the space provided the *number* of the word or expression that, of those given, best completes the statement or answers the question.

1 Which activity is an example of a decomposer recycling organic compounds back into the environment?

  (1) A tree synthesizes starch from simpler molecules.
  (2) A bacterial cell performs photosynthesis.
  (3) A bird digests proteins from its food.
  (4) A fungus breaks down the body of a dead animal.   1 _____

2 Itching and other skin problems are signs that a cat or dog may have fleas. Fleas are parasites known for their biting and blood-sucking abilities. When they bite, flea saliva enters the pet's circulatory system, sometimes causing an allergic response commonly seen as a "hot spot" on the pet's neck or the base of its tail.

Source: https://www.planetnatural.com/
pest-problem-solver/household-pests/flea-control/

These observations are best explained by the fact that

(1) flea saliva may stimulate an immune response in cats and dogs
(2) fleas are microbes whose bites cause a decreased blood flow
(3) flea saliva is a toxic substance that is released when fleas prey on cats and dogs
(4) fleas are host organisms whose saliva digests cat and dog fur, leaving "hot spots"   2 _____

3 A German measles (rubella) epidemic during the years 1963 to 1965 resulted in approximately 30,000 babies being born with birth defects. The specific cause of these birth defects was most likely

(1) the development of rubella virus infections in embryos
(2) the failure of zygotes infected with rubella to develop
(3) mutations in the nerve cells of pregnant females at the time of the rubella epidemic
(4) an increase in the amount of time needed for healthy embryonic development   3 _____

4 Placenta previa is a medical condition that occurs in some pregnant women. Women with this condition are often placed on bed rest, which prohibits them from any strenuous activity that may cause the blood vessels in the placenta to rupture. If not diagnosed, placenta previa can be a very dangerous condition because the placenta is

(1) the primary source of oxygen for the mother
(2) where the fetus obtains milk from the mother
(3) where nutrients and wastes are exchanged
(4) the primary source of estrogen and progesterone in the mother   4 _____

5 Over time, a tree that once had a total mass of 300 g increased in mass to 3000 kg. This increase in mass comes mostly from

(1) carbon dioxide that enters through the leaf openings
(2) oxygen that enters through the leaf openings
(3) soil that all plants need to grow
(4) chloroplasts that enter the roots and move to the leaves   5 _____

6 Recently, a type of genetically modified fish has been approved for sale for human consumption. The modified fish contain a growth hormone gene from a different fish species. As a result, the modified fish grow rapidly and are ready to sell in almost half the time it normally would take. The modified fish are able to produce the new growth hormone because

(1) each of their cells contains the new gene to produce growth hormone
(2) each gene contains the code to synthesize carbohydrates
(3) the altered gene directs the mitochondria to synthesize the hormone
(4) the modified body cells are able to reproduce by meiosis   6 _____

7 Melanoma is a type of skin cancer that can spread to vital organs in the body. Doctors believe that exposure to ultraviolet (UV) radiation from the Sun is a leading cause of melanoma. One practical way governments can help prevent the harmful effects of UV radiation is to

(1) require everyone to remain indoors during daylight hours
(2) regulate the production and release of gases that damage the ozone shield
(3) encourage the building of a greater number of cancer treatment centers
(4) prohibit the use of solar panels on homes and businesses   7 _____

8 Some birds have recently modified their migratory behavior. Instead of flying to warmer climates during the winter months, the birds are remaining in northern areas where they can consume discarded food that is abundant in landfills. As a result of this change in migratory behavior, many insect populations that the birds normally feed on in the warmer climate areas are now increasing. This is an example of human activity

(1) interfering with ecological succession
(2) increasing competition for infinite resources
(3) disrupting the homeostasis of organisms
(4) altering the equilibrium of ecosystems   8 _____

9 New York State charges consumers a fee when purchasing beverages sold in aluminum cans and plastic bottles. This money is returned to purchasers when they return these items for recycling. Programs such as these are an attempt to

(1) encourage people to spend more money on their beverages
(2) conserve the resources these containers are made from
(3) reduce the amount of carbon dioxide produced by deforestation
(4) totally eliminate the use of reusable containers   9 _____

10 Recently, a human trachea (a respiratory organ) was produced by using a patient's own stem cells. The benefit of using the patient's own cells to produce a trachea instead of receiving one from a donor is that

(1) there will be more enzymes produced to help maintain homeostasis in the trachea
(2) there will be an increase in the quantity of antibodies that the patient produces in response to the new trachea
(3) there is less of a chance that the patient's immune system will attack the trachea
(4) there will be a greater response to any infectious agent that may enter the body   10 _____

11 The diagram below represents the organization of structures within an organism.

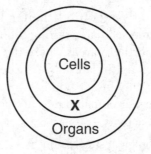

Which term best indicates the structures represented by the circle labeled X?

(1) organelles   (3) organ systems
(2) chromosomes   (4) tissues   11 _____

12 The chart below shows a comparison of the blood sugar levels for two individuals who took part in a scientific study.

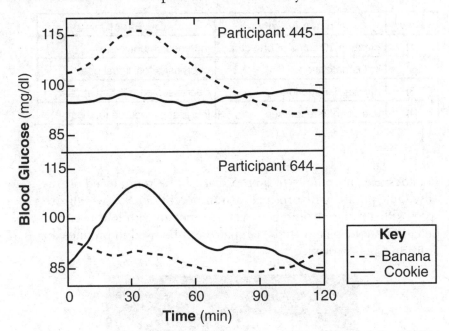

Source: Science Daily 11/19/15

Scientists have observed that blood sugar levels rose by different amounts in the two individuals even though they were given identical portions of bananas and cookies. These results were obtained because

(1) glucose is too large a molecule to be absorbed into the blood, so the researchers were only measuring the amount of glucose already present

(2) participant 445 didn't like bananas, and his body absorbed more of the food that he likes

(3) individuals have genetic differences that alter their responses to environmental factors

(4) two different foods were used; the scientists should have had only one experimental variable

12 _____

13 Which row in the chart below correctly matches the human activity with its effect?

Row	Human Activity	Effect
(1)	planting 20 acres of one crop	increases biodiversity
(2)	industrialization	decreases fossil fuel use
(3)	habitat destruction	decreases ecosystem stability
(4)	use of finite resources	increases resource renewal

13 _____

14 Potatoes are an example of a crop that can be reproduced asexually. One potato will produce a number of "eyes," which are sprouts that can grow into new plants. A potato with four eyes can be cut into four pieces, and each piece can be used to produce an individual potato plant.

"Eyes"

Source: https://www.quickcrop.ie/
blog/2014/02/growing-potatoes/

A gardener could produce a small crop of potatoes by planting the eyes from a single potato in her garden. Some of the potatoes grown in this way could be used to obtain eyes for the next season's crop.

One likely *disadvantage* of growing potatoes cloned in this way, year after year, would be that

(1) after a few years, the potatoes would stop producing eyes altogether, so no potatoes could be grown in the garden
(2) the potatoes produced each succeeding year would get larger and larger, eventually being too big for use as food
(3) the cost for growing your own potatoes in the garden would be greatly reduced
(4) a potato plant could become infected with a disease, and it could easily spread to the entire crop, killing all of the plants   14 _____

15 The back of the Namib Desert darkling beetle, shown in the photograph below, is covered in little bumps that collect water from the air. When it tilts forward, the water runs off its back into its mouth.

Source: http://myinforms.com

These specialized structures on the beetle's back allow it to

(1) locate food within the harsh desert environment
(2) obtain a substance that is required for survival
(3) reproduce asexually if mates are not available in the area
(4) increase the chances of survival by producing organic raw materials   15 _____

16 An increase in human population puts a stress on resources that can be renewed, such as

(1) trees and coal   (3) oil and natural gas
(2) water and gasoline   (4) water and trees   16 _____

17 Mitochondria provide ribosomes with

    (1) ATP for protein synthesis
    (2) amino acids for protein synthesis
    (3) oxygen for respiration
    (4) carbon dioxide for the production of sugars   17 _____

18 Mutations are most directly caused by changes in the

    (1) cell organelles of tissues
    (2) genes of chromosomes
    (3) ribosomes in gametes
    (4) receptors on membranes   18 _____

19 Animals and green plants are similar in that they

    (1) both carry out heterotrophic nutrition
    (2) all produce offspring by asexual reproduction
    (3) both use DNA to transmit hereditary information to offspring
    (4) all require oxygen to carry out photosynthesis   19 _____

20 Two organisms of different species are not likely to compete for the same

    (1) food   (3) space
    (2) mate   (4) water   20 _____

21 Some salmon have been genetically modified to grow bigger and faster than wild salmon. They are grown in fish-farming facilities. These genetically modified fish should *not* be introduced into a natural habitat because

    (1) the salmon would recycle nutrients at a rapid rate
    (2) their rapid growth rate could cause them to outcompete native salmon
    (3) they would not have enough oxygen for survival
    (4) they would reproduce asexually once they were released   21 _____

22 The diagram below represents a portion of a cell membrane.

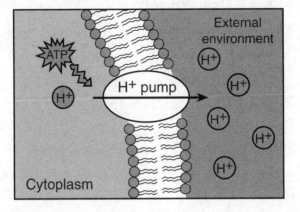

The arrow indicates that the cell membrane is carrying out the process of

(1) respiration
(2) cell recognition
(3) diffusion
(4) active transport

22 _____

23 The expression of a trait is directly dependent on the

(1) arrangement of amino acids in the protein synthesized
(2) shape of the subunits in the DNA molecule
(3) number of chromosomes present in the nucleus
(4) sequence of bases coded for by the ribosome

23 _____

24 Global warming is most closely associated with

(1) increased use of solar panels
(2) increased industrialization
(3) reducing the rate of species extinction
(4) removal of environmental wastes

24 _____

25 Which diagram below indicates that species *D* is more closely related to *C* than it is to either *A* or *B*?

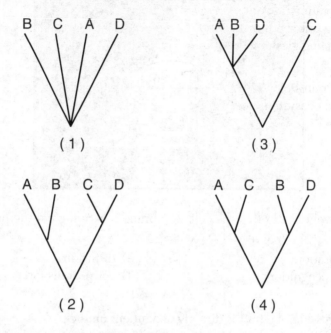

(1)   (3)

(2)   (4)

25 _____

26 As climate changes, which type of reproduction would most likely result in a greater chance of survival for a species?

(1) sexual reproduction, with a short reproductive cycle
(2) sexual reproduction, with a long reproductive cycle
(3) asexual reproduction, with a short reproductive cycle
(4) asexual reproduction, with a long reproductive cycle

26 _____

27 Adults of the *Aedes* mosquito genus are responsible for transmitting the viral diseases Zika and Dengue. Scientists have produced a modified form of male *Aedes* mosquitoes. The offspring of these male mosquitoes die before reaching adulthood. This method of reducing the spread of disease is dependent on

(1) vaccines stimulating the immune system of infected people
(2) providing medication to reduce the symptoms of disease
(3) the use of natural selection to modify the viruses so they are no longer pathogenic
(4) the use of genetic engineering to reduce the population of mosquitoes that carry the virus

27 _____

28 Humans deplete the most resources when

    (1) using wind energy as a power source
    (2) generating power by using fossil fuels
    (3) using water power to generate electricity
    (4) recycling glass and plastics   28 _____

29 The diagram below does *not* represent a sustainable energy pyramid in an ecosystem because

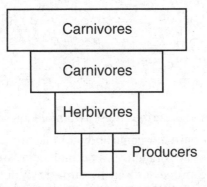

    (1) energy is never transferred between levels in ecosystems
    (2) ecosystems never have more than three levels of energy transfer
    (3) more energy must be available in the producer level than in the consumer levels
    (4) producers feed on herbivores in most ecosystems   29 _____

30 The two diagrams below represent a sugar molecule and a fat molecule that are used by living organisms.

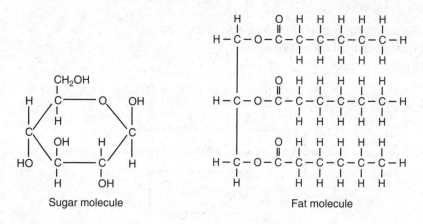

Sugar molecule   Fat molecule

Which statement best describes these two molecules?

(1) Sugar molecules are inorganic and fat molecules are organic.
(2) Sugar molecules are organic and fat molecules are inorganic.
(3) Energy for life processes can be stored within the chemical bonds of both molecules.
(4) Energy for life processes can be stored within the chemical bonds of sugar molecules, only.

30 _____

PART B–1

Answer all questions in this part.   [13]

Directions (31–43): For *each* statement or question, record in the space provided the *number* of the word or expression that, of those given, best completes the statement or answers the question.

31 A scientist analyzed a segment of DNA from a human chromosome and found that the percentage of thymine molecular bases (T) was 35%. Which row in the chart below contains the correct percentages of the other molecular bases in the DNA segment?

Row	Guanine (G)	Cytosine (C)	Adenine (A)
(1)	15%	25%	25%
(2)	25%	25%	15%
(3)	15%	15%	35%
(4)	35%	15%	15%

31 _____

32 The graph below shows changes in the populations of hares and lynx in a Canadian ecosystem.

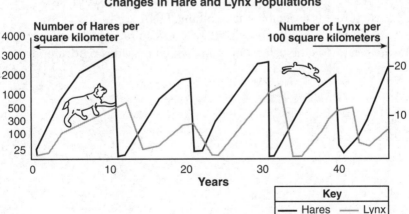

Changes in Hare and Lynx Populations

Source: Adapted from http://lbyiene-jardin-wikispaces.com

Which statement about the hares and lynx can be supported with information from the graph?

(1) The hare is the predator of the lynx because it is a larger animal.
(2) The lynx population begins to drop after the hare population drops.
(3) Both populations go through cycles due to the succession of plant species.
(4) Both populations have a carrying capacity of 3000 per square kilometer.   32 _____

33 The diagram below represents a cell in the human body.

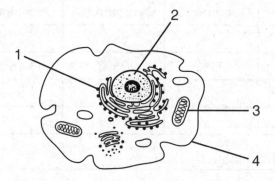

Which statement concerning the structures within this cell is accurate?

(1) Structure 1 is a chloroplast that carries out photosynthesis.
(2) Structure 2 is a vacuole that contains DNA.
(3) Structure 3 is a mitochondrion, where respiration takes place.
(4) Structure 4 is the cell membrane, which provides rigid support for the cell.   33 _____

Base your answers to questions 34 and 35 on the data table below and on your knowledge of biology. The table below indicates the amount of oxygen present at various water temperatures in a pond.

Amount of Available Oxygen in Water at Various Temperatures	
Temperature (°F)	**Dissolved Oxygen** (ppm)
68.0	9.2
71.6	8.8
78.8	8.2
82.4	7.9
86.0	7.6

34 An aquatic ecosystem experiences an increase in temperature. Which row in the chart below shows the effect of this increased temperature on the available oxygen and ecosystem?

Row	Amount of Available Oxygen	Effect on Ecosystem
(1)	decreases	greater stability of the ecosystem
(2)	increases	lessens competition between predatory organisms
(3)	decreases	reduces carrying capacity for fish
(4)	increases	increases genetic mutations in bacteria

34 _____

35 Which process performed by organisms produces oxygen for the aquatic ecosystem?

(1) respiration   (3) active transport
(2) replication   (4) autotrophic nutrition

35 _____

Base your answers to questions 36 and 37 on the diagram below and on your knowledge of biology. The diagram represents a food web illustrating some relationships in a tidal marsh ecosystem.

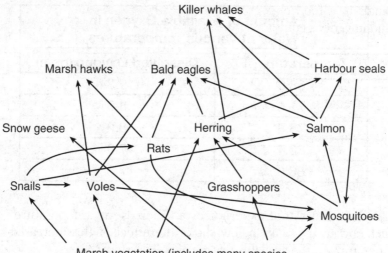

Adapted from: http://www.physicalgeography.net/fundamental/9o.html

36 Examples of autotrophs in this food web are
  (1) killer whales and grasses
  (2) sedges and bulrushes
  (3) mosquitoes and grasshoppers
  (4) snails and seals

  36 _____

37 In addition to grasshoppers, herring may also get energy from
  (1) algae   (3) snails
  (2) bald eagles   (4) voles

  37 _____

Base your answers to questions 38 and 39 on the information below and on your knowledge of biology.

Mercury is a toxic chemical that accumulates in the tissues of animals in a food chain. The chart below shows mercury levels found in various commercial fish and shellfish.

Mercury Concentration		
Species	Average Mercury Concentration (ppm)	Number of Samples
king mackerel	0.730	213
shark	0.979	356
swordfish	0.995	636
tilefish (Gulf of Mexico)	1.450	60
catfish	0.025	57
haddock	0.055	50
lobster (spiny)	0.093	13

Source: www.fda.gov/food/foodborneillnesscontaminants/metals/ucm115644.html

38 Each species listed is a predator. If the prey organisms that each predator consumes were tested, they would most likely contain

(1) the same amount of mercury as the predator species
(2) less mercury than the predator species
(3) more mercury than the predator species
(4) no mercury, since the predators probably get it from the polluted water   38 _____

39 Which statement is best supported by the data in the chart?

(1) Any fish caught in the Gulf of Mexico would have low levels of mercury.
(2) Eating catfish or haddock would be most likely to cause deadly mercury poisoning.
(3) Spiny lobsters may have more or less mercury than indicated because only a few were sampled.
(4) Tilefish are the most nutritious of all the species listed.   39 _____

40 The diagram below represents a laboratory process.

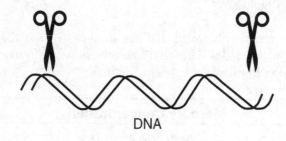

DNA

The substance represented by the scissors shown cutting the DNA is

(1) an enzyme   (3) a carbohydrate
(2) a starch molecule   (4) a fat molecule   40 _____

41 The human body has many cells that are deep inside the body. For this reason, the human body requires

(1) a transport system and other organs
(2) carbon dioxide from the air
(3) the synthesis of many inorganic compounds
(4) the breakdown of glucose by the digestive system   41 _____

Base your answers to questions 42 and 43 on the information below and on your knowledge of biology.

Bird Flu

Researchers are not sure when the H7N9 virus, referred to as bird flu, hit the China poultry markets. In February of 2012, the virus was found to have spread from birds to humans. All cases resulted from direct contact with infected poultry.

The bird flu can cause severe respiratory illness in humans. Since flu viruses constantly mutate, it would be difficult to develop a vaccine ahead of time. Scientists are worried that the virus could spread easily among people, causing a worldwide outbreak of the disease.

42 Based on the information, one danger of the new Bird Flu H7N9 strain is that it

  (1) causes death in over 75% of the individuals who become infected
  (2) is transferred to humans through consuming cooked poultry
  (3) can spread from humans to birds, such as crows and pigeons
  (4) mutates rapidly, making it hard to produce an effective vaccine   42 _____

43 The fact that the H7N9 virus has only recently infected humans helps explain why

  (1) it is highly transmissible through both the air and water
  (2) it is found only in the U.S.
  (3) humans have little or no immunity to the virus
  (4) the human population has formed antibodies against the virus   43 _____

PART B–2

Answer all questions in this part.   [12]

Directions (44–55): For those questions that are multiple choice, record your answer in the space provided. For all other questions in this part, record your answer in accordance with the directions.

Base your answers to questions 44 through 47 on the information below and on your knowledge of biology.

As part of an experiment, a bacterial culture was grown in a lab for two days. No additional nutrients were added to the culture after the initial set-up. As the bacteria reproduced asexually, the population of the culture was measured every six hours. Some of the data related to the bacterial growth are shown in the data table below.

Bacterial Growth	
Time (hrs)	**Population** (millions)
0	2.0
6	4.5
18	16.0
30	28.0
48	37.0

Directions (44–45): Using the information in the data table, construct a line graph on the grid below, following the directions below.

44 Mark an appropriate scale, without any breaks in the data, on each labeled axis.   [1]

45 Plot the data on the grid provided. Connect the points and surround each point with a small circle.   [1]

Example: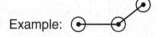

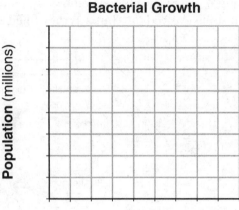

46 If data for the growth of this bacterial population continued to
be recorded, would the data point at 60 hours be above or below
37 million? Support your answer.   [1]

_____

_____

**Note: The answer to question 47 should be recorded in the
space provided.**

47 One likely reason bacteria would be grown in laboratory cultures
would be to

(1) increase the number of antibiotics produced by human cells
(2) eliminate the cloning of cells that can fight disease
(3) increase the production of specialized proteins by using genetic
engineering
(4) decrease the amount of bacteria naturally present in organisms   47 _____

Base your answers to questions 48 and 49 on the information and diagram below and on your knowledge of biology. The diagram represents a biological process.

Fossil evidence has demonstrated that birds evolved from a group of small carnivorous dinosaurs. Scientists have hypothesized that some evolved into birds as they filled available niches.

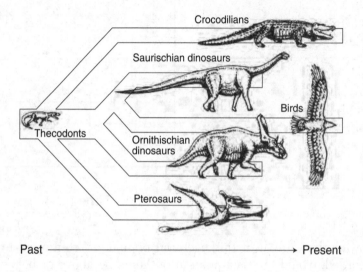

Past ——————————————————————→ Present

48 Identify *two* groups of organisms from the diagram that still exist on Earth today. Describe how they may have been able to survive to the present. [1]

Organisms: _____ and _____

_____

_____

_____

Note: The answer to question 49 should be recorded in the space provided.

49 The most recent fossil discoveries have filled in many of the gaps in the evolution of birds from dinosaurs. Before the latest fossils were found, there were some scientists who questioned this idea that birds evolved from dinosaurs. In general, scientists constantly work to

(1) clarify scientific explanations so they can be made into a law that never changes
(2) develop theories based on the data and evidence from a few experiments with inconclusive results
(3) provide enough evidence and accurate predictions to allow for widespread acceptance
(4) develop explanations that are permanent and do not change over time   49 _____

Base your answers to questions 50 through 52 on the diagram below and on your knowledge of biology. The diagram indicates some parts of the human female reproductive system.

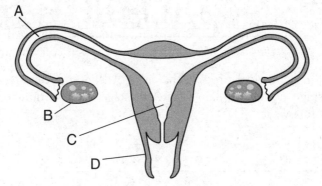

Note: The answer to question 50 should be recorded in the space provided.

50 The structure in which fertilization normally takes place is

(1) A   (3) C
(2) B   (4) D   50 _____

51 State *one* function of organ *B*.   [1]

_____

52 State *one* advantage of internal development for the human embryo.   [1]

_____

_____

_____

Base your answers to questions 53 through 55 on the information below and on your knowledge of biology. The diagram represents an ecological process that occurs in New York State over a long period of time.

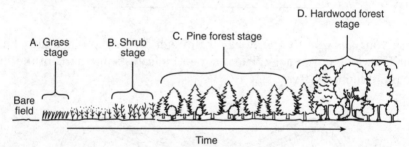

53 Identify the ecological process that is represented from stage *A* through stage *D*, and explain why each stage is important to the stage that follows it.   [1]

Process: _____

_____

_____

54 Identify *two* abiotic factors that can determine which types of organisms can inhabit an ecosystem.   [1]

_____ and _____

55 Identify the short-term effect that a forest fire during stage *D* would have on the biodiversity of the area.   [1]

_____

_____

PART C

Answer all questions in this part.   [17]

Directions (56–72): Record your answers in the spaces provided.

Base your answers to questions 56 and 57 on the information below and on your knowledge of biology.

Turtle Cells and Human Skin

New research has demonstrated that turtles and humans may have had a common ancestor 310 million years ago. A recent study looked at the genes responsible for the skin layers of turtle shells compared to the genes for human skin. The findings of the study suggest that about 250 million years ago, when turtle evolution split from other reptiles, a mutation in a specific group of genes occurred. The basic organization of this group of genes is similar in turtles and humans, and they produce the important skin proteins that produce shells in turtles and protect against infection in the skin of humans.

56 Identify the molecule that contains the hereditary material and the organelle in which it is found in turtle cells.   [1]

Molecule: _____

Organelle in turtle cells: _____

57 Describe how the mutation in the genes of a turtle ancestor turned out to be a beneficial evolutionary adaptation.   [1]

_____

_____

Base your answers to questions 58 through 60 on the illustration and information below and on your knowledge of biology.

The Little Brown Bat

Source: http://knatolee.blogspot.com/2011/09/not-ducklings.html

The illustration is of a species commonly called the little brown bat. It has 38 teeth and usually lives near bodies of water. The animal is considered beneficial by many people because it eats mosquitoes and many types of garden pests. They feed at night, detecting their prey by echolocation—a form of sonar similar to what is used on ships. They can determine the location and size of their prey by listening to the return echo.

58 The little brown bat eats mainly mosquitoes and night-flying insects. State *one* way in which the animal is adapted to prey on these organisms. [1]

_____

_____

_____

59 If a mutation occurs in some of these bats, it may result in a new inheritable trait that makes them better able to catch insects than other bats in the population. Describe what will most likely happen to the frequency of the *original* trait in the population. Support your answer. [1]

_____

_____

_____

60 Coevolution occurs when the evolution of an adaptation by one species affects the evolution of an adaptation in a second species. Some species of moths have evolved the ability to emit high frequency sounds that can block the little brown bat's echolocation. Based on the information provided, explain how this relationship between moths and bats is an example of coevolution. [1]

_____

_____

_____

Base your answers to questions 61 through 64 on the information below and on your knowledge of biology.

Kaolin as a Spray to Control a Bean Pest

Spraying kaolin, a clay-like material, on the leaves of plants has been effective in reducing insect damage to plants that grow in temperate regions, but has not been tried in tropical areas.

Researchers in the tropical Andean region of South America have recently conducted experiments to see if kaolin can be used there to control the greenhouse whitefly, a significant pest of the region's bean crops.

In the study, four groups of bean plants were used with the following treatments:

Group	Treatment	Whiteflies Killed (%)
1 (control)	No insecticide or other substance applied to the plants	0
2	Synthetic chemical insecticide applied to leaves	90
3	Leaves treated with 2.5% concentration of kaolin spray	80
4*	Leaves treated with 5% concentration of kaolin spray	80

* Note: In group 4 the plants lost 40% less water and showed a 45% increase in chlorophyll content in the leaves.

61 State *one* likely effect of the whiteflies on the bean plants in the control group (group 1) by the end of the study. Support your answer. [1]

_____

_____

62 Should the group 3 kaolin treatments be considered as an acceptable alternative control method to the group 2 insecticide treatment for whiteflies? Support your answer with data from the chart. [1]

_____

_____

63 Based on the results of groups 3 and 4, identify the kaolin treatment that would be best for bean plants grown in areas where low rainfall is a common occurrence. Support your answer. [1]

_____

_____

64 State *one* reason why the scientists are interested in reducing whitefly populations in the Andean region. [1]

_____

_____

Base your answers to questions 65 through 68 on the passage below and on your knowledge of biology.

Medical Mystery

Recently, an elderly man went to a hospital. He felt tired and was coughing and dehydrated. At first, the doctor thought he had pneumonia, but an x ray showed a spot on his lung. Because the man was a smoker, the doctor expected to find a tumor.

Instead, the surgeon discovered a pea seed growing inside the man's lung. When the pea seedling was removed, the patient quickly regained his health.

65 When he first arrived at the hospital, the man reported feeling unusually tired. Explain why damage to the man's lung caused fatigue.   [1]

_____

_____

66 In this case, the pea seed entered into the man's lung, but the immune system was not able to defend against it. Describe *one* specific way the cells of the immune system usually protect the body against certain molecules or microbes that are breathed into the lungs.   [1]

_____

_____

67 Identify *two* environmental factors inside a human lung that would help the pea begin to germinate.   [1]

_____

_____

68 State whether the pea seedling could have continued to grow and develop in the lung over a long period of time. Support your answer.   [1]

_____

_____

Base your answers to questions 69 and 70 on the information below and on your knowledge of biology.

Scientists Reprogram Plants for Drought Tolerance

Source: Lancaster Farming 2/21/15/AAAS

Arabidopsis plants respond to drought conditions by producing a stress hormone called ABA. This hormone slows down plant growth and leads to a decrease in the plant's use of water.

ABA binds to specific receptors in the plant that cause the guard cells on the leaf surfaces to close the stomatal openings through which water vapor can normally pass. This reduces water loss during the drought conditions.

Although it has been suggested that spraying plants with ABA during a drought could be beneficial, it is not practical. The chemical is expensive to produce and quickly loses its ability to bind to cell receptors in the plant cells.

Recently, however, scientists have found a way to modify the ABA receptors in *Arabidopsis* plants so they can be activated by another chemical that is both stable and inexpensive.

69 Describe how the shape of molecules, such as the hormone ABA, is critical to their function in the *Arabidopsis* plant. [1]

_____

_____

_____

70 Explain how the response of the guard cells to a drought is part of a feedback mechanism. [1]

_____

_____

_____

Base your answers to questions 71 and 72 on the passage and graph below and on your knowledge of biology.

Atmospheric Carbon Dioxide

Records from polar ice cores show that the natural range of atmospheric carbon dioxide (CO_2) over the past 800,000 years was 170 to 300 parts per million (ppm) by volume. In the early 20th century, scientists began to suspect that CO_2 in the atmosphere might be increasing beyond this range due to human activities, but there were no clear measurements of this trend. In 1958, Charles David Keeling began measuring atmospheric CO_2 at the Mauna Loa observatory on the big island of Hawaii.

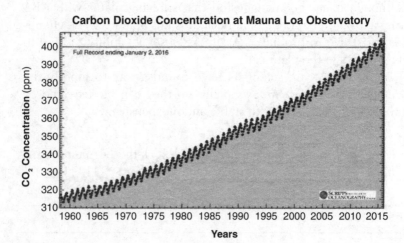

Carbon Dioxide Concentration at Mauna Loa Observatory

71 Record the approximate concentration of carbon dioxide at the start of the study and describe how it compares to the concentration in 2015.   [1]

_____ **ppm CO$_2$**

Description: _____

_____

72 Identify *one* likely reason for the overall change in CO$_2$ concentration observed between 1958 and 2015.   [1]

_____

_____

PART D

Answer all questions in this part. [13]

Directions (73–85): For those questions that are multiple choice, record your answer in the space provided. For all other questions in this part, record your answer in accordance with the directions.

Note: The answer to question 73 should be recorded in the space provided.

73 Which group of materials would be most useful to a student planning to separate a mixture of leaf pigments using paper chromatography?

(1) filter paper, dropper, solvent, beaker
(2) enzymes, beaker, goggles, compound microscope
(3) compound microscope, filter paper, coverslip, glass slide
(4) meterstick, thermometer, solvent, enzymes   73 _____

Note: The answer to question 74 should be recorded in the space provided.

74 In many parts of the world, plants are used as a source of medicine. Many of these plants are in danger of becoming extinct. It is therefore important for researchers to

(1) collect and dry all the medicinal plants to preserve them for future use
(2) search for other plant species that could be used as a new source of that medicine
(3) use the plants now while we still have them
(4) apply fertilizer to reduce the numbers of the plants that grow in the wild   74 _____

Note: The answer to question 75 should be recorded in the space provided.

75 In the lab activity *Making Connections,* an experiment was designed to test the effect of exercise on the ability to squeeze a clothespin. The number of times the clothespin was squeezed served as the

(1) independent variable   (3) hypothesis
(2) dependent variable   (4) control   75 _____

Base your answer to question 76 on the Universal Genetic Code Chart below and on your knowledge of biology.

Universal Genetic Code Chart

		SECOND BASE				
		U	**C**	**A**	**G**	
F I R S T B A S E	**U**	UUU ⎫ PHE UUC ⎭ UUA ⎫ LEU UUG ⎭	UCU ⎫ UCC ⎬ SER UCA UCG ⎭	UAU ⎫ TYR UAC ⎭ UAA ⎫ STOP UAG ⎭	UGU ⎫ CYS UGC ⎭ UGA ⎭ STOP UGG ⎭ TRP	U C A G
	C	CUU ⎫ CUC ⎬ LEU CUA CUG ⎭	CCU ⎫ CCC ⎬ PRO CCA CCG ⎭	CAU ⎫ HIS CAC ⎭ CAA ⎫ GLN CAG ⎭	CGU ⎫ CGC ⎬ ARG CGA CGG ⎭	U C A G
	A	AUU ⎫ AUC ⎬ ILE AUA ⎭ AUG ⎭ MET or START	ACU ⎫ ACC ⎬ THR ACA ACG ⎭	AAU ⎫ ASN AAC ⎭ AAA ⎫ LYS AAG ⎭	AGU ⎫ SER AGC ⎭ AGA ⎫ ARG AGG ⎭	U C A G
	G	GUU ⎫ GUC ⎬ VAL GUA GUG ⎭	GCU ⎫ GCC ⎬ ALA GCA GCG ⎭	GAU ⎫ ASP GAC ⎭ GAA ⎫ GLU GAG ⎭	GGU ⎫ GGC ⎬ GLY GGA GGG ⎭	U C A G

(Right side: **T H I R D   B A S E**)

Note: The answer to question 76 should be recorded in the space provided.

76 When provided with a sequence of bases in one segment of mRNA, the Universal Genetic Code Chart is used to

(1) directly identify the DNA from an animal cell
(2) determine the sequence of amino acids in a protein
(3) change the RNA sequence of a protein into DNA
(4) identify the specific mutations in the genetic material in a cell   76 _____

77 A student was setting up beakers that contained different solutions in order to conduct a laboratory investigation, but the next day he could not tell which beaker contained the starch and water mixture. In order to find out which beaker contained starch, he took a small sample from each of the beakers and conducted a test for starch on each of them.

Describe the test for starch that the student should use and the result that would indicate the presence of starch.   [1]

_____

_____

78 In order to survive in its environment, a single-celled organism uses a contractile vacuole to remove excess water that diffuses into its cell. Another species, a hydra, also excretes excess water. Both processes involve the use of energy.

Based on this information, state whether these two organisms live in fresh water or salt water. Support your answer.   [1]

_____

_____

79 The diagram below represents two types of carbohydrate molecules, glucose and sucrose.

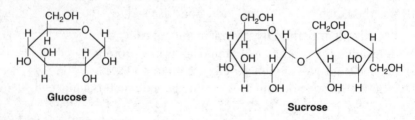

State *one* reason why a glucose molecule is more likely than a sucrose molecule to diffuse through an artificial membrane.   [1]

_____

Base your answers to questions 80 through 82 on the information below and on your knowledge of biology. The diagram represents some of the various types of giant tortoises that live on the Galapagos Islands. The chart provides information about some individual island environments.

Giant Tortoises of the Galapagos Islands

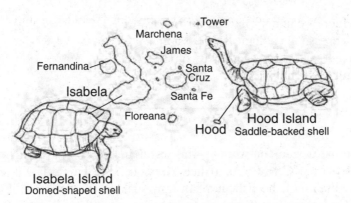

Source: Adapted from http://slideplayer.com/slide/7372273

Environmental Conditions on Certain Galapagos Islands	
Galapagos Island	**Island Characteristics**
Hood Island	sparse vegetation located high off of the ground; hot, dry, arid
Isabela Island	rich variety of vegetation located low to the ground; much rainfall; humid

80 Explain why specific Galapagos tortoise species are able to live only on certain islands. [1]

Note: The answer to question 81 should be recorded in the space provided.

81 The role that the environment plays in determining which species survive is referred to as

(1) a trade-off

(2) a gene mutation

(3) an ecological niche

(4) a selecting agent   81 _____

Note: The answer to question 82 should be recorded in the space provided.

82 Over the years, human activity introduced organisms such as goats and other herbivores to the Galapagos Islands. The addition of these invasive organisms caused the tortoise species to be threatened because there was

(1) an increase in competition for food sources
(2) a decrease in ecological succession
(3) an increase in the availability of vegetation
(4) a decrease in direct harvesting   82 _____

83 As fish are frozen for storage, the water in the cells expands as it cools from 4°C to 0°C and may cause cells to burst. This lowers the quality of the fish. Explain how soaking the fish briefly in salt water before freezing them might prevent this damage to the cells. [1]

_____

_____

Base your answers to questions 84 and 85 on the diagram below and on your knowledge of biology.

Variations in Beaks of Galapagos Islands Finches

Source: *Galapagos: A Natural History Guide*

84 Identify *one* finch population that would be *negatively* affected if the birth rate of small tree finches increased significantly. Support your answer. [1]

Finch: _____

Support: _____

_____

_____

85 A student completed two trials of the *Beaks of Finches* lab, each time picking up eleven seeds, as shown in the table below. If the student needs to collect an average of thirteen seeds to survive, how many seeds must he pick up in round 3? Record your answer in the space provided in the table below. [1]

Trial Number	Seeds Picked Up
1	11
2	11
3	————
Average	13

Answers
June 2019
Living Environment

Answer Key

PART A

1. 4	**6.** 1	**11.** 4	**16.** 4	**21.** 2	**26.** 1
2. 1	**7.** 2	**12.** 3	**17.** 1	**22.** 4	**27.** 4
3. 1	**8.** 4	**13.** 3	**18.** 2	**23.** 1	**28.** 2
4. 3	**9.** 2	**14.** 4	**19.** 3	**24.** 2	**29.** 3
5. 1	**10.** 3	**15.** 2	**20.** 2	**25.** 2	**30.** 3

PART B–1

31. 3	**34.** 3	**36.** 2	**38.** 2	**40.** 1	**42.** 4
32. 2	**35.** 4	**37.** 1	**39.** 3	**41.** 1	**43.** 3
33. 3					

PART B–2

44. *See* Answers Explained.
45. *See* Answers Explained.
46. *See* Answers Explained.
47. 3
48. *See* Answers Explained.
49. 3

50. 1
51. *See* Answers Explained.
52. *See* Answers Explained.
53. *See* Answers Explained.
54. *See* Answers Explained.
55. *See* Answers Explained.

PART C. *See Answers Explained.*

PART D

73. 1
74. 2
75. 2
76. 2
77. *See* Answers Explained.
78. *See* Answers Explained.
79. *See* Answers Explained.

80. *See* Answers Explained.
81. 4
82. 1
83. *See* Answers Explained.
84. *See* Answers Explained.
85. *See* Answers Explained.

Answers Explained

PART A

1. **4** *A fungus breaks down the body of a dead animal* is the activity that is an example of a decomposer recycling organic compounds back into the environment. A fungus is one of several species known as decomposers that derive their nutrition by digesting the tissues of dead animals or plants and absorbing the simpler digestive products. Through this activity, the fungus recycles the complex compounds that make up living things into simpler materials that can be taken up and used by living plants.

WRONG CHOICES EXPLAINED:

(1) *A tree synthesizes starch from simpler molecules* is *not* the activity that is an example of a decomposer recycling organic compounds back into the environment. This activity is an example of a producer (tree) manufacturing complex organic molecules (starch) from simpler molecules (glucose).

(2) *A bacterial cell performs photosynthesis* is *not* the activity that is an example of a decomposer recycling organic compounds back into the environment. Bacteria generally do not carry out the process of photosynthesis. This is a nonsense distracter.

(3) *A bird digests proteins from its food* is *not* the activity that is an example of a decomposer recycling organic compounds back into the environment. This activity is an example of a consumer (bird) breaking down complex organic foods (proteins) into simpler molecules for use within the body of the consumer.

2. **1** These observations are best explained by the fact that *flea saliva may stimulate an immune response in cats and dogs*. Antigens in the flea saliva enter the tissues of the animal and stimulate the animal's immune response, including the release of histamines. These histamines cause an itching sensation at the area of the insect bite and/or at secondary sites such as the neck or the base of the tail.

WRONG CHOICES EXPLAINED:

(2) It is *not* true that these observations are best explained by the fact that *fleas are microbes whose bites cause a decreased blood flow*. Fleas are complex multicellular insects, not simple microbes such as bacteria or viruses. This is a nonsense distracter.

(3) It is *not* true that these observations are best explained by the fact that *flea saliva is a toxic substance that is released when fleas prey on cats and dogs*. While flea saliva is not toxic, it does contain antigens that stimulate the immune response in the tissues of cats and dogs.

(4) It is *not* true that these observations are best explained by the fact that *fleas are host organisms whose saliva digests cat and dog fur, leaving hot spots*. Fleas are parasites, not hosts. No information is provided in the question that would lead to the conclusion that flea saliva digests the fur of cats and dogs.

3. **1** The specific cause of these birth defects was most likely *the development of rubella virus infections in embryos*. It is possible for a virus to breech the placental barrier and enter the blood stream of an unborn embryo. A viral infection of an embryo can do significant damage to the normal development of that embryo and may lead to birth defects as a result.

WRONG CHOICES EXPLAINED:

(2) The specific cause of these birth defects was *not* most likely *the failure of zygotes infected with rubella to develop*. If a zygote does not develop, then no embryo will form, so no embryonic defect will be in evidence.

(3) The specific cause of these birth defects was *not* most likely *mutations in the nerve cells of pregnant females at the time of the rubella epidemic*. No information is provided in the question that would lead to the conclusion that mutations occurred in pregnant females to cause this condition. This is a nonsense distracter.

(4) The specific cause of these birth defects was *not* most likely *an increase in the amount of time needed for healthy embryonic development*. The process of normal human embryonic development requires approximately nine months to complete. A viral infection caused by rubella or any other pathogen would have no effect on this development period.

4. **3** If not diagnosed, placenta previa can be a very dangerous condition because the placenta is *where nutrients and wastes are exchanged*. This exchange occurs between mother and fetus and is critical to the survival of the fetus. Without efficient exchange, the fetus may easily die.

WRONG CHOICES EXPLAINED:

(1) It is *not* true that, if not diagnosed, placenta previa can be a very dangerous condition because the placenta is *the primary source of oxygen for the mother*. Adult female humans obtain oxygen from the atmosphere via the lungs, not the placenta.

(2) It is *not* true that, if not diagnosed, placenta previa can be a very danger-ous condition because the placenta is *where the fetus obtains milk from the mother*. The human fetus does not utilize maternal milk as a source of nutrition, but instead depends on obtaining nutrients from the mother via the placental connection.

(4) It is *not* true that, if not diagnosed, placenta previa can be a very danger-ous condition because the placenta is *the primary source of estrogen and pro-gesterone in the mother*. Adult female humans obtain estrogen and progesterone primarily from the ovaries, not the placenta.

5. **1** This increase in mass comes mostly from *carbon dioxide that enters through the leaf openings*. Trees, like all green plants, absorb carbon dioxide from the atmosphere and convert it by the process of photosynthesis into simple sugar. This sugar is then used in the synthesis of complex carbohydrates such as cellu-lose, which is the primary component of plant fiber and wood. Over time, a single tree can convert many tons of atmospheric carbon dioxide into cellulose by this process.

WRONG CHOICES EXPLAINED:

(2) It is *not* true that this increase in mass comes mostly from *oxygen that enters through the leaf openings*. Oxygen is a product of plant photosynthesis whose net movement is out of, not into, leaf openings.

(3) It is *not* true that this increase in mass comes mostly from *soil that all plants need to grow*. Soil provides minerals and water for growth and physical support for many plant species. However, soil is not the primary source of mate-rials adding to the mass of these plants.

(4) It is *not* true that this increase in mass comes mostly from *chloroplasts that enter the roots and move to the leaves*. Chloroplasts are organelles found in and produced by the leaf cells, the photosynthesizing cells of all green plants. Chloroplasts are not absorbed by roots for transport to leaves.

6. **1** The modified fish are able to produce the new growth hormone because *each of their cells contains the new gene to produce growth hormone*. The genetic engineering procedure that was used to create this new fish variety inserted a gene into the genome of a single fish egg cell. This egg cell was then fertilized to produce a zygote containing the desired gene. As the zygote developed, each new cell received a copy of this gene via mitosis until an adult fish was hatched. Breed-ing this fish with other similarly modified fish eventually resulted in the develop-ment of a viable farm population of modified fish, all containing the desired gene for rapid growth.

WRONG CHOICES EXPLAINED:

(2) It is *not* true that the modified fish are able to produce the new growth hormone because *each gene contains the code to synthesize carbohydrates*. Of the millions of genes in the genome of this fish species, some, not all, of them are able to control the synthesis of carbohydrates. Hormones are proteins, not carbohydrates.

(3) It is *not* true that the modified fish are able to produce the new growth hormone because *the altered gene directs the mitochondria to synthesize the hormone*. Mitochondria are cell organelles responsible for the conversion of chemical bond energy in glucose to the chemical bond energy of ATP. Mitochondria do not contain the genes necessary to synthesize this growth hormone.

(4) It is *not* true that the modified fish are able to produce the new growth hormone because *the modified body cells are able to reproduce by meiosis*. Meiosis is a process in which homologous chromosome pairs and the genes they carry are separated into haploid gametes during gametogenesis. Meiosis is not the process by which new diploid cells are formed.

7. **2** One practical way governments can help prevent the harmful effects of UV radiation is to *regulate the production and release of gases that damage the ozone shield*. Chemicals known as CFCs are used as propellants in many aerosol (spray) products. CFCs released into the atmosphere are known to adversely affect the Earth's ozone layer. The ozone layer protects the Earth's living species from excessive UV solar radiation. Governmental regulations that limit the use of CFCs help to preserve the ozone layer and thereby help protect living species from harmful UV radiation.

WRONG CHOICES EXPLAINED:

(1) One practical way governments can help prevent the harmful effects of UV radiation is *not* to *require everyone to remain indoors during daylight hours*. Although this action would be effective in reducing the harmful effects of UV radiation, it is not practical to limit human behavior in this way.

(3) One practical way governments can help prevent the harmful effects of UV radiation is *not* to *encourage the building of a greater number of cancer treatment centers*. Although additional cancer treatment centers would help cancer patients receive treatments, it is not practical to address this problem by allowing additional cancers to be produced in the human population.

(4) One practical way governments can help prevent the harmful effects of UV radiation is *not* to *prohibit the use of solar panels on homes and businesses*. Solar panels are used to convert solar energy into electrical energy. Their use represents a positive effect on the environment, and one totally unrelated to the problem of UV-induced melanoma.

8. **4** This is an example of human activity *altering the equilibrium of ecosystems*. The presence of landfills represents a human activity that attracts bird populations that normally migrate to warmer climates in the winter where food is more abundant. The birds' changed migratory patterns have the effect of disrupting long-established patterns and thus the delicate balance (equilibrium) of the birds' environment.

WRONG CHOICES EXPLAINED:

(1) This is *not* an example of human activity *interfering with ecological succession*. Ecological succession refers to the series of changes that occur to the dominant plant community of an area that has been significantly affected by an environmental change. This example involves animal, not plant, populations.

(2) This is *not* an example of human activity *increasing competition for infinite resources*. Competition in a balanced ecosystem normally occurs between species that attempt to inhabit the same ecological niche and complete for limited, not infinite, resources. This is a nonsense distracter.

(3) This is *not* an example of human activity *disrupting the homeostasis of organisms*. Homeostasis relates to the dynamic equilibrium, or steady state, that normally exists within the body of a living organism. Homeostasis allows an organism to carry out its life processes in an efficient manner that promotes its survival. This example does not relate to the concept of homeostasis.

9. **2** Programs such as these are an attempt to *conserve the resources these containers are made from*. Aluminum is a metal that is in short supply for industrial and commercial uses. Recycling aluminum cans allows already refined aluminum to be reused for new aluminum products over and over without the need for mining and refining additional aluminum. This saves money, energy, and scarce natural resources.

WRONG CHOICES EXPLAINED:

(1) It is *not* true that programs such as these are an attempt to *encourage people to spend more money on their beverages*. The cost of beverages is subject to market pressures that are actually reduced as a result of recycling efforts.

(3) It is *not* true that programs such as these are an attempt to *reduce the amount of carbon dioxide produced by deforestation*. Deforestation decreases the volume of plant matter available to reduce atmospheric carbon dioxide via the photosynthetic reactions. Deforestation does not directly produce carbon dioxide; it only reduces nature's ability to absorb it.

(4) It is *not* true that programs such as these are an attempt to *totally eliminate the use of reusable containers*. Aluminum recycling makes economic and environmental sense and in no way affects the recycling or reuse of other containers or the materials from which they are made.

10. **3** The benefit of using the patient's own cells to produce a trachea instead of receiving one from a donor is that *there is less of a chance that the patient's immune system will attack the trachea*. The immune system functions in humans by detecting the presence of foreign antigens in the body and by producing antibodies to neutralize them. Any cells carrying these antigens are destroyed and removed from the body. This benefits the body by eliminating potential pathogens but works against the body when the cells attacked are transplanted cells meant to assist the patient. By using a structure constructed of the patient's own cells, the antigen-antibody reaction does not occur because the antigens in the printed trachea are not foreign to the body.

WRONG CHOICES EXPLAINED:
(1) The benefit of using the patient's own cells to produce a trachea instead of receiving one from a donor is *not* that *there will be more enzymes produced to help maintain homeostasis in the trachea*. The use of the patient's own cells for this procedure will have no effect on the production of enzymes in the trachea.

(2) The benefit of using the patient's own cells to produce a trachea instead of receiving one from a donor is *not* that *there will be an increase in the quantity of antibodies that the patient produces in response to the new trachea*. Because the new trachea is constructed with the cells of the patient, no foreign antigens will be introduced, so there will be no increase in the quantity of antibodies produced.

(4) The benefit of using the patient's own cells to produce a trachea instead of receiving one from a donor is *not* that *there will be a greater response to any infectious agent that may enter the body*. The use of the patient's own cells for this procedure will have no effect on the body's ability to protect itself through the immune response.

11. **4** *Tissues* is the term that best indicates the structures represented by the circle labeled X. In the organizational patterns of multicellular organisms, cells are the basic living units of structure and function (e.g., heart muscle cells). Groups of similar cells that perform discrete functions in the body are known as tissues (e.g., heart muscle tissue). Groups of tissues that function together to perform a life function (e.g., transport) are known to make up the structures of organs (e.g., heart).

WRONG CHOICES EXPLAINED:
(1) *Organelles* is *not* the term that best indicates the structures represented by the circle labeled X. Organelles are sub-cellular structures that are specialized for carrying out the essential life processes needed to maintain homeostatic balance in cells. In this diagram, organelles might be indicated as a circle labeled "Organelles" within the circle labeled "Cells."

(2) *Chromosomes* is *not* the term that best indicates the structures represented by the circle labeled X. Chromosomes are sub-organellar structures located within nuclei that carry the genetic codes for living things. In this diagram, chromosomes might be indicated as a circle labeled "Sub-organellar structures" within a circle labeled "Organelles."

(3) *Organ systems* is *not* the term that best indicates the structures represented by the circle labeled X. An organism is a grouping of organs functioning together to perform a life function in a living thing. In this diagram, organ systems might be indicated as a circle labeled "Organ systems" surrounding the circle labeled "Organs."

12. **3** These results were obtained because *individuals have genetic differences that alter their responses to environmental factors*. In this case, the environmental factors are represented by the bananas and cookies. It is possible that the individuals tested had genetic differences enabling different enzyme synthesis reactions such that participant 445 was able to digest the bananas but not the cookies, while participant 644 was able to digest the cookies but not the bananas.

WRONG CHOICES EXPLAINED:

(1) It is *not* true that these results were obtained because *glucose is too large a molecule to be absorbed into the blood, so the researchers were only measuring the amount of glucose already present*. Glucose is among the smallest of organic substances and can pass readily through cell membranes.

(2) It is *not* true that these results were obtained because *participant 445 didn't like bananas, and his body absorbed more of the food that he likes*. The body does not alter its biological processes in response to individual preferences.

(4) It is *not* true that these results were obtained because *two different foods were used; the scientists should have had only one experimental variable*. Although details of this study are not described in the question, it is likely that the researchers administered the cookies and bananas to participants at different times. In this manner, the researchers are able to isolate the effects of different experimental variables from each other.

13. **3** Row *3* in the chart correctly matches the human activity with its effect. When natural environments (habitats) are destroyed, living species populations are eliminated or greatly reduced. Food webs are interrupted when species are removed, causing surviving species to leave the environment or face starvation and death. This has the effect of destabilizing the ecosystem and accelerating its decline.

WRONG CHOICES EXPLAINED:

(1) Row *1* in the chart does *not* correctly match the human activity with its effect. Biodiversity is increased when multiple compatible species migrate into and establish themselves in a healthy environment. Planting a single plant crop in the area actually has the effect of reducing, not increasing, biodiversity.

(2) Row *2* in the chart does *not* correctly match the human activity with its effect. Industrialization normally requires the use of energy, which commonly includes the burning of fossil fuels such as coal, oil, and natural gas. Industrialization increases, not decreases, fossil fuel use.

(4) Row *4* in the chart does *not* correctly match the human activity with its effect. Finite resources such as fossil fuels, metals, and rare earth elements are used in many industrial processes. Once these finite resources are used up, they cannot be replaced. Use of finite resources decreases, not increases, resource renewal.

14. **4** One likely *disadvantage* of growing potatoes cloned in this way, year after year, would be that *a potato plant could become infected with a disease, and it could easily spread to the entire crop, killing all the plants*. Cloning and other methods of asexual plant propagation are used to ensure that particular fruit and vegetable varieties are maintained. However, plants that are propagated in this way lack the important gene reshuffling that occurs in the process of sexual reproduction. Such genetic reshuffling is essential to provide the species with new gene combinations, some of which could provide resistance to blights and other plant diseases.

WRONG CHOICES EXPLAINED:

(1), (2) It is *not* true that one likely *disadvantage* of growing potatoes cloned in this way, year after year, would be that *after a few years, the potatoes would stop producing eyes altogether, so no potatoes could be grown in the garden* or that *the potatoes produced each succeeding year would get larger and larger, eventually being too big for use as food*. The genetic characteristics of potato plants propagated asexually would not change over time, so eyes would still be produced, and the relative size of the potatoes would not change.

(3) It is *not* true that one likely *disadvantage* of growing potatoes cloned in this way, year after year, would be that *the cost for growing your own potatoes in the garden would be greatly reduced*. Home gardening is generally considered to be less expensive than buying foods commercially. This is an advantage, not a disadvantage, of this type of plant propagation.

15. **2** These specialized structures on the beetle's back allow it to *obtain a substance that is required for survival*. Water is an essential material that is needed by all living things for survival. This adaptation provides this beetle species with an advantage in its environment.

WRONG CHOICES EXPLAINED:
(1), (3), (4) It is *not* true that these specialized structures on the beetle's back allow it to *locate food within the harsh desert environment, reproduce asexually if mates are not available in the area,* or *increase the chances of survival by producing organic raw materials.* The information provided in the paragraph clearly states that these adaptations enable the beetle to collect water, not food, mates, or organic material.

16. **4** An increase in the human population puts stress on resources that can be renewed, such as *water and trees.* Renewable resources are those that can be replaced to provide energy or materials for human uses. As long as the environment remains healthy, trees can be renewed for material supplies by replanting new trees, and water can be replaced for power generation through the natural water cycle.

WRONG CHOICES EXPLAINED:
(1), (2), (3) It is *not* true that an increase in the human population puts stress on resources that can be renewed, such as *trees and coal, water and gasoline,* or *oil and natural gas.* Coal, gasoline, oil, and natural gas are all examples of fossil fuels that cannot be renewed or replaced.

17. **1** Mitochondria provide ribosomes with *ATP for protein synthesis.* Mitochondria are cell organelles responsible for the conversion of chemical bond energy in glucose to the chemical bond energy of ATP. This ATP releases its energy to cell processes, including protein synthesis, as needed.

WRONG CHOICES EXPLAINED:
(2), (3), (4) Mitochondria do *not* provide ribosomes with *amino acids for protein synthesis, oxygen for respiration,* or *carbon dioxide for the production of sugars.* Amino acids, oxygen, and carbon dioxide are materials that are absorbed into living things from their environment. They are not provided by mitochondria.

18. **2** Mutations are most directly caused by changes in the *genes of chromosomes.* Mutations are random events in which genetic material (DNA) is altered. DNA is located primarily in the genes that are located on chromosomes within the nucleus of the cell.

WRONG CHOICES EXPLAINED:
(1), (3), (4) Mutations are *not* most directly caused by changes in the *cell organelles of tissues, ribosomes in gametes,* or *receptors on membranes.* None of these structures contain DNA, so mutations cannot occur within them.

19. **3** Animals and green plants are similar in that they *both use DNA to transmit hereditary information to offspring*. DNA located in the cell nuclei of all living cells carries the genetic information needed to produce a living thing including all of its species characteristics. This is true of both animals and green plants.

WRONG CHOICES EXPLAINED:

(1) It is *not* true that animals and green plants are similar in that they *both carry out heterotrophic nutrition*. Animals typically carry on heterotrophic (other feeding) nutrition, while green plants typically carry on autotrophic (self-feeding) nutrition.

(2) It is *not* true that animals and green plants are similar in that they *all produce offspring by asexual reproduction*. Although some animals and some green plants reproduce by asexual means, most reproduce by sexual means.

(4) It is *not* true that animals and green plants are similar in that they *all require oxygen to carry out photosynthesis*. The photosynthetic reactions carried on by green plants require water and carbon dioxide as raw materials, not oxygen. Oxygen is a byproduct of photosynthesis.

20. **2** Two organisms of different species are *not* likely to compete for the same *mate*. By definition, a species is considered separate from other species when its members cannot mate with members of a similar species to produce viable, fertile offspring. This distinction would most likely extend to the selection of mating partners, since only mating activities involving members of the same species would be successful in producing such offspring.

WRONG CHOICES EXPLAINED:

(1), (3), (4) It is *not* true that two organisms of different species are *not* likely to compete for the same *food, space,* or *water*. Competition in a balanced ecosystem normally occurs between species that attempt to inhabit the same ecological niche (role) and compete for the same limited resources. To the extent that food, space, and water become limited in the environment, it is possible that organisms of different species may compete for them.

21. **2** These genetically modified fish should *not* be introduced into a natural habitat because *their rapid growth rate could cause them to outcompete native salmon*. A species' competitive success in its natural environment is partially determined by its rate of reproduction and growth relative to similar species. If these modified salmon were released into the natural environment of native salmon, it might compete successfully with the native salmon for food, space, and similar limited resources. This could lead to the elimination of native salmon from their natural environment.

WRONG CHOICES EXPLAINED:

(1) It is *not* true that these genetically modified fish should *not* be introduced into a natural habitat because *the salmon would recycle nutrients at a rapid rate*. Like most animal species, salmon consume organic foods and release the undigested material into the environment for use by bacteria and other decomposers. There is no information presented to suggest that this characteristic would change under these circumstances.

(3) It is *not* true that these genetically modified fish should *not* be introduced into a natural habitat because *they would not have enough oxygen for survival*. If oxygen concentration became a limited resource in the salmons' natural habitat, then it is possible that this situation could lead to competition between the native and modified salmon.
[NOTE: This may be considered a second correct answer to the question as written, although it is not the best answer.]

(4) It is *not* true that these genetically modified fish should *not* be introduced into a natural habitat because *they would reproduce asexually once they were released*. Like most animal species, salmon reproduce by sexual, not asexual, means. There is no information presented to suggest that this characteristic would change under these circumstances.

22. **4** The arrow indicates that the cell membrane is carrying out the process of *active transport*. Active transport is a process by which substances are transported across cell membranes from an area of low relative concentration to an area of high relative concentration, requiring the expenditure of cell energy. The diagram illustrates hydrogen ions being moved out of the cell from low to high concentration by means of a "proton pump." This mechanism requires the input of energy from ATP molecules.

WRONG CHOICES EXPLAINED:

(1), (2) It is *not* true that the arrow indicates that the cell membrane is carrying out the process of *respiration* or *cell recognition*. Neither of these processes requires hydrogen ions to be removed from the cell in this manner.

(3) It is *not* true that the arrow indicates that the cell membrane is carrying out the process of *diffusion*. Diffusion is a process by which substances are transported across cell membranes from an area of high relative concentration to an area of low relative concentration, *not* requiring the expenditure of cell energy.

23. **1** The expression of a trait is directly dependent on the *arrangement of amino acids in the protein synthesized*. The arrangement of amino acids defines the nature of the protein, including the biochemical reactions it may catalyze in the cell. This arrangement is ultimately determined by the sequence of bases in the DNA of the individual, which provides a template for the process of protein synthesis.

WRONG CHOICES EXPLAINED:

(2) The expression of a trait is *not* directly dependent on the *shape of the subunits in the DNA molecule*. The subunits of DNA are the nitrogenous bases A, T, C, and G, whose chemical shape is critical to the coding function of DNA. The role of the shapes of A, T, C, and G in determining the expression of a trait is indirect, not direct.

(3) The expression of a trait is *not* directly dependent on the *number of chromosomes present in the nucleus*. The number of chromosomes in the nucleus provides an indication of the species of the individual, but role of that number in determining the expression of a trait is indirect, not direct.

(4) The expression of a trait is *not* directly dependent on the *sequence of bases coded for by the ribosome*. The role of the ribosome in protein synthesis is to capture mRNA molecules, attract tRNA molecules, and link amino acids in the arrangement determined by the genetic code. The ribosome does not code for a sequence of bases, so its role in determining the expression of a trait is indirect, not direct.

24. **2** Global warming is most closely associated with *increased industrialization*. Competent scientific evidence around the issue of climate change points to increases in Earth's average temperature sufficient to cause major disruption to long-stable climate patterns and associated habitat destruction. This temperature increase is known to be caused by the release of "greenhouse gases" such as carbon dioxide and methane into the atmosphere. A major source of these gases over the past 300 years has been the rapid growth of human industrial, commercial, and agricultural practices.

WRONG CHOICES EXPLAINED:

(1) Global warming is *not* most closely associated with *increased use of solar panels*. Solar panels represent an alternative energy source that, if widely implemented, would help to reduce our dependence on fossil fuels (a major source of carbon dioxide) to provide energy for industrial processes.

(3) Global warming is *not* most closely associated with *reducing the rate of species extinction*. Global warming and the associated climate change are thought by competent scientists to be increasing, not reducing, the rate of extinction, which has resulted in the loss of thousands of Earth's living species over the past 300 years. The pace of such extinctions is known to be increasing.

(4) Global warming is *not* most closely associated with *removal of environmental wastes*. Global warming is known to be enhanced by the increase, not removal, of environmental wastes such as carbon dioxide in the atmosphere.

25. **2** Diagram 2 indicates that species *D* is more closely related to *C* than it is to either *A* or *B*. In this diagram, two lines are shown to have diverged from a distant common ancestor (unidentified). Both of these lines diverged again from more recent common ancestors (unidentified), leading to modern species *A* and *B* in one line and species *C* and *D* in the other line. In this depiction, species *A* and *B* are closely related to each other and species *C* and *D* are closely related to each other. It also shows that species *D* is more closely related to *C* than it is to either *A* or *B*.

WRONG CHOICES EXPLAINED:

(1) Diagram *1* does *not* indicate that species *D* is more closely related to *C* than it is to either *A* or *B*. This diagram illustrates that modern species *A*, *B*, *C*, and *D* all diverged from a distant common ancestor at about the same time and are equally related to each other.

(3) Diagram *3* does *not* indicate that species *D* is more closely related to *C* than it is to either *A* or *B*. In this diagram, two lines are shown to have diverged from a distant common ancestor. One of these lines diverged again from a more recent common ancestor, leading to modern species *A*, *B*, and *D*. The other line resulted in species *C*. In this depiction, species *A*, *B*, and *D* are closely related to each other but are only distantly related to species *C*.

(4) Diagram *4* does *not* indicate that species *D* is more closely related to *C* than it is to either *A* or *B*. In this diagram, two lines are shown to have diverged from a distant common ancestor. Both of these lines diverged again from more recent common ancestors, leading to modern species *A* and *C* in one line and species *B* and *D* in the other line. In this depiction, species *A* and *C* are closely related to each other and species *B* and *D* are closely related to each other. It also shows that species *D* is more closely related to *B* than it is to either *A* or *C*.

26. **1** As climate changes, *sexual reproduction, with a short reproductive cycle,* is the type of reproduction that would most likely result in a greater chance of survival for a species. Sexual reproduction provides species with the greatest adaptability to changing environmental conditions because sexual reproduction maximizes the opportunity to produce new genetic combinations. A short reproductive cycle provides species with the best opportunity to adapt quickly to such changes.

WRONG CHOICES EXPLAINED:

(2) As climate changes, *sexual reproduction, with a long reproductive cycle,* is *not* the type of reproduction that would most likely result in a greater chance of survival for a species. Although sexual reproduction provides species with the greatest adaptability to changing environmental conditions, a long reproductive cycle does not provide those species with the best opportunity to adapt quickly to such changes.

(3), (4) As climate changes, *asexual reproduction, with a short reproductive cycle,* and *asexual reproduction, with a long reproductive cycle,* are *not* the types of reproduction that would most likely result in a greater chance of survival for a species. Asexual reproduction, whether coupled with long or short reproductive cycles, provides species with only limited ability to adapt to changing environmental conditions due to the limited opportunity to produce new genetic combinations. That being said, bacteria, which reproduce asexually and have extremely short reproductive cycles, are among the most adaptable and successful of Earth's species.

27. **4** This method of reducing the spread of disease is dependent on *the use of genetic engineering to reduce the population of mosquitoes that carry the virus.* The method described in the paragraph involves scientists using a laboratory technique to alter the genome of the male *Ades* mosquito such that its sex cells carry a gene that proves lethal to its offspring. The described method meets the definition of genetic engineering.

WRONG CHOICES EXPLAINED:
(1), (2), (3) This method of reducing the spread of disease is *not* dependent on *vaccines stimulating the immune system of infected people, providing medication to reduce the symptoms of disease,* or *the use of natural selection to modify the viruses so they are no longer pathogenic.* None of these situations are described in the paragraph.

28. **2** Humans deplete the most resources when *generating power by using fossil fuels.* Fossil fuels such as coal, oil, and natural gas are nonrenewable resources formed millions of years ago under unique environmental and geological conditions. When fossil fuels are burned as a source of energy, the global supply of these resources is reduced and made unavailable for other potential uses in the future.

WRONG CHOICES EXPLAINED:
(1), (3), (4) It is *not* true that humans deplete the most resources when *using wind energy as a power source, using water power to generate electricity,* or *recycling glass and plastics.* Each of these activities provides a positive impact either by using renewable resources for energy (wind and water) or reusing (recycling) materials to reduce the energy needed to manufacture new products from raw materials.

29. **3** The diagram does *not* represent a sustainable energy pyramid in an ecosystem because *more energy must be available in the producer level than in the consumer levels.* In any traditional energy pyramid, the base of the pyramid

(producer level) is presumed to contain the greatest quantity of energy. As producers are consumed by herbivores, some of this energy ($\sim$10%) is retained in the chemical bonds of the herbivore tissues. The remainder ($\sim$90%) is dissipated as heat to the surrounding environment. A similar energy transfer/dissipation occurs at each trophic level transition.

WRONG CHOICES EXPLAINED:

(1) It is *not* true that the diagram does *not* represent a sustainable energy pyramid in an ecosystem because *energy is never transferred between levels in ecosystems*. Energy is always transferred between trophic levels in ecosystems to provide higher trophic levels with the energy captured by green plants via photosynthesis.

(2) It is *not* true that the diagram does *not* represent a sustainable energy pyramid in an ecosystem because *ecosystems never have more than three levels of energy transfer*. Ecosystems may have multiple interwoven trophic levels that support the energy needs of many biodiverse populations in a stable ecological community.

(4) It is *not* true that the diagram does *not* represent a sustainable energy pyramid in an ecosystem because *producers feed on herbivores in most ecosystems*. Producers are self-feeders that do not feed on herbivores in any known ecosystem (insectivorous plants provide possible exceptions).

30. **3** *Energy for life processes can be stored within the chemical bonds of both molecules* is the statement that best describes these two molecules. The carbon-carbon (C-C) and the carbon-hydrogen (C-H) bonds of these molecules contain considerable energy that can be released during the respiratory reactions of living cells. The chemical bond energy released in this manner is transferred to molecules of ATP for use by the cells in carrying out their physiological functions. [Note that fat molecules must undergo chemical digestion and other enzymatic alterations before they can enter the respiratory reactions.]

WRONG CHOICES EXPLAINED:

(1), (2) *Sugar molecules are inorganic and fat molecules are organic* and *Sugar molecules are organic and fat molecules are inorganic* are *not* the statements that best describe these two molecules. Organic molecules contain the elements carbon (C), hydrogen (H), and oxygen (O). Both sugar and fat are organic molecules.

(4) *Energy for life processes can be stored within the chemical bonds of sugar molecules, only* is *not* the statement that best describes these two molecules. Both sugar molecules and fat molecules contain chemical bond energy that can be used by living things for their life processes. See correct answer above.

PART B–1

31. **3** Row 3 in the chart contains the correct percentages of the other molecular bases in the DNA segment. The percentage of thymine (T) at 35% must match the percentage of adenine (A) in the sample, since these two bases are complementary in a DNA strand, for a combined total of 70%. The percentages of guanine (G) and cytosine (C), as complementary bases, must also be equal to each other, and their combined total must make up the balance of the 100% of all bases in the sample. G and C therefore must make up a combined total of 30% of the sample, or 15% each. Only Row 3 contains the required percentages of these molecular bases as described.

WRONG CHOICES EXPLAINED:
(1), (2), (4) Rows 1, 2, and 4 in the chart do *not* contain the correct percentages of the other molecular bases in the DNA segment. None of these rows contains values that are consistent with the complementarity and mathematical conditions described in the correct answer above.

32. **2** *The lynx population begins to drop after the hare population drops* is the statement about the hares and lynx that can be supported with information from the graph. This phenomenon can be observed by tracking the graphed lines for both species simultaneously over time (from left to right). As the hare population rises, so does the lynx population. When the hare population drops (due to food scarcity and/or excessive predation), the lynx population drops soon after (due to food scarcity).

WRONG CHOICES EXPLAINED:
(1) *The hare is the predator of the lynx because it is a larger animal* is *not* the statement about the hares and lynx that can be supported with information from the graph. No information is presented in the graph that would support this conclusion. In fact, the lynx is the predator and the hare is the prey in this predator-prey relationship.

(3) *Both populations go through cycles due to the succession of plant species* is *not* the statement about the hares and lynx that can be supported with information from the graph. No information is presented in the graph concerning plant succession.

(4) *Both populations have a carrying capacity of 3000 per square kilometer* is *not* the statement about the hares and lynx that can be supported with information from the graph. According to the information presented in the graph (left vertical axis), it appears that the carrying capacity of hares is approximately

3000 per square kilometer. However, the right vertical axis indicates that the carrying capacity for lynx is approximately 20 per 100 square kilometers, or approximately 0.2 per square kilometer.

33. **3** *Structure 3 is a mitochondrion, where respiration takes place* is the statement concerning the structures within this cell that is accurate. Although it is difficult to determine accurately given the scale of this diagram, structure 3 resembles a mitochondrion and is correctly linked to the mitochondrial function in housing the human respiratory reactions.

WRONG CHOICES EXPLAINED:

(1) *Structure 1 is a chloroplast that carries out photosynthesis* is *not* the statement concerning the structures within this cell that is accurate. Human cells do not contain chloroplasts or carry on photosynthesis. This is a nonsense distracter.

(2) *Structure 2 is a vacuole that contains DNA* is *not* the statement concerning the structures within this cell that is accurate. Structure 2 is likely the nucleus, not a vacuole. Human vacuoles do not contain DNA.

(4) *Structure 4 is the cell membrane, which provides rigid support for the cell* is *not* the statement concerning the structures within this cell that is accurate. Structure 4 is likely the cell membrane, but cell membranes do not provide "rigid support" for human cells.

34. **3** Row *3* in the chart shows the effect of this increased temperature on the available oxygen and ecosystem. The solubility of oxygen gas in water is partially dependent on the temperature of the water medium. Warm water holds less oxygen gas in solution than cold water. As water temperature increases and the dissolved oxygen level in the aquatic ecosystem decreases, less oxygen is available to support the respiratory requirements of aquatic species such as fish. This phenomenon has the effect of reducing the carrying capacity for the fish in a warm water environment compared to a cold water environment.

WRONG CHOICES EXPLAINED:

(1) Row *1* in the chart does *not* show the effect of this increased temperature on the available oxygen and ecosystem. The amount of available oxygen in the aquatic ecosystem will decrease under these circumstances, but this will result in lesser, not greater, stability of the ecosystem.

(2), (4) Rows *2* and *4* in the chart do *not* show the effect of this increased temperature on the available oxygen and ecosystem. The amount of available oxygen in the aquatic ecosystem will decrease under these circumstances, not increase.

35. **4** *Autotrophic nutrition* is the process performed by organisms that produces oxygen for the aquatic ecosystem. The most common type of autotrophic (self-feeding) nutrition is photosynthesis. Photosynthesis is performed by green plants and algae whenever a source of light energy is available to those plants. Aquatic plants and algae directly add gaseous oxygen, a byproduct of photosynthesis, to their water environment for use by aquatic animals such as fish.

WRONG CHOICES EXPLAINED:

(1) *Respiration* is *not* the process performed by organisms that produces oxygen for the aquatic ecosystem. Respiration is a process that uses, not produces, oxygen to assist in the oxidation of glucose and the release of chemical bond energy in plant and animal cells. Oxygen is not released to the aquatic ecosystem by this process.

(2) *Replication* is *not* the process performed by organisms that produces oxygen for the aquatic ecosystem. Replication is the process by which molecules of DNA exactly self-duplicate to produce new DNA molecules during cell division. Oxygen is not released to the aquatic ecosystem by this process.

(3) *Active transport* is *not* the process performed by organisms that produces oxygen for the aquatic ecosystem. Active transport is a process by which substances are transported across cell membranes from an area of low relative concentration to an area of high relative concentration, requiring the expenditure of cell energy. Oxygen is not released to the aquatic ecosystem by this process.

36. **2** Examples of autotrophs in this food web are *sedges and bulrushes*. Autotrophs (self-feeders) are green plants (vegetation) capable of performing the process of photosynthesis. In this process, carbon dioxide and water are combined to produce glucose and oxygen gas. Photosynthesis and the organisms that perform it constitute the foundation of any stable food web.

WRONG CHOICES EXPLAINED:

(1) It is *not* true that examples of autotrophs in this food web are *killer whales and grasses*. Grasses are autotrophs, but killer whales function as heterotrophs (other feeders) in this food web.

(3), (4) It is *not* true that examples of autotrophs in this food web are *mosquitoes and grasshoppers* or *snails and seals*. These organisms all function as heterotrophs (other feeders) in this food web.

37. **1** In addition to grasshoppers, herring may also get energy from *algae*. Based on the information provided in the diagram, which uses arrows to trace the transfer of energy among organisms, herring are known to consume algae, grasshoppers, and mosquitoes.

WRONG CHOICES EXPLAINED:

(2), (3), (4) It is *not* true that in addition to grasshoppers, herring may also get energy from *bald eagles, snails,* or *voles*. The diagram shows no arrows leading from these organisms to the herring, so the herring are not known to use them as sources of energy.

38. **2** If the prey organisms that each predator consumes were tested, they would most likely contain *less mercury than the predator species*. Prey fish and shellfish absorb trace amounts of mercury (and other metal contaminants) in sea water and store them in their body tissues. Mercury is not readily removed from these tissues by normal metabolic activities. As predator fish consume many prey fish, the mercury contaminants become progressively more concentrated in the tissues of the predator. This phenomenon is known as bioaccumulation.

WRONG CHOICES EXPLAINED:

(1), (3), (4) If the prey organisms that each predator consumes were tested, they would *not* most likely contain *the same amount of mercury as the predator species; more mercury than the predator species;* or *no mercury, since the predators probably get it from the polluted water*. The prey fish would likely contain less mercury in their tissues than predator fish, as described above.

39. **3** *Spiny lobsters may have more or less mercury than indicated because only a few were sampled* is the statement that is best supported by the data in the chart. The chart indicates that only 13 spiny lobsters were sampled. It is possible that these lobsters are not representative of the larger population in regard to their levels of mercury concentration. To verify these results, a larger number of samples should be taken.

WRONG CHOICES EXPLAINED:

(1) *Any fish caught in the Gulf of Mexico would have low levels of mercury* is *not* the statement that is best supported by the data in the chart. Insufficient information is presented in the chart concerning the locations from which these samples were taken. Only tilefish are identified as being collected from the Gulf of Mexico, and their contaminant level was high relative to other organisms tested.

(2) *Eating catfish or haddock would be most likely to cause deadly mercury poisoning* is *not* the statement that is best supported by the data in the chart. No information is presented in the chart concerning the relative lethality to humans of the mercury concentrations measured.

(4) *Tilefish are the most nutritious of all the species listed* is *not* the statement that is best supported by the data in the chart. No information is presented in the chart concerning the relative nutritional value of the species sampled.

40. **1** The substance represented by the scissors shown cutting the DNA is *an enzyme*. Enzymes are proteins coded by genes to perform specific actions in the cell. The enzymes used by researchers to cut DNA molecules at specific points are known as restriction enzymes.

WRONG CHOICES EXPLAINED:
(2), (3), (4) It is *not* true that the substance represented by the scissors shown cutting the DNA is *a starch molecule, a carbohydrate,* or *a fat molecule*. None of these substances is a protein and none has the ability to cut DNA in the manner described.

41. **1** For this reason, the human body requires *a transport system and other organs*. The human transport system provides a mechanism to carry essential substances to all tissues of the body, as well as to carry metabolic wastes away from these tissues. Other organs and organ systems perform discrete functions including coordination and control, gas exchange, waste removal, nutrient digestion, support and locomotion, and many other essential tasks to help the body maintain homeostasis.

WRONG CHOICES EXPLAINED:
(2) It is *not* true that for this reason, the human body requires *carbon dioxide from the air*. Carbon dioxide is a metabolic waste eliminated from, not needed by, the human body.

(3) It is *not* true that for this reason, the human body requires *the synthesis of many inorganic compounds*. The human body can produce essential organic, not inorganic, substances.

(4) It is *not* true that for this reason, the human body requires *the breakdown of glucose by the digestive system*. In the human body, glucose is broken down in the mitochondria during the respiratory reactions, not by the digestive system.

42. **4** Based on the information, one danger of the new Bird Flu H7N9 strain is that it *mutates rapidly, making it hard to produce an effective vaccine*. This danger, is expressed in the second sentence of the second paragraph, is true of many viral infections, which make effective vaccination programs essential to human health.

WRONG CHOICES EXPLAINED:
(1), (2), (3) It is *not* true that, based on the information, one danger of the new Bird Flu H7N9 strain is that it *causes death in over 75% of the individuals who become infected;* that it *is transferred to humans through consuming cooked poultry;* or that it *can spread from humans to birds, such as crows and pigeons*. No information is presented in the passage concerning any of these claims.

43. **3** The fact that the H7N9 virus has only recently infected humans helps to explain why *humans have little or no immunity to it*. The human immune system responds to the presence of a foreign antigen by producing antibodies specifically designed to attack it. Until H7N9 began infecting humans in 2012, the immune system had had no opportunity to build individual resistance to the disease. In addition, without direct knowledge of the presence of the virus, scientists could not begin researching the origins, effects, and treatments of the disease. This research started only after H7N9 became known to infect humans, so no vaccines could have been developed prior to 2012.

WRONG CHOICES EXPLAINED:

(1), (2), (4) The fact that the H7N9 virus has only recently infected humans does *not* help to explain why *it is highly transmissible through both air and water; it is found only in the U.S.;* or *the human population has formed antibodies against the virus.* No information is presented in the passage concerning any of these claims.

PART B–2

44. One credit is allowed for correctly marking an appropriate scale, without any breaks in the data, on each labeled axis.   [1]

45. One credit is allowed for correctly plotting the data on the grid, connecting the points, and surrounding each point with a small circle.   [1]

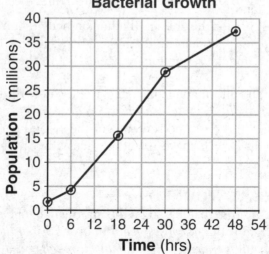

Bacterial Growth

46. One credit is allowed for correctly stating that the data point at 60 hours would be either above 37 million or below 37 million and supporting the answer. Acceptable responses include but are not limited to: [1]

- *Below: Since there are no more nutrients being added to the culture, the bacterial population will begin to drop below 37 million.*
- *Above: The bacteria will continue to increase above 37 million as long as nutrients are available in the culture.*
- *Below: Without additional nutrients being added to the culture, the bacteria population will begin to drop below 37 million.*
- *Below 37 million: The population would decrease as wastes build up.*
- *Above 37 million: If the trend continues because, as time increased, the population increased.*

47. **3** One likely reason bacteria would be grown in laboratory cultures would be to *increase the production of specialized proteins by using genetic engineering*. Genetic engineering is a laboratory technique in which a gene for a desired trait is snipped from the DNA of a donor cell and inserted into the genome of a recipient bacterial cell. This technique may be used effectively by scientists to produce specialized proteins such as insulin for human medical use.

WRONG CHOICES EXPLAINED:

(1) It is *not* true that one likely reason bacteria would be grown in laboratory cultures would be to *increase the number of antibiotics produced by human cells*. Antibiotics are biochemical compounds produced naturally by certain molds that are lethal to specific bacterial species. Human cells are not capable of producing antibiotics.

(2) It is *not* true that one likely reason bacteria would be grown in laboratory cultures would be to *eliminate the cloning of cells that can fight disease*. Cloning is a laboratory technique that can be controlled by the researchers employing it. There would be no scientific purpose in eliminating cloning of disease fighting cells. This is a nonsense distracter.

(4) It is *not* true that one likely reason bacteria would be grown in laboratory cultures would be to *decrease the amount of bacteria naturally present in organisms*. Beneficial bacteria are present in the intestinal tracts of most animal species and provide beneficial effects in those animals. There would be no scientific purpose in reducing the number of such bacteria. This is a nonsense distracter.

48. One credit is allowed for correctly identifying crocodilians and birds and for describing how they may have been able to survive to the present. Acceptable responses include but are not limited to: [1]

- *The crocodilians/crocodiles and birds: They may have survived because they had certain adaptations that allowed them to be successful in their environment then and now.*
- *Crocodilians and birds: They had characteristics that allowed them to survive, reproduce, and pass their traits on to their offspring.*
- *Birds and crocodiles: They had adaptations to their environments that allowed them to fill available niches effectively.*
- *Crocs and birds: They expressed transferrable genetic traits that provided them with an adaptive advantage in a dramatically changing environment.*

49. **3** In general, scientists constantly work to *provide enough evidence and accurate predictions to allow for widespread acceptance*. In doing so, competent scientists carefully design controlled experiments, collect and interpret data, and develop inferences based on data analyses. When an experiment has been replicated multiple times by others with the same results, general scientific predictions may be developed that can be applied to similar situations occurring in nature or other laboratory investigations. When a sufficient body of experimental data has been accumulated by multiple scientists, these inferences and predictions may gain widespread acceptance in the scientific community.

WRONG CHOICES EXPLAINED:
(1), (4) It is *not* true that in general, scientists constantly work to *clarify scientific explanations so they can be made into a law that never changes* or to *develop explanations that are permanent and do not change over time*. Competent scientists understand that experimental results and the inferences that are drawn from them are always subject to change as new information becomes available. A tremendous quantity of experimental work must be accomplished, replicated, verified, and reviewed before any science concept can be considered scientific law.

(2) It is *not* true that in general, scientists constantly work to *develop theories based on the data and evidence from a few experiments with inconclusive results*. To the contrary, competent scientists develop theories only after many experiments are conducted with conclusive results that can be replicated and verified by others. Also necessary to be recognized as scientific theory, experimental work must be subjected to extensive peer review and gain widespread acceptance by the scientific community.

50. **1** The structure in which fertilization normally takes place is *A*. Structure *A* is the oviduct (Fallopian tube) through which the unfertilized egg passes on its way from the ovary to the uterus. Sperm cells introduced into the female tract in the vagina swim through the uterus and up the oviduct until they encounter the egg and fertilization is accomplished.

WRONG CHOICES EXPLAINED:

(2) It is *not* true that the structure in which fertilization normally takes place is *B*. Structure *B* is the ovary, which is responsible for the production and periodic release of mature unfertilized egg cells. Fertilization does not normally occur in the ovary.

(3) It is *not* true that the structure in which fertilization normally takes place is *C*. Structure *C* is the uterus, which is responsible for the protection and development of fertilized egg cells. Fertilization does not normally occur in the uterus.

(4) It is *not* true that the structure in which fertilization normally takes place is *D*. Structure *D* is the vagina, which is responsible for receiving sperm cells provided by the male parent. Fertilization does not normally occur in the vagina.

51. One credit is allowed for correctly stating *one* function of organ *B*. Acceptable responses include but are not limited to: [1]

- *Production of egg cells*
- *Production of female gametes*
- *Structure B functions to produce female sex hormones.*
- *It can produce estrogen and progesterone needed for fetal development.*
- *Structure B is the ovary, which is responsible for the production and periodic release of mature unfertilized egg cells.*

52. One credit is allowed for correctly stating *one* advantage of internal development for the human embryo. Acceptable responses include but are not limited to: [1]

- *Embryos that grow inside the body of the female parent get greater protection from harsh environmental conditions than embryos that develop externally.*
- *The survival rate of offspring is higher for internal development than external development.*
- *It is an efficient way to provide nutrients to the fetus.*

- *The moist, warm conditions inside the mother's body protect the developing embryo from extreme variations of the external environment.*
- *The mother's body and the amniotic fluid protect the fetus from mechanical shock and tissue damage.*

53. One credit is allowed for correctly identifying the ecological process that is represented from stage *A* through stage *D* and explaining why each stage is important to the stage that follows. Acceptable responses include but are not limited to:　[1]

- *Ecological succession: Each stage modifies the environment for the next stage.*
- *Succession: Each stage makes the environment more suitable for the replacement community.*
- *Ecological succession: Each stage provides additional plant matter that mixes with the soil to provide a richer organic mix to support the growth of larger plant species in later stages of succession.*
- *Succession: As each successive stage occurs the soil is increasingly shaded from direct sunlight, allowing shade-tolerant shrubs and trees to root and grow effectively in the later stages.*
- *Ecological succession: In the final/climax stage, the combination of organic-rich soil and depth of shade provided by earlier stages favor the perpetuation of climax hardwood tree species such as oak, beech, and maple.*

54. One credit is allowed for correctly identifying *two* abiotic factors that can determine which types of organisms can inhabit an ecosystem. Acceptable responses include but are not limited to:　[1]

- *Light intensity and temperature*
- *Soil composition and soil chemistry*
- *Soil depth and land topography*
- *Average rainfall and light availability*
- *Wind conditions and rock substrata*
- *Ground slope and water table*
- *Surface elevation and length of growing season*

55. One credit is allowed for correctly identifying the short-term effect that a forest fire during stage *D* would have on the biodiversity of the area. Acceptable responses include but are not limited to:　[1]

- *Biodiversity would decrease.*
- *Many trees would no longer be present, reducing biodiversity.*

- *Mature trees would be damaged/lost but their seeds would survive to begin the process of ecosystem recovery, so the reduction of biodiversity would be minimal.*
- *The loss of biodiversity would be temporary because shade-producing forest canopy would be destroyed, allowing the growth of light-tolerant herbs and shrubs in open areas.*
- *Species of forest plants and animals would be eliminated, but species of field plants and animals would quickly take their place.*

PART C

56. One credit is allowed for correctly identifying the molecule that contains the hereditary material and the organelle in which it is found in turtle cells. Acceptable responses include: [1]

- *Molecule: DNA/deoxyribonucleic acid; Organelle in turtle cells: nucleus*

57. One credit is allowed for correctly describing how the mutation in the genes of a turtle ancestor turned out to be a beneficial evolutionary adaptation. Acceptable responses include but are not limited to: [1]

- *The turtles produced proteins that strengthened skin, resulting in a tough shell for a defense mechanism.*
- *In turtles, the mutation resulted in a shell with better protection than skin.*
- *They resulted in the production of skin proteins in humans that protect against infection.*
- *The turtles were better protected from predators.*
- *The humans were better protected from pathogens.*

58. One credit is allowed for correctly stating *one* way in which the animal is adapted to prey on these organisms. Acceptable responses include but are not limited to: [1]

- *The bat can catch these insects because its modified forelimbs enable it to fly.*
- *It uses echolocation and can hear the return echo to find prey.*
- *It has large ears so it can receive the echo as it bounces from the prey.*
- *Because it feeds at night, the bat can find insects when they are most active.*
- *The little brown bat has teeth with which to catch and consume its prey.*

59. One credit is allowed for correctly describing what will most likely happen to the frequency of the *original* trait in the population. Acceptable responses include but are not limited to: [1]

- *The frequency of the original trait will decrease because these bats will not be as successful at obtaining food. They will be less likely to produce off-spring than the bats with the new mutation.*

- *Since they will not be able to compete successfully with the bats with the mutation, they will produce fewer offspring and the trait will decrease.*
- *The trait would decrease in that population because bats would be less successful.*
- *The original trait will decrease due to natural selection for the new trait, which is beneficial.*
- *There could be no change if there's plenty of food.*

60. One credit is allowed for correctly explaining how this relationship between moths and bats is an example of coevolution. Acceptable responses include but are not limited to: [1]

- *After bats evolved echolocation, some moths with an adaptive advantage to emit high frequency sound survived to produce offspring able to block the bats.*
- *Bats that had adapted to find and eat moths by using echolocation eliminated moths that had no ability to produce high frequency sound, leaving moths with that ability to produce future generations inheriting this adaptation.*
- *The bats evolved echolocation because this random adaptation made those with the trait better at finding food, surviving, and producing offspring with the same adaptation. Some moths with the random adaptation to emit sounds that blocked the echolocation of the bats were able to survive and produce offspring with that same adaptation.*
- *Bat echolocation served as a selecting agent in the moth's environment that eliminated the non-sound-emitting moths and favored the survival of the sound-emitting moths.*

61. One credit is allowed for correctly stating *one* likely effect of the whiteflies on the bean plants in the control group (group 1) by the end of the study and supporting the answer. Acceptable responses include but are not limited to: [1]

- *Since no pest control method was used in the control group, the bean plants were probably eaten to a much greater extent than in other groups.*
- *Without pesticide or kaolin, the whiteflies probably caused serious damage to the plants in the control group.*
- *With no protective spray/treatment added to kill the whiteflies, these plants were probably eaten.*

62. One credit is allowed for stating your opinion about whether the group 3 kaolin treatments should be considered as an acceptable alternative control method to the group 2 insecticide treatment for whiteflies and supporting the answer. Acceptable responses include but are not limited to: [1]

- *Group 3 should be used. The insecticide was about 90% effective, while the kaolin treatments were only about 80% effective. However, 80% effectiveness is still quite effective, so it would be a good alternative to the synthetic spray.*

- *The kaolin treatment was only 10% less effective than the insecticide, and when kaolin's other benefits are considered, it is a good alternative to insecticide.*
- *10% more whiteflies will survive with the kaolin treatments, which can lead to more damage to the bean plants, so the insecticide should be used, not the kaolin.*
- *Kaolin (group 3): Even though the insecticide was more effective (90%), it could damage beneficial insects such as bees, so I think kaolin should be used instead.*

63. One credit is allowed for correctly stating that group 4 kaolin treatments would be best for bean plants grown in areas where low rainfall is a common occurrence and supporting the answer. Acceptable responses include but are not limited to: [1]

- *Treat the plants with 5% spray, since the plants will survive better in drought conditions.*
- *Use the group 4 treatment, since the plants will function better in a drought and will grow better with more chlorophyll.*
- *Group 4: 5% kaolin would give protection against insects and help the plants survive dry conditions, too.*
- *The plants treated with 5% kaolin spray (group 4) will lose 40% less water, so it should be used under these conditions.*

64. One credit is allowed for correctly stating *one* reason why the scientists are interested in reducing whitefly populations in the Andean region. Acceptable responses include but are not limited to: [1]

- *Whiteflies are a significant pest of the region's bean crops.*
- *The whitefly is a major problem affecting bean plant farming in the Andes.*
- *The whiteflies decrease bean crop yields.*
- *They cause damage to the plants/eat the crops grown in the Andean region.*
- *Scientists have identified the whitefly as an impediment to growing beans and other crops in this area.*

65. One credit is allowed for correctly explaining why damage to the man's lung caused fatigue. Acceptable responses include but are not limited to: [1]

- *If the lungs do not function well, less oxygen is available to release energy in his cells.*
- *If his lungs aren't working efficiently, he wouldn't get as much oxygen into his blood.*

- *The damage to the man's lung resulted in a decrease in his ability to breathe.*
- *The inefficiency would cause the man to retain too much carbon dioxide in his tissues.*

66. One credit is allowed for correctly describing *one* specific way the cells of the immune system usually protect the body against certain molecules or microbes that are breathed into the lungs. Acceptable responses include but are not limited to: [1]

- *White blood cells engulf and devour the pathogen.*
- *The body makes killer T-cells that destroy the microbes.*
- *The cells make specific antibodies to neutralize foreign antigens.*
- *Certain white blood cells would mark the foreign invader for destruction.*
- *The cells release histamines that help flush pathogens out of the body.*

67. One credit is allowed for correctly identifying *two* environmental factors inside a human lung that would help the pea begin to germinate. Acceptable responses include but are not limited to: [1]

- *Moisture/water and oxygen*
- *Warmth and moisture*
- *Water and air*
- *Temperature and pH*

68. One credit is allowed for correctly stating that the pea seedling could *not* have continued to grow and develop in the lung over a long period of time and supporting the answer. Acceptable responses include but are not limited to: [1]

- *No, because the lungs have no source of light for photosynthesis.*
- *No, since the plant would die after the nutrients in the seed had been used up.*

69. One credit is allowed for correctly describing how the shape of molecules, such as the hormone ABA, is critical to their function in the *Arabidopsis* plant. Acceptable responses include but are not limited to: [1]

- *Only molecules with a specific shape can fit into the receptor.*
- *If the molecule had a different shape, it would not fit into the receptor and the response would not occur.*
- *If the active site of the hormone is different, it won't regulate/run/operate/cause the reaction.*
- *If the molecular shape of ABA changes, it won't match the shape of the other molecule it reacts with.*
- *The shape of the molecule allows it to bind to the correct receptor.*

70. One credit is allowed for correctly explaining how the response of the guard cells to a drought is part of a feedback mechanism. Acceptable responses include but are not limited to: [1]

- *The plant responds to a stimulus. When there is a drought, ABA is produced and the stomates are closed. When the drought is over, no ABA is produced and the stomates are opened.*
- *The plant maintains homeostasis by opening or closing leaf openings according to the amount of water available.*
- *The plant closes the stomates/slows growth when there is a drought or opens the stomates/resumes growth when it rains.*
- *The stomates open or close in response to changes in the amount of water available.*
- *The guard cells close the stomates when there is less water available.*

71. One credit is allowed for correctly recording the appropriate concentration of carbon dioxide at the start of the study and describing how it compares to the concentration in 2015. Acceptable responses include but are not limited to: [1]

- *315 ppm—in 2015, the concentration was approximately 400 ppm, while in 1958 it was only 315 ppm*
- *~318 ppm—the concentration in 2015 is much more than in 1958*
- *313 ppm—the concentration was much lower when the study began*
- *About 316 ppm—it has increased by about 85 ppm since 1958*

72. One credit is allowed for correctly identifying *one* likely reason for the overall change in CO_2 concentration observed between 1958 and 2015. Acceptable responses include but are not limited to: [1]

- *The concentration of CO_2 increased between 1958 and 2015 due to more CO_2 produced by burning fossil fuels.*
- *There was an increase in CO_2 due to more people driving more cars.*
- *There has been an increase in the use of fossil fuels due to increased industrialization.*
- *CO_2 in the atmosphere increased due to human activities.*
- *Reduction of plants/deforestation has eliminated organisms capable of absorbing CO_2 from the atmosphere.*

PART D

73. **1** *Filter paper, dropper, solvent, beaker* is the group of materials that would be most useful to a student planning to separate a mixture of leaf pigments using paper chromatography. Paper chromatography is a laboratory technique that is used to help researchers study mixtures of organic chemicals such as plant pigments. In this technique, a concentrated mixture of pigments is placed on a strip of filter paper with a dropper. Solvent in a breaker carries the pigments through the filter paper when it is drawn by capillary action. Banding of the pigments helps the researcher to identify individual pigments in the mixture.

WRONG CHOICES EXPLAINED:

(2), (3), (4) *Enzymes, beaker, goggles, compound microscope; compound microscope, filter paper, coverslip, glass slide;* and *meterstick, thermometer, solvent enzymes* are *not* the groups of materials that would be most useful to a student planning to separate a mixture of leaf pigments using paper chromatography. Each of these groups contains one or more materials that would not be useful in a study of this type,

74. **2** It is therefore important for researchers to *search for other plant species that could be used as a new source of that medicine.* Many such plant species produce their unique biochemical compounds important to human medicine as a result of mutant genes that they, alone, carry. It is often impractical to search for another species displaying the same identical characteristic as the endangered or extinct species. It may be more feasible to find a way to synthesize these compounds in the laboratory. Genetic engineering to insert the genes controlling production of the biochemical into the genome of a more viable plant species is another possibility. A better long-term solution to this problem is to stop the human behaviors that are leading to the extinction of so many of the Earth's living species.

WRONG CHOICES EXPLAINED:

(1), (3) It is *not* true that it is therefore important for researchers to *collect and dry all the medicinal plants to preserve them for future use* or *to use the plants now while we still have them.* These are poor, short-term actions that do not address the real problem of species extinction. The medicines derived from these techniques would last for only a few months or years, would be extremely expensive to obtain, and would be available to only a few people.

(4) It is *not* true that it is therefore important for researchers to *apply fertilizer to reduce the numbers of the plants that grow in the wild.* This does not

resemble any practical technique that could be employed to resolve this problem. This is a nonsense distracter.

75. **2** The number of times the clothespin was squeezed served as the *dependent variable*. The dependent variable in a scientific experiment is the variable that is determined by the independent variable. In this case, the time limit (one minute) was the independent variable (controlled by the researcher) and the number of times the clothespin was squeezed in one minute was the dependent variable that was measured.

[NOTE: In order to arrive at this answer, we must assume that this version of *Making Connections* was set up to measure the number of times the subjects could squeeze a clothespin in one minute.]

WRONG CHOICES EXPLAINED:

(1) It is *not* true that the number of times the clothespin was squeezed served as the *independent variable*. The independent variable in a scientific experiment is the variable that is controlled by the investigator. In this case, the independent variable set up by the researcher is the time limit of one minute for each experimental trial and/or the number of trials to be run consecutively by each experimental subject.

(3) It is *not* true that the number of times the clothespin was squeezed served as the *hypothesis*. The hypothesis in a scientific experiment is the investigator's educated guess as to the outcome of an experiment, not a variable.

(4) It is *not* true that the number of times the clothespin was squeezed served as the *control*. The control in a scientific experiment is set up in exactly the same manner as the experimental group except for the independent variable that is manipulated by the researcher, not a variable.

76. **2** When provided with a sequence of bases in one segment of mRNA, the Universal Genetic Code Chart is used to *determine the sequence of amino acids in a protein*. Since the base sequence of the mRNA and the complementarity of bases between mRNA and tRNA (A complementary to U and G complementary to C) are known, it is possible to use a tool such as the Code Chart to work out the order in which amino acids will be linked together in the formation of a new polypeptide (protein).

WRONG CHOICES EXPLAINED:

(1) It is *not* true that, when provided with a sequence of bases in one segment of mRNA, the Universal Genetic Code Chart is used to *directly identify the DNA from an animal cell*. Although this is not the common use of the Code Chart, it might be possible to work backward through the protein amino acid

sequence and the mRNA base sequence to indirectly identify a DNA segment. Such a use would not be restricted to animal cells.

(3) It is *not* true that, when provided with a sequence of bases in one segment of mRNA, the Universal Genetic Code Chart is used to *change the RNA sequence of a protein into DNA*. RNA and proteins are fundamentally different biochemicals compared to each other and to DNA. Such a conversion would not be chemically possible.

(4) It is *not* true that, when provided with a sequence of bases in one segment of mRNA, the Universal Genetic Code Chart is used to *identify the specific mutations in the genetic material in a cell*. Mutations are random events that leave telltale signs in the DNA of a cell. Mutations in DNA may be transferred to molecules of mRNA. Although this is not the common use of the Code Chart, it might be possible to work backward through the protein amino acid sequence and the mRNA base sequence to indirectly identify a DNA mutation.

77. One credit is allowed for correctly describing the test for starch that the student should use and the result that would indicate the presence of starch. Acceptable responses include but are not limited to: [1]

- *Add starch indicator to the samples. The one that turns black/blue-black would be the one that contains starch.*
- *Add iodine to each sample. If one of the samples turns blue-black, it is the one that contains starch.*
- *To test for starch, add Lugol's solution and look for a color change to black if starch is present.*
- *Add starch indicator, and if starch is present, it will change color.*

78. One credit is allowed for correctly stating that these organisms live in fresh water and supporting the answer. Acceptable responses include but are not limited to: [1]

- *Fresh water—water would move into their cells and they would need to remove it.*
- *Fresh water—in a fresh water environment, the water would be diffusing into the cells all the time and would have to be pumped out of the cells in order to maintain homeostasis.*
- *Fresh water—if they are pumping out excess water, it is because water is diffusing into them from an environment with a higher concentration of water molecules than inside the cells. This means that they live in fresh (not salt) water.*
- *Fresh water—if they were living in salt water, they would be losing water all the time to the environment, so pumping out more would have no benefit.*

79. One credit is allowed for correctly stating *one* reason why a glucose molecule is more likely than a sucrose molecule to diffuse through an artificial membrane. Acceptable responses include but are not limited to: [1]

- *The sucrose molecule may be too large to diffuse through the membrane.*
- *Glucose is a smaller molecule than sucrose.*
- *Glucose is less complex than sucrose.*
- *The pores on the artificial membrane may be too small to accommodate the sucrose molecule.*

80. One credit is allowed for correctly explaining why specific Galapagos tortoise species are able to live only on certain islands. Acceptable responses include but are not limited to: [1]

- *The tortoises have beneficial structures/adaptations that allow them to feed on specific food/vegetation located on the islands they inhabit.*
- *Tortoises have inherited genetic adaptations favorable to their survival in certain island environments.*
- *Different islands have different climates/types of food that favor the different tortoise species' physical characteristics.*
- *They are adapted to the food, temperature, and climate on certain islands.*

81. **4** The role that the environment plays in determining which species survive is referred to as *a selecting agent*. Natural selection, sometimes referred to as survival of the fittest, postulates that certain members of a species are better adapted to their environment than others and that these organisms are more likely to survive and pass their favorable adaptations on to future generations. A selecting agent is any aspect of the natural environment that puts pressure on the species population and favors certain variations over others.

WRONG CHOICES EXPLAINED:
(1) The role that the environment plays in determining which species survive is *not* referred to as *a trade-off*. A trade-off is any decision that is made that takes into consideration the positive and negative aspects of that decision.

(2) The role that the environment plays in determining which species survive is *not* referred to as *a gene mutation*. Gene mutations are random events in which genetic material (DNA) is altered.

(3) The role that the environment plays in determining which species survive is *not* referred to as *an ecological niche*. An ecological niche is the environmental role played by species that inhabit a particular ecosystem.

82. **1** The addition of these invasive organisms caused the tortoise species to be threatened because there was *an increase in competition for food sources*. Invasive organisms are organisms native to one ecosystem that are introduced into a foreign ecosystem and that often have the effect of crowding out and eliminating native species from their natural habitats. The goats and other herbivores that were brought to the Galapagos Islands by Europeans had the effect of competing with the native Galapagos tortoises for their available food sources.

WRONG CHOICES EXPLAINED:
(2) It is *not* true that the addition of these invasive organisms caused the tortoise species to be threatened because there was *a decrease in ecological succession*. Ecological succession refers to the series of changes that occur to the dominant plant community of an area that has been significantly affected by an environmental change. No evidence is presented that leads to this conclusion.

(3) It is *not* true that the addition of these invasive organisms caused the tortoise species to be threatened because there was *an increase in the availability of vegetation*. The situation as described usually results in a decrease, not an increase, in the availability of vegetation. No evidence is presented that leads to this conclusion.

(4) It is *not* true that the addition of these invasive organisms caused the tortoise species to be threatened because there was *a decrease in direct harvesting*. Direct harvesting is a human practice in which plants or animals are removed (harvested) from a natural environment for their economic value. No evidence is presented that leads to this conclusion.

83. One credit is allowed for correctly explaining how soaking the fish briefly in salt water before freezing them might prevent this damage to the cells. Acceptable responses include but are not limited to: [1]

- *Soaking the fish briefly in salt water would reduce the water content of the cells, allowing space for the remaining water to expand when it freezes.*
- *The fish cells will lose water, so they won't burst.*
- *The salt soak would draw water out of the fish cells. When these cells are frozen, the expansion of the remaining water will only refill the cell volume, not burst the cell membranes.*

84. One credit is allowed for correctly identifying *one* finch population that would be *negatively* affected if the birth rate of small tree finches increased significantly and supporting the answer. Acceptable responses include but are not limited to: [1]

- *Large tree finch—They eat mostly animal food/have biting tips on their beaks/have grasping bills and competition for food will increase.*

- *Woodpecker finch—They eat mostly animal food/both have biting beak tips, so they will compete for food.*
- *Warbler finch—They will compete for food because the small tree finch eats the warbler finch's only food source.*

85. One credit is allowed for correctly calculating and recording the number of seeds that must be picked up in the last trial to average 13 seeds per trial. Acceptable responses include: [1]

- 17/seventeen seeds

[NOTE: This is a simple math problem that can be solved using the algebraic equation:

$$A = (N_1 + N_2 + N_3) \div 3$$
$$3A = (N_1 + N_2 + N_3)$$
$$3A - N_3 = (N_1 + N_2)$$
$$N_3 = (3A) - (N_1 + N_2)$$
$$N_3 = (3 \times 13) - (11 + 11)$$
$$N_3 = 39 - 22$$
$$N_3 = 17$$

Where: N_1 = number of seeds in trial 1 (11)
N_2 = number of seeds in trial 2 (11)
N_3 = number of seeds in trial 3 (unknown)
A = average number of seeds/trial (13)]

STUDENT SELF-APPRAISAL GUIDE
Living Environment June 2019

Standards/Key Ideas	June 2019 Question Numbers	Number of Correct Responses
Standard 1		
Key Idea 1: The central purpose of scientific inquiry is to develop explanations of natural phenomena in a continuing and creative process.	49	
Key Idea 2: Beyond the use of reasoning and consensus, scientific inquiry involves the testing of proposed explanations involving the use of conventional techniques and procedures and usually requiring considerable ingenuity.		
Key Idea 3: The observations made while testing proposed explanations, when analyzed using conventional and invented methods, provide new insights into natural phenomena.	32, 44, 45, 46, 62, 63, 71	
Laboratory Checklist	39, 61, 64	
Standard 4		
Key Idea 1: Living things are both similar to and different from each other and from nonliving things.	11, 12, 17, 19, 20, 22, 29, 33, 36, 37, 41	
Key Idea 2: Organisms inherit genetic information in a variety of ways that result in continuity of structure and function between parents and offspring.	6, 18, 23, 27, 31, 40, 47, 56	
Key Idea 3: Individual organisms and species change over time.	25, 26, 48, 57, 58, 59, 60	
Key Idea 4: The continuity of life is sustained through reproduction and development.	3, 4, 50, 51, 52	
Key Idea 5: Organisms maintain a dynamic equilibrium that sustains life.	2, 5, 10, 30, 35, 42, 43, 65, 66, 68, 69, 70	
Key Idea 6: Plants and animals depend on each other and their physical environment.	1, 15, 21, 34, 38, 53, 54, 55, 67	
Key Idea 7: Human decisions and activities have a profound impact on the physical and living environment.	7, 8, 9, 13, 14, 16, 24, 28, 72	
Required Laboratories		
Lab 1: "Relationships and Biodiversity"	73, 74, 76	
Lab 2: "Making Connections"	75	
Lab 3: "The Beaks of Finches"	80, 81, 82, 84, 85	
Lab 5: "Diffusion Through a Membrane"	77, 78, 79, 83	

Examination
August 2019
Living Environment

PART A
Answer all questions in this part. [30]

Directions (1–30): For *each* statement or question, record in the space provided the *number* of the word or expression that, of those given, best completes the statement or answers the question.

1 The diagram below represents an energy pyramid.

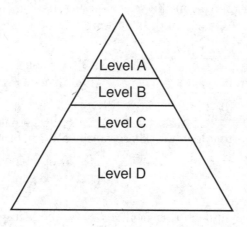

In this pyramid, the greatest amount of stored energy is found at level

(1) A   (3) C

(2) B   (4) D   1 ____

2 A certain species of plant serves as the only food for the young larvae of a particular species of butterfly. In a large field, a disease kills all the members of this plant species. As a result of the plant disease, the butterfly population will most likely

(1) quickly adapt to eat other plants
(2) disappear from the area
(3) evolve to form a new species
(4) enter the adult stage more quickly   2 _____

3 When handling cat litter, humans can potentially be exposed to a harmful single-celled protozoan. Its primary host is the common domestic cat, but it can also live in humans. This protozoan is an example of a

(1) predator   (3) parasite
(2) producer   (4) scavenger   3 _____

4 Certain seaweeds contain a greater concentration of iodine inside their cells than there is in the seawater surrounding them. The energy required to maintain this concentration difference is most closely associated with the action of

(1) ribosomes   (3) vacuoles
(2) mitochondria   (4) nuclei   4 _____

5 Doctors sometimes use a vaccine to prepare the body to defend itself against future infections. These vaccines most often contain

(1) antibodies   (3) white blood cells
(2) antibiotics   (4) weakened pathogens   5 _____

6 Building large manufacturing facilities can affect ecosystems by increasing the

(1) atmospheric quality
(2) biodiversity in the area
(3) demand for resources such as fossil fuels
(4) availability of space and resources for organisms   6 _____

7 An ameba is a single-celled organism. It uses its cell membrane to obtain food from its environment, digests the food with the help of organelles called lysosomes, and uses other organelles to process the digested food. From this, we can best infer that

  (1) all single-celled organisms have lysosomes to digest food
  (2) amebas are capable of digesting any type of food molecule
  (3) single-celled organisms are as complex as multicellular organisms
  (4) structures in amebas have functions similar to organs in multicellular organisms   7 _____

8 White blood cells are most closely associated with which two body systems?

  (1) circulatory and digestive
  (2) immune and circulatory
  (3) digestive and excretory
  (4) excretory and immune   8 _____

9 Carnivorous plants, such as pitcher plants and sundews, live in bogs where many other organisms cannot. Due to the high rate of decomposition occurring in bogs, the environment is acidic and contains very little oxygen and nutrients. The bogs only support certain types of organisms because

  (1) organisms in an environment are not limited by available energy and resources
  (2) the growth and survival of organisms depends upon specific physical conditions
  (3) favorable gene mutations only occur when organisms live in harsh environments
  (4) photosynthetic organisms can only inhabit environments that have a low acidity   9 _____

10 Anhidrosis is the inability to sweat normally. If the human body cannot sweat properly, it cannot cool itself, which is potentially harmful. Anhidrosis most directly interferes with

  (1) a feedback mechanism that maintains homeostasis
  (2) an immune system response to harmless antigens
  (3) the synthesis of hormones in the circulatory system
  (4) the enzymatic breakdown of water in cells   10 _____

11 The hair colors of the members of a family are listed below.

> mother – brown hair
> father – blond hair
> older son – brown hair
> younger son – blond hair

The hair colors of the sons are most likely a direct result of

(1) natural selection in males
(2) heredity
(3) evolution
(4) environmental influences   11 _____

12 A sample of DNA from a human skin cell contains 32% cytosine (C) bases. Approximately what percentage of the bases in this sample will be thymine (T)?

(1) 18   (3) 32
(2) 24   (4) 36   12 _____

13 Carmine, a compound that comes from the cochineal beetle, shown below, is used as a food coloring.

Source: https://alibi.com/events/256770/
Cochineal-Empire-making-Insect.html

The food coloring is not harmful to most people, but in a small number of individuals, it causes a reaction and affects their ability to breathe. This response to carmine is known as

(1) a stimulus   (3) natural selection
(2) an allergy   (4) an adaptation   13 _____

14 As a way to reduce the number of cases of malaria, a human tropical disease, a specific DNA sequence is inserted into the reproductive cells of *Anopheles* mosquitoes. Which process was most
   likely used to alter these mosquitoes?

   (1) cloning studies   (3) natural selection
   (2) genetic engineering   (4) random mutations   14 _____

15 Which row in the chart below accurately identifies two causes of
   mutations and the cells that must be affected in order for the
   mutations to be passed on to offspring?

Row	Cause of Mutations	Cells Affected
(1)	infections and antigens	body cells
(2)	meiosis and mitosis	body cells
(3)	disease and differentiation	sex cells
(4)	chemicals and radiation	sex cells

15 _____

16 Many tiny plants can be seen developing asexually along the edge of the mother-of-thousands plant leaf, as shown in the photo below. The tiny plants eventually drop to the ground and grow into new plants of the same species.

Source: http://www.plantamundo.com/produto_completo.asp?IDProduto=255

One way this form of reproduction differs from sexual reproduction is

(1) more genetic variations are seen in the offspring
(2) there is a greater chance for mutations to occur
(3) the offspring and the parents are genetically identical
(4) the new plants possess the combined genes of both parents   16 _____

17 A food web is represented in the diagram below.

Ocean Food Web

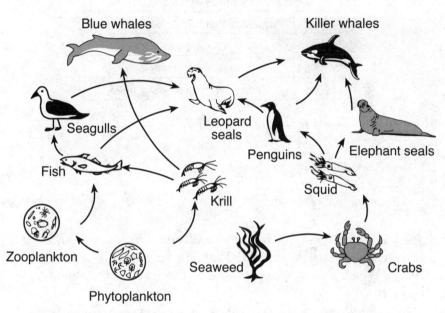

Adapted from: www.siyavula.com/gr7-9-websites/natural-sciences/gr8/gr8-11-02.html

If the fish population *decreases*, what is the most direct effect this will have on the aquatic ecosystem?

(1) The leopard seals will all die from lack of food.
(2) The krill population will only be consumed by seagulls.
(3) The zooplankton population will increase in size.
(4) The phytoplankton population will increase in size.   17 _____

18 The chart below shows a sequence of events that was observed at an abandoned ski center over a period of years.

Changes in Plant Species Over Time	
Year	Dominant Plant Species Observed
1985	grasses
1995	shrubs and bushes
2005	cherry, birch, and poplar trees

This sequence of changes is the result of

(1) ecological succession
(2) decreased biodiversity
(3) biological evolution
(4) environmental trade-offs   18 _____

19 Some salmon have been genetically modified to grow bigger and mature faster than wild salmon. They are kept in fish-farming facilities. Which statement regarding genetically modified salmon is correct?

(1) Genetically modified salmon produce more of some proteins than wild salmon.
(2) Genetically modified salmon and wild salmon would have identical DNA.
(3) Wild salmon reproduce asexually while genetically modified salmon reproduce sexually.
(4) Wild salmon have an altered protein sequence, but genetically modified salmon do not.   19 _____

20 Which group of organisms in an ecosystem fills the niche of recycling organic matter back to the environment?

(1) carnivores   (3) producers
(2) decomposers   (4) predators   20 _____

21 The use of solar panels has increased in the last ten years. A benefit of using solar energy would include

(1) adding more carbon dioxide to the atmosphere
(2) using less fossil fuel to meet energy needs
(3) using a nonrenewable source of energy
(4) releasing more gases for photosynthesis   21 _____

22 In a sewage treatment facility, an optimal environment is maintained for the survival of naturally occurring species of microorganisms. These organisms can then break the sewage down into relatively harmless wastewater. For these microorganisms, the wastewater facility serves as

(1) its carrying capacity   (3) an ecosystem
(2) a food chain   (4) an energy pyramid   22 _____

23 The diagram below represents a process taking place in a cell.

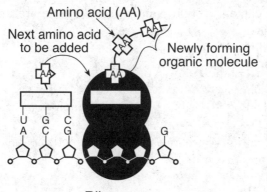

Ribosome

The type of organic molecule that is being synthesized is

(1) DNA   (3) protein
(2) starch   (4) fat   23 _____

24 The governments of many countries have regulations that are designed to prevent the accidental introduction of nonnative insects into their countries. This is because, in these new habitats, the nonnative insects might

(1) become food for birds
(2) not survive a cold winter
(3) not have natural predators
(4) add to the biodiversity   24 _____

25 The process of transferring energy during respiration occurs in a series of steps. This prevents too much heat from being released at one time. Maintaining an appropriate temperature is beneficial to an organism because

(1) enzymes need a proper range of temperatures to catalyze vital reactions
(2) cellular waste products can only be excreted in cooler temperatures
(3) hormones can only produce antibodies if temperatures are not excessive
(4) nutrients diffuse faster into cells when temperatures are lower   25 _____

26 The diagram below represents some structures in the human
female reproductive system.

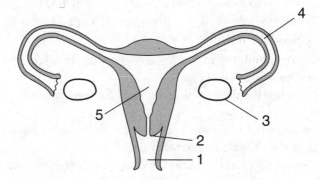

The processes of meiosis and fertilization are essential in human
reproduction. Which row in the chart correctly identifies where in
the female reproductive system these two processes occur?

Row	Meiosis	Fertilization
(1)	1	3
(2)	2	5
(3)	3	4
(4)	4	5

26 _____

27 Fruits and vegetables exposed to air begin to brown because of a
chemical reaction in their cells. This may result in these foods
being thrown out. Some people have found that adding lemon
juice (citric acid) to apple slices keeps them from turning brown.
The prevention of browning is likely the result of

(1) increasing the concentration of enzymes
(2) increasing the temperature
(3) slowing the rate of enzyme action
(4) maintaining the pH

27 _____

28 Scientists monitoring frog populations have noticed that the ratio of male frogs to female frogs varies when certain chemicals are present in the environment. The influence of estrogen, for example, has a noticeable effect. In the presence of a higher amount of estrogen, it would be most likely that

  (1) fewer males would be found because they are much larger and fewer are produced

  (2) fewer females would be found because they are more sensitive to pesticides

  (3) more males would be found because estrogen promotes the development of male characteristics

  (4) more females would be found because estrogen promotes the development of female characteristics   28 _____

29 Which action could humans take to slow the rate of global warming?

  (1) Cut down trees for more efficient land use.

  (2) Increase the consumption of petroleum products.

  (3) Use alternate sources of energy such as wind.

  (4) Reduce the use of fuel-efficient automobiles.   29 _____

30 The role of antibodies in the human body is to

  (1) stimulate pathogen reproduction to produce additional white blood cells

  (2) increase the production of guard cells to defend against pathogens

  (3) promote the production of antigens to stimulate an immune response

  (4) recognize foreign antigens and mark them for destruction   30 _____

PART B–1

Answer all questions in this part.   [13]

Directions (31–43): For *each* statement or question, record in the space provided the *number* of the word or expression that, of those given, best completes the statement or answers the question.

Base your answers to questions 31 and 32 on the information below and on your knowledge of biology.

Fossil Footprints

Scientists examined a trail of fossil footprints left by early humans in soft, volcanic ash in Eastern Africa. A drawing of the trail of footprints is shown below. Each footprint is represented as a series of lines indicating the depth that different parts of the foot sank into the volcanic ash.

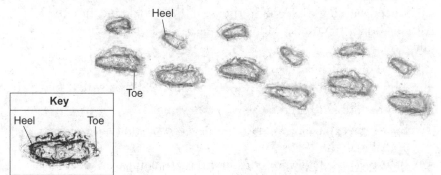

Source: http://www.indiana.edu/~eniweb/lessons/foot-topo-10inch.pdf

31 Which statement is an accurate observation that can be made based on this trail of footprints?

  (1) The individuals were running from a predator.
  (2) The volcano was about to erupt again.
  (3) One individual was much taller than the other.
  (4) One individual had larger feet than the other.   31 _____

32 The type of information directly provided by these fossil footprints is useful because it

    (1) offers details about how these individuals changed during their lifetime

    (2) offers data regarding their exposure to ultraviolet (UV) radiation

    (3) is a record of information about what these individuals ate during their lifetime

    (4) is a record of some similarities and differences they share with present-day species   32 _____

33 Since the early 1990s, proton pump inhibitors (PPIs) have been widely used to treat acid reflux disease. Although clinical tests in the 1980s deemed PPIs to be safe for humans, in 2012 the FDA announced warnings that long-term use of PPIs could increase the risk of bone fractures, kidney disease, and some intestinal infections.

Which statement best explains why the safety of PPIs is now in question when clinical experiments in the 1980s provided evidence that they were safe?

    (1) Researchers have been able to collect more data than were available in the 1980s.

    (2) Fewer people had acid reflux in the 1980s compared to today.

    (3) The medication containing PPIs has changed since the 1980s when tests were done.

    (4) The original experiments in the 1980s used only test animals and did not use human subjects.   33 _____

34 The process of embryonic development is represented in the diagram below.

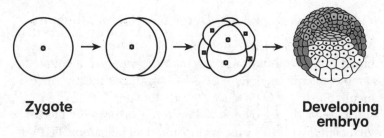

Zygote   **Developing embryo**

The three arrows in the diagram each represent a process known as

(1) mitotic cell division
(2) meiotic cell division
(3) fertilization of gamete cells
(4) differentiation of tissues   34 _____

35 A cell with receptors for two different hormones is represented below.

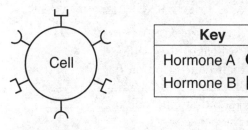

Which chemical would most likely interfere with the activity of hormone *A*, but *not* hormone *B*?

       ◆   ◗

(1)   (2)   (3)   (4)   35 _____

Base your answers to questions 36 through 39 on the information and photograph below and on your knowledge of biology.

Scientists Investigate Sex Determination in Alligators

The sex of some reptiles, including the American alligator, is determined by the temperature at which the eggs are incubated. For example, incubating them at 33°C produces mostly males, while incubation at 30°C produces mostly females.

Scientists recently discovered a thermosensor protein, TRPV4, that is associated with this process in American alligators. TRPV4 is activated by temperatures near the mid-30s, and increases the movement of calcium ions into certain cells involved with sex determination.

A baby alligator emerges from its egg shell during hatching

Source: http://www.dailymail.
co.uk/news/article-2190839/

36 The results of this scientific investigation will most likely lead other scientists to hypothesize that

(1) human sex cells also contain the TRPV4 protein
(2) other reptiles may have the TRPV4 protein in their eggs
(3) the TRPV4 protein affects the growth of plants
(4) the TRPV4 protein is present in all of the foods eaten by alligators

36 _____

37 Which information was most essential in preparing to carry out this scientific investigation?

(1) a knowledge of the variety of mutations found in American alligator populations
(2) the arrangement of the DNA bases found in the TRPV4 protein
(3) the effects of temperature on the incubation of alligator eggs
(4) a knowledge of previous cloning experiments conducted on alligators and other reptiles 37 _____

38 The movement of the calcium ions into certain cells is most likely due to

(1) the destruction of the TRPV4 when it contacts the cell membrane
(2) the action of TRPV4 proteins on the cells involved with sex determination
(3) the sex of the alligator embryo present in that particular egg
(4) the action of receptor proteins attached to the mitochondria in alligator sex cells 38 _____

39 Environmental changes, such as global warming, could affect species such as the American alligator because even slight increases in environmental temperature could

(1) lead to an overabundance of females and few, if any, males
(2) lead to an overabundance of males and few, if any, females
(3) cause the breakdown of the TRPV4 protein in female alligators
(4) increase the rate at which calcium ions exit the sex cells 39 _____

40 Which human activity can have a *negative* impact on the stability of a mature ecosystem?

(1) replanting trees in areas where forests have been cut down for lumber
(2) building dams to control the flow of water in rivers, in order to produce electricity
(3) preserving natural wetlands, such as swamps, to reduce flooding after heavy rainfalls
(4) passing laws that limit the dumping of pollutants in forests 40 _____

Base your answer to question 41 on the diagram below and on your knowledge of biology. The diagram represents a pond ecosystem.

Source: freshwaterecosystemswebquest.wikispaces.com/
ponds,+lakes,+and+inland+seas

41 Energy in this ecosystem passes directly from the Sun to

(1) herbivores   (3) heterotrophs

(2) consumers   (4) autotrophs   41 _____

Base your answers to questions 42 and 43 on the information and photograph below and on your knowledge of biology.

Wild Horse Roundup

Source: http://tuesdayshorse.wordpress.com/2012/10/31outrage-over-advisory-board-proposal-to-sterilize-wild-mustangs/

Wild horses called mustangs roam acres of federally owned land in the western United States. These horses have overgrazed the local vegetation to the extent that plants and soils are being lost entirely.

When the number of mustangs that roam the land exceeds the number of horses that the land can sustain, the government organizes helicopter-driven roundups. The horses are guided into a roped-off area and then are sold to the public or brought to pastures in the Midwest. About one percent of the horses captured die from injuries or accidents that occur during roundups.

42 The risk to the horses during the roundups compared to the entire loss of plants and soils is considered

(1) selective breeding   (3) direct harvesting

(2) a technological fix   (4) a trade-off   42 _____

43 The number of organisms that an area of land can sustain over a long period of time is known as

(1) ecological succession   (3) its carrying capacity

(2) its finite resources   (4) evolutionary change   43 _____

PART B–2

Answer all questions in this part. [12]

Directions (44–55): For those questions that are multiple choice, record your answer in the space provided. For all other questions in this part, record your answer in accordance with the directions.

Base your answers to questions 44 through 48 on the information and data table below and on your knowledge of biology.

White Nose Syndrome Found in Bats

White nose syndrome (WNS) is a disease found in bats. The disease, first detected in bats during the winter of 2006, is characterized by the appearance of a white fungus on the nose, skin, and wings of some bats, which live in and around caves and mines. It affects the cycle of hibernation and is responsible for the deaths of large numbers of bats of certain species. In some areas, 80–90% of bats have died. Not all bats in an area are affected, and certain bats that are susceptible in one area are not affected in other areas.

The roles of temperature and humidity in the environment of the bats are two of the many factors being investigated to help control the disease. Over the past few years, the Conserve Wildlife Foundation of New Jersey conducted summer bat counts of two bat species at 22 different sites, totaled the number, and reported the results. The approximate numbers of bats counted (to the nearest hundred) are listed in the table below.

Summer Bat Count (Total Number of Bats)		
Year	Big Brown Bats (*Eptesicus fuscus*)	Little Brown Bats (*Myotis lucifugus*)
2009	900	6100
2010	1000	1700
2011	1000	500
2012	1000	400
2013	1300	300

Directions (44–46): Using the information in the data table, construct a line graph on the grid below, following the directions below.

44 Mark an appropriate scale, without any breaks in the data, on the axis labeled "Number of Bats."   [1]

45 Plot the data for big brown bats on the grid, connect the points, and surround each point with a small circle.   [1]

Example:

46 Plot the data for little brown bats on the grid, connect the points, and surround each point with a small triangle.   [1]

Example:

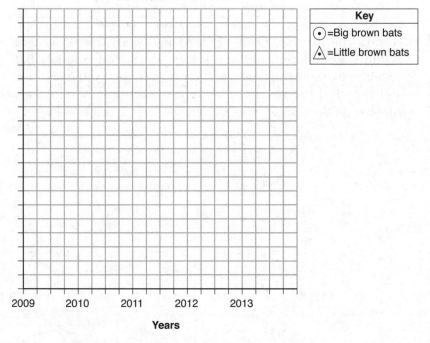

Summer Brown Bat Count

Key
⊙ =Big brown bats
△ =Little brown bats

Number of Bats

2009   2010   2011   2012   2013

Years

Note: The answer to question 47 should be recorded in the space provided.

47 Biologists in New York and Vermont have noted that, in recent years, a higher percentage of the little brown bats are now surviving. Which statement best explains this increased survival rate?

(1) A few of the bats possessed an immunity to the WNS disease and produced offspring that were immune.

(2) The bats needed to reproduce in greater numbers, otherwise they would have died out completely.

(3) The people that performed the recent counts did not identify the bats correctly and were counting bats of a different species.

(4) The original decline in the bat population due to WNS was a natural occurrence and is part of a natural cycle.   47 _____

48 Conservation groups have promoted the building and placing of bat houses in areas thought to be most suitable for bat populations. Explain why this might have a positive effect on the control of WNS in bats. [1]

_____

_____

Note: The answer to question 49 should be recorded in the space provided.

49 In the coastal waters off western North America, there is a starfish species that feeds primarily on mussels, another marine organism. In an experimental area, the starfish were removed from the waters. The effect of this removal is shown in the graph below.

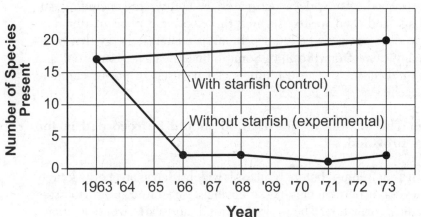

Influence of Starfish

Source: *Biology, 8th Ed.,* Campbell, Reese, et al. Pearson, San Francisco, CA, 2009, p. 1208.

What conclusion can be made regarding the role of the starfish in this ecosystem?

(1) The biodiversity of this ecosystem increased within ten years as organisms adjusted to the loss of the starfish.
(2) The starfish is important in maintaining the biodiversity of this ecosystem.
(3) When the starfish were removed, the ecosystem decreased in stability and increased in biodiversity.
(4) Biodiversity in this ecosystem is not dependent on the presence of starfish.

49 _____

Base your answers to questions 50 through 52 on the information below and on your knowledge of biology.

Biomass Energy

Biomass is the term for all living, or recently living, materials coming from plants and animals that can be used as a source of energy. Biomass can be burned to produce heat and used to make electricity. The most common materials used for biomass energy are wood, plants, decaying materials, and wastes, including garbage and food waste. Burning the wood and plant matter does produce some air pollutants. Biomass contains energy that originally came from the Sun. Some biomass can be converted into liquid biofuels. These biofuels can be used to power cars and machinery.

Note: The answer to question 50 should be recorded in the space provided.

50 In a community, before biomass is widely used as an energy source, several experts, including an ecologist, are hired to provide specific information. The ecologist would most likely be asked about

(1) the cost of producing the fuel compared with the profit when the fuel is sold
(2) whether the fuel will be widely accepted by consumers
(3) what effect the production of the fuel will have on the environment
(4) the time it will take to produce large amounts of the fuel   50 _____

51 Explain why biomass is considered a renewable energy source. [1]

_____

_____

_____

52 State *one* specific advantage and *one* specific disadvantage of the use of biofuels as an energy source. [1]

Advantage: _____

_____

Disadvantage: _____

_____

Base your answers to questions 53 and 54 on the information below and on your knowledge of biology.

Photosynthesis is a process that is important to the survival of many organisms on Earth.

53 Identify *two* raw materials necessary for photosynthesis. [1]

_____ and _____

54 State *one* reason why photosynthesis is necessary for animals to survive. [1]

_____

_____

55 The diagram below represents a pond ecosystem.

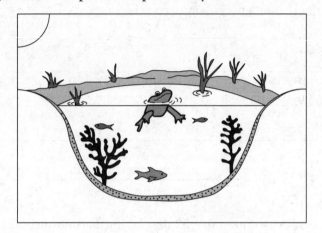

Identify *one* abiotic factor present in the pond ecosystem and explain how this abiotic factor would affect the frogs in the pond. [1]

Abiotic factor: _____

_____

_____

PART C

Answer all questions in this part. [17]

Directions (56–72): Record your answers in the spaces provided.

Base your answers to questions 56 through 58 on the information below and on your knowledge of biology.

Bye–Bye Bananas?

The world's most popular type of banana is facing a major health crisis. According to a new study, a disease caused by a powerful fungus is killing the Cavendish banana, which accounts for 99% of the banana market around the globe. The disease, called tropical race 4 (TR4), has affected banana crops in southeast Asia for decades. In recent years, it has spread to the Middle East and the African nation of Mozambique. Now experts fear the disease will show up in Latin America, where the majority of the world's bananas are grown. ...

...Once a banana plant is infected with TR4, it cannot get nourishment from water and nutrients, and basically dies of thirst. TR4 lives in soil, and can easily end up on a person's boots. If the contaminated boots are then worn on a field where Cavendish bananas are grown, the disease could be transferred. "Once a field has been contaminated with the disease, you can't grow Cavendish bananas there anymore," Randy Ploetz [scientist] says. "The disease lasts a long time in the soil."...

...But Cavendish [banana] is also particularly vulnerable to TR4. The banana is grown in what is called monoculture. "You see a big field of bananas and each one is genetically identical to its neighbor" Ploetz says. "And they are all uniformly susceptible to this disease. So once one plant gets infected, it just runs like wildfire throughout that entire plantation."...

Source: http://www.timeforkids.com/new/bye-bye-bananas/3311666

56 State how the TR4 fungus threatens homeostasis within the banana plant. [1]

_____

_____

57 Explain why the entire Cavendish banana crop worldwide is particularly vulnerable to the TR4 fungus. [1]

_____

_____

58 If the fungus cannot be stopped by chemical treatment of the soil, describe *one* other possible way that the growers may be able to combat the disease. [1]

_____

_____

Base your answers to questions 59 through 61 on the passage below and on your knowledge of biology.

Lead Poisoning

Two pathways by which lead can enter the human body are ingestion and inhalation. Once in the bloodstream, lead is distributed to parts of the body including the brain, bones, and teeth.

One reason that lead is toxic is that it interferes with the functioning of a variety of enzymes. It acts like metals such as calcium and iron and replaces them, changing the molecular structure of these enzymes. In the case of calcium, lead is absorbed through the same cell membrane channels that take in calcium.

Lead affects children and adults in different ways. Even low lead levels in children can cause many different problems, including nervous system damage, learning disabilities, decreased intelligence, poor bone growth, and death. In adults, high levels of lead can cause hearing problems, memory and concentration problems, muscle and joint pain, brain damage, and death. It wasn't until 1971 that steps were taken against the use of lead with the passage of the Lead Poisoning Prevention Act. However, lead is still a public health risk today.

59 It is recommended that children eat foods high in calcium and iron as a way to reduce the accumulation of lead in their cells and enzymes. Explain why this is a scientifically valid recommendation.   [1]

_____

_____

60 Describe how the presence of lead in body cells could interfere with the ability of enzymes to function.   [1]

_____

_____

_____

61 Based on the parts of the body that are most affected by lead intake, identify _one_ type of cell that would be expected to have numerous calcium channels. Support your answer.   [1]

Type of cell: _____

Support: _____

_____

_____

Base your answers to questions 62 and 63 on the illustration and passage below and on your knowledge of biology.

The Telltale DNA of Manx Cats

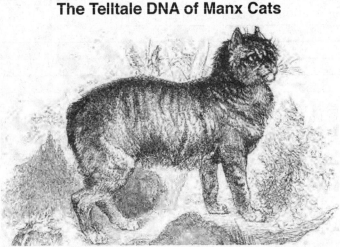

Source: http://commons.wikimedia.org/wiki/File: Manx cat (stylizes) 1885.jpg

A few breeds of cat have no tails. Manx cats have extremely short tails and may even appear to have no tail at all. Manx cats were first discovered several hundred years ago.

Scientists have determined that a certain mutation in a group of genes (called T-box genes) interferes with the development of the spine in the cat embryo. Mutations in these T-box genes can cause abnormalities in the number, shape, and/or size of bones in the spines of Manx cats, which results in smaller spines and shorter tails.

If a Manx cat inherits one copy of the mutated T-box gene and one copy of the normal gene, it will have a very short tail or no tail at all.

If the cat embryo inherits two copies of these mutated genes, it will stop developing and die. Therefore, all surviving Manx cats have only one copy of the mutated gene.

62 State *one* reason why the mutation in Manx cat embryos causes them to have such very short tails.   [1]

_____

_____

63 Two Manx cats have several litters of offspring. Explain how the genes that the kittens inherit determine whether they will have a normal tail or a short tail. [1]

_____

_____

_____

Base your answer to questions 64–66 on the passage below and on your knowledge of biology.

Hummingbirds Are Sugar Junkies

Source: http://bug-bird.com/hummingbirds-large-images/

Most humans enjoy candy, cake, and ice cream. As a result of evolutionary history, we have a wide variety of tastes. This is not true of all animals. Cats do not seek sweets. Over the course of their evolutionary history, the cat family tree lost a gene to detect sweet flavors. Most birds also lack this gene, with a few exceptions. Hummingbirds are sugar junkies.

Hummingbirds evolved from an insect-eating ancestor. The genes that detect the savory flavor of insects underwent changes, making hummingbirds more sensitive to sugars. These new sweet-sensing genes give hummingbirds a preference for high-calorie flower nectar. Hummingbirds actually reject certain flowers whose nectar is not sweet enough!

64–66 Discuss how sweet sensitivity in hummingbirds has developed. In your answer, be sure to:

- identify the initial event responsible for the new sweet-sensing gene   [1]
- explain how the presence of the sweet-sensing gene increased in the hummingbird population over time   [1]
- describe how the fossil record of hummingbird ancestors might be used to learn more about the evolution of food preferences in hummingbirds   [1]

_____

_____

_____

_____

_____

_____

_____

_____

_____

Base your answers to questions 67 through 70 on the information below and on your knowledge of biology.

Folic acid is a type of vitamin that is essential for the normal growth and development of cells in the body. If a woman consumes folic acid in her diet before and during the earliest stages of pregnancy, it can help to reduce her baby's risk for developing a type of birth defect called a neural tube defect. Early in pregnancy, the neural tube forms the brain and spinal cord. If the neural tube does not form properly, serious birth defects may result.

67 Explain why taking folic acid early in pregnancy is important to the prevention of neural tube defects. [1]

_____

_____

_____

68 Describe how a fetus receives folic acid and other essential materials directly from its mother for its development. [1]

_____

_____

69 Identify *one* factor, other than a lack of folic acid, that may interfere with the proper development of essential organs during pregnancy. [1]

_____

_____

70 Many foods, such as breads, cereals, pastas, and rice, are fortified or enriched with folic acid. Explain why adding folic acid to foods is an advantage to people other than pregnant women. [1]

_____

_____

_____

Base your answers to questions 71 and 72 on the diagram below and on your knowledge of biology. The diagram represents an evolutionary tree.

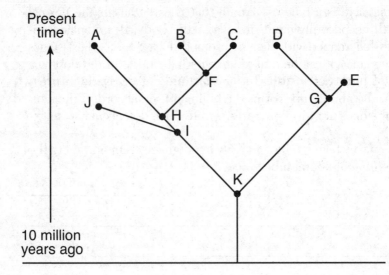

71 Are species A and B more closely related than A and D? Circle yes *or* no and support your answer with information from the diagram.   [1]

Circle one: Yes   *or*   No

_____

_____

72 State *one* possible cause for the extinction of species E.   [1]

_____

PART D

Answer all questions in this part. [13]

Directions (73–85): For those questions that are multiple choice, record your answer in the space provided. For all other questions in this part, record your answers in accordance with the directions.

Note: The answer to question 73 should be recorded in the space provided.

Base your answer to question 73 on the information and diagram below and on your knowledge of biology.

During the *Relationships and Biodiversity* lab, simulated pigments from three plant species were compared to those in *Botana curus*. The results were similar to those represented below.

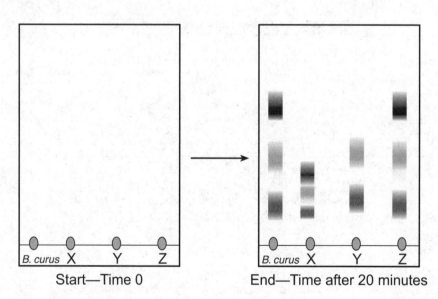

Start—Time 0   End—Time after 20 minutes

73 Based on the results of this comparison alone, is there enough information to conclude which of the other three species is most closely related to *Botana curus*?

(1) Yes. Only species X has the same bands as *Botana curus*.
(2) Yes. Species Z has only two of the bands that *Botana curus* has.
(3) No. Additional tests should be done to test for other chemical similarities.
(4) No. Other rainforest plant species should be tested.

73 _____

Note: The answer to question 74 should be recorded in the space provided.

Base your answer to question 74 on the diagram below and on your knowledge of biology.

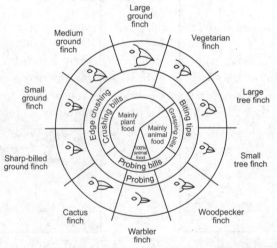

Variations in Beaks of Galapagos Islands Finches

Source: *Galapagos: A Natural History Guide*

74 Insects can get diseases just like other organisms. A deadly bacteria infected the insects on one Galapagos Island. Among the birds living there, the finches most likely to experience a drastic *decrease* in population size would be the

(1) warbler finches
(2) cactus finches
(3) large ground finches
(4) medium ground finches

74 _____

Note: The answer to question 75 should be recorded in the space provided.

75 A variety of species of Galapagos finches evolved from one original species long ago through the process of

(1) asexual reproduction   (3) natural selection

(2) ecological succession   (4) selective breeding   75 _____

Note: The answer to question 76 should be recorded in the space provided.

76 If scientists want to determine the similarities in the DNA fragments in several plant species, they should

(1) add salt water to cells from each plant

(2) analyze electrophoresis results

(3) compare seed structures of the plants

(4) examine their chromosomes with a microscope   76 _____

77 One coach of an Olympic rowing team makes his athletes warm up by doing 30 minutes of stretching and jogging in place before practicing each day. Another coach suggests that resting before practicing will result in better performance by her team. They decide to conduct an experiment to see which practice is correct. One team rests before practice, the other team warms up for thirty minutes, and they then record the time that it takes each team to row a specific distance. Identify the dependent variable in this experiment. [1]

_____

78 A student squeezes a clothespin 82 times in a minute. Then, using the same hand and the same clothespin, he squeezes the clothespin 68 times in a minute. State *one* biological reason for the decrease in the number of squeezes during the second trial. [1]

_____

_____

79 During rest, an adult's heart rate averages 60–100 beats per minute. When exercising, an adult's heart rate may increase to 100–170 beats per minute. State *one* reason why the heart rate increased during exercise. [1]

_____

_____

80 In some single-celled protozoans living in fresh water, such as the paramecium, contractile vacuoles are organelles used to pump excess water out of the cell. Explain why a paramecium would require contractile vacuoles while a similar protozoan living in salt water would *not*. [1]

_____

_____

_____

Base your answers to questions 81 through 83 on the information and diagram below and on your knowledge of biology.

A cube cut from a potato is placed in a beaker of distilled water. The potato cells have a relatively high concentration of starch and a relatively low concentration of water. The diagram represents the water and starch molecules in and around one of the potato cells in contact with the water in the beaker.

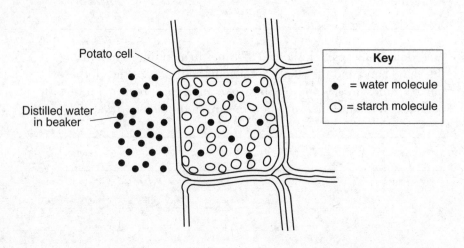

Note: The answer to question 81 should be recorded in the space provided.

81 Which row in the chart correctly describes what would be expected to occur in the potato cells, with regard to both the starch and water molecules?

Row	Water	Starch
(1)	More water will move into the cell than will leave the cell.	Starch will remain inside the potato cell.
(2)	More water will leave the cell than will enter.	Starch will move out of the cell.
(3)	Water content of the cell will not change.	Starch will move out of the cell.
(4)	More water will leave the cell than will enter.	Starch will remain inside the potato cell.

81 _____

Note: The answer to question 82 should be recorded in the space provided.

82 Which statement correctly describes a possible result if starch indicator is added to the water in the beaker one hour after the potato cube was added?

(1) The indicator solution would turn to an amber color in the water if starch molecules were present in the water in the beaker.
(2) The indicator would remain amber in color if starch molecules were not present in the water in the beaker.
(3) The indicator would change to a black color if starch molecules were not present in the water in the beaker.
(4) The indicator solution would remain black in color if starch molecules were present in the water in the beaker.

82 _____

83 Before placing the potato in the beaker, the student used an electronic balance to determine the mass of the potato cube. The mass of the cube was determined again after it was in the beaker for an hour. Describe how this information could specifically be used to determine if water moved during the investigation. [1]

_____

_____

_____

84 Select *one* row in the chart below and explain how the systems in that row work together during exercise. [1]

	System	System	System
Row 1	Respiratory	Circulatory	Muscular
Row 2	Muscular	Circulatory	Excretory
Row 3	Digestive	Circulatory	Muscular

Row: _____

_____

_____

_____

Base your answer to question 85 on the information below and on your knowledge of biology.

In an experiment, a membrane bag containing 95% water and 5% salt was placed in a beaker containing 80% water and 20% salt, as shown below. The setup was put aside until the next day.

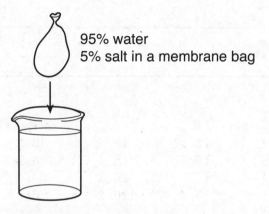

95% water
5% salt in a membrane bag

80% water
20% salt in beaker

85 Based on the information given, state *one* way the bag or its contents will have changed by the next day. Support your answer with an explanation. [1]

_____

_____

_____

Answers
August 2019
Living Environment

Answer Key

PART A

1. 4	6. 3	11. 2	16. 3	21. 2	26. 3
2. 2	7. 4	12. 1	17. 3	22. 3	27. 3
3. 3	8. 2	13. 2	18. 1	23. 3	28. 4
4. 2	9. 2	14. 2	19. 1	24. 3	29. 3
5. 4	10. 1	15. 4	20. 2	25. 1	30. 4

PART B–1

31. 4	34. 1	36. 2	38. 2	40. 2	42. 4
32. 4	35. 1	37. 3	39. 2	41. 4	43. 3
33. 1					

PART B–2

44. *See* Answers Explained.
45. *See* Answers Explained.
46. *See* Answers Explained.
47. 1
48. *See* Answers Explained.
49. 2
50. 3
51. *See* Answers Explained.
52. *See* Answers Explained.
53. *See* Answers Explained.
54. *See* Answers Explained.
55. *See* Answers Explained.

PART C. *See* **Answers Explained.**

PART D

73. 3
74. 1
75. 3
76. 2
77. *See* Answers Explained.
78. *See* Answers Explained.
79. *See* Answers Explained.
80. *See* Answers Explained.
81. 1
82. 2
83. *See* Answers Explained.
84. *See* Answers Explained.
85. *See* Answers Explained.

Answers Explained

PART A

1. **4** In this pyramid, the greatest amount of stored energy is found at level *D*. In any traditional energy pyramid, the base of the pyramid (producer level) is presumed to contain the greatest quantity of energy. This level represents green plants that absorb solar energy and convert it to the chemical bond energy of glucose.

WRONG CHOICES EXPLAINED:
(1) In this pyramid, the greatest amount of stored energy is *not* found at level *A*. Level *A*, which represents consumers such as top predators, contains the smallest amount of energy of the levels shown in the diagram.
(2) In this pyramid, the greatest amount of stored energy is *not* found at level *B*. Level *B*, which represents carnivores (meat eaters), contains less energy than the levels below it in the diagram.
(3) In this pyramid, the greatest amount of stored energy is *not* found at level *C*. Level *C*, which represents herbivores (plant eaters), contains less energy than the level below it in the diagram.

2. **2** As a result of the plant disease, the butterfly population will most likely *disappear from the area*. Absent their main food source, the butterfly larvae will likely die of starvation before they can pupate. Adult butterflies of this particular species will lay their eggs on members of this certain plant species that may be surviving in nearby fields.

WRONG CHOICES EXPLAINED:
(1), (3), (4) As a result of the plant disease, the butterfly population will *not* most likely *quickly adapt to eat other plants, evolve to form a new species,* or *enter the adult stage more quickly*. These responses all imply evolutionary modifications based on need rather than on natural selection. These changes are unlikely to occur as a result of natural selection, which requires random genetic mutations coupled with environmental selection pressures over substantial time.

3. **3** This protozoan is an example of a *parasite*. Parasites are other feeders that feed on the bodies of living plants and animals in their ecosystems. This protozoan may enter the bodies of humans and feed on their living tissues.

WRONG CHOICES EXPLAINED:

(1) This protozoan is *not* an example of a *predator*. Predators are other feeders that hunt and feed on the killed bodies of herbivores and carnivores in their ecosystems. The protozoan described in the question does not fit the definition of a predator.

(2) This protozoan is *not* an example of a *producer*. Producers are photosynthetic self-feeders that generally do not feed on any other organisms in their ecosystems. The protozoan described in the question does not fit the definition of a producer.

(4) This protozoan is *not* an example of a *scavenger*. Scavengers are other feeders that feed on the dead bodies of animals in their ecosystems. The protozoan described in the question does not fit the definition of a scavenger.

4. **2** The energy required to maintain this concentration difference is most likely associated with the action of *mitochondria*. Mitochondria are cell organelles responsible for the conversion of chemical bond energy in glucose to the chemical bond energy of ATP. ATP is used in the cell as a source of energy to operate cell functions such as active transport. Active transport is a process by which substances are transported across cell membranes from an area of low relative concentration to an area of high relative concentration, requiring the expenditure of cell energy. The seaweed's ability to concentrate iodine in its cells is an example of the energy-consuming action of active transport in nature.

WRONG CHOICES EXPLAINED:

(1) The energy required to maintain this concentration difference is *not* most likely associated with the action of *ribosomes*. Ribosomes function in protein synthesis by capturing mRNA molecules, attracting tRNA molecules, and linking amino acids in the arrangement determined by the genetic code. Ribosomes are not directly associated with the process of active transport of iodine in cells.

(3) The energy required to maintain this concentration difference is *not* most likely associated with the action of *vacuoles*. Vacuoles are cell organelles that store food or wastes in the cell. Such vacuoles are not directly associated with the process of active transport of iodine in cells.

[NOTE: The action of the contractile vacuole, found in certain protozoa, pumps water from inside the cell (an area of low relative concentration or water) to the cell's aquatic environment (an area of high relative concentration of water) with the expenditure of ATP energy. This process is similar to but distinct from the process of active transport that allows seaweed to concentrate iodine in its cells.]

(4) The energy required to maintain this concentration difference is *not* most likely associated with the action of *nuclei*. Nuclei are cell organelles that store the genetic codes of the cell. Nuclei are not directly associated with the process of active transport of iodine in cells.

5. **4** These vaccines most often contain *weakened pathogens*. Proteins known as antigens are embedded in the cell membrane of pathogenic organisms such as bacteria and viruses. By adding weakened or dead pathogens to a vaccine and injecting the vaccine into the tissues of a human, the body's immune system is stimulated to produce specific antibodies to neutralize the antigen. This immune response protects the body against future attacks by live pathogens of the same type as are found in the vaccine.

WRONG CHOICES EXPLAINED:
(1) These vaccines do *not* most often contain *antibodies*. Antibodies are substances produced by the immune system in response to the presence of a foreign antigen. Antibodies are not found in vaccines.
(2) These vaccines do *not* most often contain *antibiotics*. Antibiotics are biochemical compounds produced naturally by certain molds that are lethal to specific bacterial species. Antibiotics are not found in vaccines.
(3) These vaccines do *not* most often contain *white blood cells*. White blood cells are structures produced in the circulatory system that engulf and devour pathogens as a means of protecting the body against infectious disease. White blood cells are not found in vaccines.

6. **3** Building large manufacturing facilities can affect ecosystems by increasing the *demand for resources such as fossil fuels*. Human manufacturing activities require the expenditure of large quantities of electrical energy. Much of the electrical energy produced in the United States comes from power plants that burn fossil fuels such as coal, oil, and natural gas. Producing electrical energy in this manner uses the limited carbon-based resources, pollutes the air and water, and produces greenhouse gases that are known to contribute to climate change.

WRONG CHOICES EXPLAINED:
(1), (2), (4) It is *not* true that building large manufacturing facilities can affect ecosystems by increasing the *atmospheric quality*, the *biodiversity in the area*, or the *availability of space and resources for organisms*. Manufacturing activities typically decrease, not increase, these environmental factors in the area in which factories are constructed.

7. **4** From this, we can best infer that *structures in amebas have functions similar to organs in multicellular organisms*. In order to survive in their environment, all organisms must perform the same basic life functions. In multicellular organisms such as the human, these life functions are performed by specialized tissues, organs, and organ systems. In single-celled organisms such as the ameba, these life functions are performed by specialized organelles.

WRONG CHOICES EXPLAINED:

(1) It is *not* true that from this, we can best infer that *all single-celled organisms have lysosomes to digest food*. The information presented in the question provides information about only amebas, not all single-celled organisms.

(2) It is *not* true that from this, we can best infer that *amebas are capable of digesting any type of food molecule*. No information presented is in the question concerning the types of food molecules the ameba is capable of digesting.

(3) It is *not* true that from this, we can best infer that *single-celled organisms are as complex as multicellular organisms*. Single-celled organisms such as the ameba are among the simplest life-forms that inhabit Earth. By definition, multicellular organisms are more complex than single-celled organisms.

8. **2** White blood cells are most closely associated with the *immune and circulatory* systems. White blood cells are immune cells that travel through the circulatory system to engulf and devour pathogens as a means of protecting the body against infectious disease. In this way, white blood cells function in both the immune and circulatory systems of the human body.

WRONG CHOICES EXPLAINED:

(1), (3), (4) White blood cells are *not* most closely associated with the *circulatory and digestive* systems, the *digestive and excretory* systems, or the *excretory and immune* systems. The digestive system processes food for the body's metabolic activities, while the excretory system eliminates waste materials resulting from those activities. White blood cells do not directly function in either of these systems.

9. **2** The bogs only support certain types of organisms because *the growth and survival of organisms depends on specific physical conditions*. Each of Earth's species functions best under a specific set of abiotic (nonliving or physical) conditions unique to that species. Carnivorous bog plant species have evolved in a manner that allows them to thrive under conditions of low oxygen, low mineral nutrients, and high acidity.

WRONG CHOICES EXPLAINED:

(1) It is *not* true that the bogs support only certain types of organisms because *organisms in an environment are not limited by available energy and resources*. All organisms in all environments are subject to limitations imposed by the amounts of energy and other resources available in those environments.

(3) It is *not* true that the bogs support only certain types of organisms because *favorable gene mutations only occur when organisms live in harsh environments*. Mutations are random events in which genetic material (DNA) is altered, sometimes in favorable ways and sometimes in unfavorable ways. Mutations occur in the cells of organisms irrespective of the environments those organisms inhabit.

(4) It is *not* true that the bogs support only certain types of organisms because *photosynthetic organisms can only inhabit environments that have low acidity*. Bogs and other acidic environments support photosynthetic organisms adapted to survive and thrive in those environments.

10. **1** Anhidrosis most directly interferes with *a feedback mechanism that maintains homeostasis*. Under normal circumstances, the human body maintains its internal temperature within a narrow range around 98.68° F (37° C). The release and evaporation of perspiration (sweat) from tiny sweat glands embedded in the skin is a key factor in lowering body temperature when it gets too high from exercise or environmental heating. A feedback mechanism regulates this cooling effect by opening sweat glands when blood temperature increases and closing them when the blood returns to normal temperature. This feedback mechanism helps to maintain the body's homeostatic balance. Anhidrosis prevents the sweat glands from operating properly and threatens to upset this homeostatic balance.

WRONG CHOICES EXPLAINED:

(2) It is *not* true that anhidrosis most directly interferes with *an immune system response to harmless antigens*. Such responses are known as allergies. The action of anhidrosis is not known to affect the body's allergic response.

(3) It is *not* true that anhidrosis most directly interferes with *the synthesis of hormones in the circulatory system*. Hormones are synthesized in discrete endocrine tissues embedded in certain glands and body organs, not in the blood. The action of anhidrosis is not known to affect the body's synthesis of hormones.

(4) It is *not* true that anhidrosis most directly interferes with *the enzymatic breakdown of water in cells*. Water molecules cannot be broken down by means of any known enzymatic reaction. This is a nonsense distracter.

11. **2** The hair colors of the sons are most likely a direct result of *heredity*. During sexual reproduction, chromosomes carrying parental genes are transmitted

from the parents to their offspring. This transmittal, sometimes referred to as heredity, allows the offspring of the parents to inherit and display traits that may be observed in the parents. Hair color is known to be an observable inheritable trait.

WRONG CHOICES EXPLAINED:
(1), (3) The hair colors of the sons are *not* most likely a direct result of *natural selection in males* or *evolution*. Natural selection is thought to be the major biological mechanism that drives the evolution of Earth's species. These processes occur over vast periods of geological time and are not observable in a single generational change. That being said, hair color in humans is almost certainly a species trait resulting from natural selection and evolution that has occurred over time.

[NOTE: This ambiguity makes these responses possible correct answers to the question as written, although they are not the best answers.]

(4) The hair colors of the sons are *not* most likely a direct result of *environmental influences*. Environmental factors can and often do modify the expression of genetic traits in humans and other organisms. Because the question does not explicitly state that the identified hair colors of the parents and sons are natural, not dyed (an environmental effect), it leaves open the question of whether genetic inheritance or environmental factors are responsible for the observed traits.

[NOTE: This ambiguity makes this response a possible correct answer to the question as written, although it is not the best answer.]

12. **1** The percentage of the bases that will be thymine (T) is approximately *18*. The percentage of cytosine (C) at 32% must match the percentage of guanine (G) in the sample since these two bases are complementary in a DNA strand. The combined total of C and G is 64%. The percentages of thymine (T) and adenine (A), as complementary bases, must also be equal to each other. Their combined total must make up the balance of the 100% of all bases in the sample. T and A, therefore, must make up a combined total of 36% of the sample, or approximately 18% each.

WRONG CHOICES EXPLAINED:
(2), (3), (4) The percentage of the bases that will be thymine (T) is *not* approximately *24, 32,* or *36*. None of these responses can result from using the appropriate calculation method shown in the correct answer above.

13. **2** This response to carmine is known as *an allergy*. An allergy is an immune response to a usually harmless environmental substance. Common environmental substances, including plant and animal pigments such as carmine,

carry foreign proteins known as antigens. When these substances are ingested, the body's immune system recognizes these antigens and may respond to them by producing histamines. Histamines, in turn, cause symptoms such as congestion and swelling of mucous membranes that can interfere with breathing. The symptoms caused by the immune response are known collectively as an allergic reaction.

WRONG CHOICES EXPLAINED:

(1) This response to carmine is *not* known as *a stimulus*. A stimulus is any change in the environment to which an organism responds. The introduction of carmine into the body can be thought of as an environmental change (stimulus) that triggers the immune response in an allergic person. However, the body's allergic reaction to carmine itself is not a stimulus.

(3) This response to carmine is *not* known as *natural selection*. The theory of natural selection postulates that certain members of a species are better adapted to their environment than others and that these organisms are more likely to survive and pass on their favorable adaptations to future generations. The human body's reaction to carmine does not constitute natural selection.

(4) This response to carmine is *not* known as *an adaptation*. An adaptation is any physical, chemical, or behavioral characteristic of an organism that helps it to survive potentially harsh environmental conditions. The allergic reaction (an immune system function) in response to carmine can be thought of as a behavioral characteristic that aids the survival of the allergic person.

[NOTE: This ambiguity makes this response a possible correct answer to the question as written, although it is not the best answer.]

14. **2** *Genetic engineering* is the process that was most likely used to alter these mosquitoes. Genetic engineering is a laboratory technique in which a gene for a desired trait is inserted into the genome of a recipient reproductive cell. When the recipient reproductive cell participates in successful fertilization, the resulting offspring will carry the altered gene and pass it on to future generations. In this case, the *Anopheles* mosquito is the recipient organism and the desired trait is the inability to transmit malaria.

WRONG CHOICES EXPLAINED:

(1) *Cloning studies* is *not* the process that was most likely used to alter these mosquitoes. Cloning is a laboratory technique that involves the removal of a diploid nucleus from a donor organism and inserting it into a denucleated fertilized egg cell of the same or related species. The resulting offspring is a genetic duplicate (clone) of the donor organism, not an artificially altered organism.

(3) *Natural selection* is *not* the process that was most likely used to alter these mosquitoes. The theory of natural selection postulates that certain members of a species are better adapted to their environment than others and that these organisms are more likely to survive and pass on their favorable adaptations to future generations. The question describes a laboratory technique, not a natural process.

(4) *Random mutations* is *not* the process that was most likely used to alter these mosquitoes. Mutations are random events in which genetic material (DNA) is altered through the action of natural mutagenic agents, sometimes in favorable ways and sometimes in unfavorable ways. The question describes a laboratory technique, not a natural process.

15. **4** Row *4* in the chart accurately identifies two causes of mutations and the cells that must be affected in order for the mutations to be passed on to offspring. Mutations are random events in which genetic material (DNA) is altered through the action of mutagenic agents, including chemicals and radiation. When mutations occur in sex cells, they may be passed on to future generations by sexual reproduction, thus perpetuating the mutations and their effects.

WRONG CHOICES EXPLAINED:

(1), (2) Rows *1* and 2 in the chart do *not* accurately identify two causes of mutations and the cells that must be affected in order for the mutations to be passed on to offspring. Mutations occurring in body cells cannot be passed on to future generations. Mutagenic agents do not include infections, antigens, meiosis, or mitosis.

(3) Row *3* in the chart does *not* accurately identify two causes of mutations and the cells that must be affected in order for the mutations to be passed on to offspring. Mutagenic agents, even in sex cells, do not include disease and differentiation.

16. **3** One way this form of reproduction differs from sexual reproduction is *the offspring and the parents are genetically identical*. The reproductive process illustrated in the question is asexual reproduction, in which new organisms are formed from a single parent organism. This process involves mitotic cell division, which does not provide opportunities for the creation of new genetic combinations. For this reason, the offspring of asexual reproduction are genetic duplicates of the parent.

WRONG CHOICES EXPLAINED:

(1) It is *not* true that one way this form of reproduction differs from sexual reproduction is *more genetic variations are seen in the offspring*. In fact, the offspring of asexual reproduction show no genetic variation from each other whatsoever.

(2) It is *not* true that one way this form of reproduction differs from sexual reproduction is *there is a greater chance for mutations to occur*. In fact, the chance of mutations occurring is the same in both sexual and asexual reproduction.

(4) It is *not* true that one way this form of reproduction differs from sexual reproduction is *the new plants possess the combined genes of both parents*. In fact, the offspring of asexual reproduction have only one parent with which they share identical genetic information.

17. **3** If the fish population *decreases*, the most direct effect this will have on the aquatic ecosystem is that *the zooplankton population will increase in size*. According to the diagram, zooplankton serve as a food source for fish. If fewer fish are present in the environment, the zooplankton population will likely increase until some limiting factor (such as a reduction of the phytoplankton population) acts to balance it.

WRONG CHOICES EXPLAINED:
(1) It is *not* true that if the fish population *decreases*, the most direct effect this will have on the aquatic ecosystem is that *the leopard seals will all die from lack of food*. According to the diagram, leopard seals feed on seagulls and penguins in addition to fish. If fewer fish are present in the environment, the leopard seals will likely move to consume more of their alternate food sources.

(2) It is *not* true that if the fish population *decreases*, the most direct effect this will have on the aquatic ecosystem is that *the krill population will only be consumed by seagulls*. According to the diagram, there is no direct nutritional link between krill and seagulls.

(4) It is *not* true that if the fish population *decreases*, the most direct effect this will have on the aquatic ecosystem is that *the phytoplankton population will increase in size*. According to the diagram, phytoplankton serve as a food source for zooplankton. If fewer fish are present in the environment, the zooplankton population will increase and consume more phytoplankton. So, the phytoplankton population will likely decrease, not increase.

18. **1** This sequence of changes is the result of *ecological succession*. Ecological succession refers to the series of changes that occur to the dominant plant community of an area that has been significantly affected by an environmental disruption. The sequence shown in the chart is typical of ecological succession that occurs in abandoned fields in the northeastern United States. Such succession typically begins with grasses, which are then replaced by shrubs, which are replaced in turn by hardwood forest trees.

WRONG CHOICES EXPLAINED:

(2) This sequence of changes is *not* the result of *decreased biodiversity*. Biodiversity is a measure of the number of different, compatible species that inhabit an area. Ecological succession is typically characterized by increased, not decreased, biodiversity.

(3) This sequence of changes is *not* the result of *biological evolution*. Biological evolution is a process characterized by gradual change of existing species over time that results in new or modified species. In this example, new species are not being developed. Rather, the sequence of changes listed in the chart indicates that existing species are replacing other existing species in a modified environment.

(4) This sequence of changes is *not* the result of *environmental trade-offs*. An environmental trade-off is any human decision that takes into consideration the positive and negative aspects of that decision. The sequence of changes listed in the chart describe a natural, not a human-made, phenomenon.

19. **1** *Genetically modified salmon produce more of some proteins than wild salmon* is the statement regarding genetically modified salmon that is correct. Scientists working in this area of science have discovered that adding certain genes to the genome of wild salmon results in larger, faster-growing fish. By modifying the wild salmon in this way, more of certain proteins are synthesized in the modified salmon than in the wild salmon.

WRONG CHOICES EXPLAINED:

(2) *Genetically modified salmon and wild salmon would have identical DNA* is *not* the statement regarding genetically modified salmon that is correct. Although the wild and modified salmon share most genes in common, the fact that scientists added genes to the modified salmon's genome ensures that their DNA is not identical.

(3) *Wild salmon reproduce asexually while genetically modified salmon reproduce sexually* is *not* the statement regarding genetically modified salmon that is correct. Fish of all species and all types reproduce sexually, not asexually. No information is presented to suggest that this characteristic would change under these circumstances.

(4) *Wild salmon have an altered protein sequence, but genetically modified salmon do not* is *not* the statement regarding genetically modified salmon that is correct. In fact, the opposite is true. Scientists have altered the genetic sequence of genetically modified salmon. So, the protein sequence of these salmon is altered by comparison with wild salmon.

20. **2** *Decomposers* is the group of organisms in an ecosystem that fills the niche of recycling organic matter back to the environment. Decomposers, such as bacteria and fungi, secrete enzymes that digest complex organic matter and convert it to simpler materials that can be taken up by green plants. In this way, decomposers aid in the recycling of carbon, hydrogen, oxygen, nitrogen, and other elements in the natural environment.

WRONG CHOICES EXPLAINED:

(1), (4) *Carnivores* and *predators* are *not* the groups of organisms in an ecosystem that fill the niche of recycling organic matter back to the environment. Carnivores and predators represent a nutritional class of animals that consume the bodies of other animals for food. Although carnivores and predators convert organic matter to other forms and release metabolic wastes to the environment, they do not perform the recycling action of decomposers.

(3) *Producers* is *not* the group of organisms in an ecosystem that fills the niche of recycling organic matter back to the environment. Producers represent a nutritional class of organisms that convert inorganic carbon dioxide and water into organic sugars with the input of solar energy. Although producers convert inorganic matter to organic matter and release metabolic wastes to the environment, they do not perform the recycling action of decomposers.

21. **2** A benefit of using solar energy would include *using less fossil fuel to meet energy needs.* Much of the electrical energy produced in the United States comes from power plants that burn fossil fuels, such as coal, oil, and natural gas. Producing electrical energy using solar collectors instead of fossil fuels preserves limited carbon-based resources and reduces pollution of the air and water.

WRONG CHOICES EXPLAINED:

(1), (4) A benefit of using solar energy would *not* include *adding more carbon dioxide to the atmosphere* or *releasing more gases for photosynthesis.* Photosynthesis uses carbon dioxide gas in the atmosphere as a raw material for the manufacture of glucose. Solar energy technology does not produce carbon dioxide for release to the atmosphere.

(3) A benefit of using solar energy would *not* include *using a nonrenewable source of energy.* Solar energy is renewable because the Sun is a stable energy source that will not run out for billions of years into the future.

22. **3** For these microorganisms, the wastewater facility serves as *an ecosystem.* An ecosystem is the basic functional unit of the living environment, including all the communities of organisms that interact with each other and with the nonliving environment. Because the microorganisms in the wastewater facility

represent a community of living species interacting with each other and with a limited set of nonliving components, it can be said that the microorganisms inhabit an ecosystem of sorts.

WRONG CHOICES EXPLAINED:

(1) It is *not* true that for these microorganisms, the wastewater facility serves as *its carrying capacity*. The carrying capacity of an environment is the maximum number of individuals of a species that can be supported given limited food and other resources. To the degree that the wastewater facility contains a finite amount of food and other resources, it may, in fact, provide a set of factors that limits the number of microorganisms that can be supported in that facility. However, the facility itself cannot be the species' carrying capacity.

(2) It is *not* true that for these microorganisms, the wastewater facility serves as *a food chain*. Food chains illustrate nutritional relationships among species in terms of the transfer of energy from producers (green plants), which manufacture their own food via photosynthesis, to sequential levels of consumer organisms. A wastewater facility contains only decomposers. So, it cannot represent a true food chain.

(4) It is *not* true that for these microorganisms, the wastewater facility serves as *an energy pyramid*. An energy pyramid is a construct used to illustrate the fact that energy is lost at each trophic level in a food web, with producers containing the most energy. A wastewater facility contains only decomposers. So, it cannot represent a true energy pyramid.

23. **3** The type of organic molecule that is being synthesized is *protein*. The diagram illustrates the manufacture of amino acid chains at the ribosome based on the codon pattern supplied by messenger RNA (mRNA) and transcribed by transfer RNA (tRNA). This amino acid chain is the basis of a specific protein molecule produced in the cell.

WRONG CHOICES EXPLAINED:

(1) It is *not* true that the type of organic molecule that is being synthesized is *DNA*. DNA is synthesized via the process of replication (exact self-duplication) in cells, which involves only DNA, not mRNA, tRNA, or ribosomes.

(2), (4) It is *not* true that the type of organic molecule that is being synthesized is *starch* or *fat*. Starch and fat are organic molecules that are synthesized in cells via the action of enzymatically controlled reactions, not mRNA, tRNA, or ribosomes.

24. **3** This is because, in these new habitats, the nonnative insects might *not have natural predators*. Invasive organisms are organisms native to one ecosystem

that are introduced into a foreign ecosystem. Invasive organisms often have the effect of competing with and eliminating native species from their natural habitats. Because invasive species are not native to their new habitat, it is likely that no native predators will be adapted to control their growth.

WRONG CHOICES EXPLAINED:

(1) It is *not* true that this is because, in these new habitats, the nonnative insects might *become food for birds*. Birds represent one type of natural predator that might control an invasive species. However, the birds in a particular area may not be adapted to hunt and consume the nonnative insects.

(2) It is *not* true that this is because, in these new habitats, the nonnative insects might *not survive a cold winter*. Cold winters represent one type of natural abiotic pressure that might control an invasive species. However, the winters may not prove sufficiently cold to eliminate the nonnative insects.

(4) It is *not* true that this is because, in these new habitats, the nonnative insects might *add to the biodiversity*. Invasive species tend to compete with and eliminate native species. So, the typical effect of nonnative insects in their new environment is to reduce, not add to, biodiversity.

25. **1** Maintaining an appropriate temperature is beneficial to an organism because *enzymes need a proper range of temperatures to catalyze vital reactions*. Enzymes are proteins that function as biological catalysts in the cells of all living things and that help to regulate the body's metabolism. If the cell's temperature is too high, enzymes' rate of catalytic action will decline and may even cease permanently.

WRONG CHOICES EXPLAINED:

(2) It is *not* true that maintaining an appropriate temperature is beneficial to an organism because *cellular waste products can only be excreted in cooler temperatures*. Cellular wastes are normally excreted from cells via the processes of passive and active transport. These are physical processes that occur more rapidly in warmer, not cooler, temperatures.

(3) It is *not* true that maintaining an appropriate temperature is beneficial to an organism because *hormones can only produce antibodies if temperatures are not excessive*. Hormones do not produce antibodies in humans, although hormones may help to regulate certain aspects of the immune response.

(4) It is *not* true that maintaining an appropriate temperature is beneficial to an organism because *nutrients diffuse faster into cells when temperatures are lower*. Diffusion is a physical process that occurs more rapidly when temperatures are higher, not lower.

26. **3** Row 3 in the chart correctly identifies where in the female reproductive system these two processes occur. Meiosis, the process in which homologous chromosome pairs and the genes they carry are separated into haploid gametes during gametogenesis, normally occurs in the ovary (structure 3). Fertilization, the process in which the haploid sperm nucleus merges with the haploid egg nucleus, normally occurs within the oviduct (structure 4).

WRONG CHOICES EXPLAINED:

(1) Row *1* in the chart does *not* correctly identify where in the female reproductive system these two processes occur. Structure 1 is the vagina, within which haploid sperm is implanted by the male parent, not where meiosis occurs. Structure 3 is the ovary, within which eggs cells are produced by the process of meiosis, not where fertilization occurs.

(2) Row 2 in the chart does *not* correctly identify where in the female reproductive system these two processes occur. Structure 2 is the cervix, which bounds the lower extremity of the uterus and prevents premature birth of the embryo, not where meiosis occurs. Structure 5 is the uterus, within which fertilized eggs develop into embryos, not where fertilization occurs.

(4) Row *4* in the chart does *not* correctly identify where in the female reproductive system these two processes occur. Structure 4 is the oviduct, within which fertilization of the haploid egg by the haploid sperm occurs, not where meiosis occurs. Structure 5 is the uterus, within which fertilized eggs develop into embryos, not where fertilization occurs.

27. **3** The prevention of browning is likely the result of *slowing the rate of enzyme action*. Enzymes are known to be sensitive to the acidity (pH) of the environment in which they operate. Most enzymes work best in a neutral pH (7.0). Squeezing lemon juice onto the surface of the apple slices lowers the pH of the apple cells to an acidic level (pH < 7.0). This, in turn, slows the catalytic action of enzymes in the apple cells and delays browning of the apple slices.

WRONG CHOICES EXPLAINED:

(1) The prevention of browning is *not* likely the result of *increasing the concentration of enzymes*. Increasing the concentration of enzymes in the system would typically increase, not decrease, the rate of the enzyme's catalytic action.

(2) The prevention of browning is *not* likely the result of *increasing the temperature*. Increasing the temperature slightly would typically increase, not decrease, the rate of the enzyme's catalytic action. If the temperature of the system gets too high, the enzyme molecules may denature, which can end the enzyme's action altogether.

(4) The prevention of browning is *not* likely the result of *maintaining the pH*. If the pH and other factors remain unchanged in the system, the enzyme's action will be neither decreased nor increased.

28. **4** In the presence of a higher amount of estrogen, it would be most likely that *more females would be found because estrogen promotes the development of female characteristics*. Estrogen is a hormone produced in many plant and animal species and by human agricultural activities. Estrogen's effect in most species is to promote the development of female characteristics and to regulate the production of egg cells needed for sexual reproduction. Scientists studying frog populations have noticed a higher ratio of female frogs compared to male frogs in estrogen-rich freshwater environments surrounded by human suburbs than in those surrounded by natural forest environments. The scientists hypothesize that estrogens produced by human activities and released into the frogs' freshwater environments are responsible for this phenomenon.

WRONG CHOICES EXPLAINED:
(1) It is *not* true that in the presence of a higher amount of estrogen, it would be most likely that *fewer males would be found because they are much larger and fewer are produced*. Male frogs are typically smaller than females of the same species. In a balanced natural freshwater environment, male frogs are typically found in numbers nearly equal to that of females.

(2) It is *not* true that in the presence of a higher amount of estrogen, it would be most likely that *fewer females would be found because they are more sensitive to pesticides*. No information is provided in the question that would lead to this conclusion.

(3) It is *not* true that in the presence of a higher amount of estrogen, it would be most likely that *more males would be found because estrogen promotes the development of male characteristics*. Estrogen is known to be associated with the production of female, not male, characteristics.

29. **3** *Use alternate sources of energy such as wind* is an action that humans could take to slow the rate of global warming. Competent scientific evidence around the issue of climate change points to the release of greenhouse gases (such as carbon dioxide) into the atmosphere as being responsible for increases in Earth's average temperature (global warming) sufficient to cause major disruption of long-stable climate patterns. A major source of carbon dioxide has been identified as electrical plants powered by fossil fuels (coal, oil, and natural gas). By encouraging the implementation of clean alternative energy solutions such as wind turbine electrical generators, our dependence on fossil fuel electrical plants and the carbon dioxide they release will decrease.

WRONG CHOICES EXPLAINED:

(1) *Cut down trees for more efficient land use* is *not* an action that humans could take to slow the rate of global warming. Trees and other green plants absorb carbon dioxide from the atmosphere and fix it into stable carbon chain molecules such as cellulose. It is imperative that we maintain our forests, not cut them down, to help reduce the quantity of carbon dioxide in the air.

(2) *Increase the consumption of petroleum products* is *not* an action that humans could take to slow the rate of global warming. Petroleum products include coal, oil, and natural gas, which are carbon-based fuels the burning of which is thought to be responsible for the increase of carbon dioxide in the atmosphere. It is imperative that we reduce, not increase, our consumption of petroleum products to help reduce the quantity of carbon dioxide in the air.

(4) *Reduce the use of fuel-efficient automobiles* is *not* an action that humans could take to slow the rate of global warming. Cars and trucks burn gasoline and release many tons of carbon dioxide into the atmosphere each day. It is imperative that we increase, not reduce, the fuel efficiency of automobiles to help reduce the quantity of carbon dioxide in the air.

30. **4** The role of antibodies in the human body is to *recognize foreign antigens and mark them for destruction*. Antigens are proteins produced by the cells of all living things, including pathogens, and are specific to the species that produces them. The immune response in humans detects the antigens of foreign pathogens and responds to these antigens by manufacturing antibodies that specifically target and neutralize the infectious agents.

WRONG CHOICES EXPLAINED:

(1) The role of antibodies in the human body is *not* to *stimulate pathogen reproduction to produce additional white blood cells*. Pathogens are infectious agents, including viruses, bacteria, fungi, and parasites, that cause homeostatic disruption in the human body. Antibodies are specialized to slow and stop the reproduction of pathogens in the body, not stimulate them.

(2) The role of antibodies in the human body is *not* to *increase the production of guard cells to defend against pathogens*. Guard cells are specialized structures found in green plant leaves that have no role in the human immune response. This is a nonsense distracter.

(3) The role of antibodies in the human body is *not* to *promote the production of antigens to stimulate an immune response*. Antigens are proteins produced by the cells of all living things, including pathogens. Antibodies are produced in the body in response to the presence of foreign antigens, not to produce antigens.

PART B–1

31. **4** *One individual had larger feet than the other* is the statement that is an accurate observation that can be made based on this trail of footprints. Direct observation of the footprints illustrated shows a size difference of those on the top (left) from those on the bottom (right). A reasonable inference may be drawn that one of the individuals had larger feet than the other.

WRONG CHOICES EXPLAINED:

(1), (2), (3) *The individuals were running from a predator, the volcano was about to erupt again,* and *one individual was much taller than the other* are *not* the statements that are accurate observations that can be made based on this trail of footprints. Insufficient data are provided in the diagram to support any of these inferences.

32. **4** The type of information directly provided by these fossil footprints is useful because it *is a record of some similarities and differences they share with present-day species*. To the extent that these fossil footprints can be compared to the footprints of modern human beings, it is possible to point to similarities of foot structure, walking gait (pattern), and other features that these individuals share with modern humans.

WRONG CHOICES EXPLAINED:

(1), (2), (3) It is *not* true that the type of information directly provided by these fossil footprints is useful because it *offers details about how these individuals changed during their lifetime, offers data regarding their exposure to ultraviolet (UV) radiation,* or *is a record of information about what these individuals ate during their lifetime*. Insufficient data are provided in the diagram to support any of these inferences.

33. **1** *Researchers have been able to collect more data than were available in the 1980s* is the statement that best explains why the safety of PPIs is now in question when clinical experiments in the 1980s provided evidence that they were safe. A characteristic of scientific inquiry is that the results of earlier experiments form the basis of continued study and the discovery of new information. Such continued study commonly results in new information that modifies the inferences drawn from earlier experiments.

WRONG CHOICES EXPLAINED:

(2), (3), (4) *Fewer people had acid reflux in the 1980s compared to today, the medication containing PPIs has changed since the 1980s when tests were done,* and *the original experiments in the 1980s used only test animals and did not use human subjects* are *not* the statements that best explain why the safety of PPIs is now in question when clinical experiments in the 1980s provided evidence that they were safe. No information is provided in the question that would lead to any of these explanations.

34. **1** The three arrows in the diagram each represent a process known as *mitotic cell division*. The diagram illustrates a zygote (fertilized egg) undergoing the process of cleavage to form the developing embryo. Each new cell shown in the diagram is formed by the process of mitotic cell division, in which a diploid cell divides to produce two genetically identical daughter cells. When this process has occurred several times, a cell mass is formed that is the developing embryo.

WRONG CHOICES EXPLAINED:

(2) The three arrows in the diagram do *not* each represent a process known as *meiotic cell division*. Meiotic cell division is the process in which homologous chromosome pairs and the genes they carry are separated into haploid gametes during gametogenesis. Meiotic cell division results in the formation of haploid gametes, not diploid developing embryos.

(3) The three arrows in the diagram do *not* each represent a process known as *fertilization of gamete cells*. Fertilization is the process in which a haploid sperm nucleus merges with a haploid egg nucleus to produce a diploid zygote. The diagram shows one cell dividing to create two, not two cells merging to create one.

(4) The three arrows in the diagram do *not* each represent a process known as *differentiation of tissues*. Differentiation is the process by which the cells of a developing embryo undergo specialization to become specific body tissues. The cells of the developing embryo illustrated in the diagram have not yet begun to differentiate.

35. **1** Chemical *1* would most likely interfere with the activity of hormone *A*, but not hormone *B*. The cell is illustrated to have receptors that are shaped in a concave circular manner so as to accommodate the circular shape of hormone *A*. Because the shape of chemical *1* is convex circular, it may theoretically bind to the cell's receptors and block hormone *A* from binding to the receptor, thus inhibiting the activity of hormone *A* activity in the cell's metabolic reactions. Chemical *1* will not, however, be able to bind with the cell's concave square receptors. So, it will not interfere with the activities of hormone *B*.

WRONG CHOICES EXPLAINED:

(2), (3), (4) Chemicals *2*, *3*, and *4* would *not* most likely interfere with the activity of hormone *A*, but not hormone *B*. The shapes of these chemicals are illustrated to be square. As such, they would tend to bind to the cell receptors for hormone *B*, which are shaped as concave squares. Thus, these chemicals would tend to interfere with hormone *B*, but not hormone *A*.

36. **2** The results of this scientific investigation will most likely lead other scientists to hypothesize that *other reptiles may have the TRPV4 protein in their eggs*. Closely related organisms often share genetic traits in common. It would be a logical assumption that other reptile species, such as crocodiles, turtles, lizards, and snakes, would carry the genes for production of TRPV4 in their genomes.

WRONG CHOICES EXPLAINED:
(1) The results of this scientific investigation will *not* most likely lead other scientists to hypothesize that *human sex cells also contain the TRPV4 protein*. Humans and reptiles are both vertebrate animals that share many genes in common. Although it is possible that humans carry the gene to produce TRPV4, it is unlikely that the protein would perform the same function in humans since humans are warm-blooded mammals whose offspring develop in the temperature-controlled environment of the uterus.
(3) The results of this scientific investigation will *not* most likely lead other scientists to hypothesize that *the TRPV4 protein affects the growth of plants*. Plants are only distantly related to alligators and other animal species. It is unlikely, though possible, that plants have the genetic ability to synthesize the TRPV4 protein.
(4) The results of this scientific investigation will *not* most likely lead other scientists to hypothesize that *the TRPV4 protein is present in all of the foods eaten by alligators*. It is possible that some of an alligator's food contains TRPV4, but this TRPV4 would not be the source of the TRPV4 in the alligator's cells. Instead, this food would only be the source of amino acids necessary for the synthesis of TRPV4 in an alligator's cells.

37. **3** The information that was most essential in preparing to carry out this scientific investigation was *the effects of temperature on the incubation of alligator eggs*. Although the investigation carried out is not described in the passage, it is clear from the information given that scientists have been studying the effects of incubation temperature on sex determination in alligators. Scientists have also studied the biochemical basis for this phenomenon and determined that the protein TRPV4 is involved.

WRONG CHOICES EXPLAINED:
(1) It is *not* true that the information that was most essential in preparing to carry out this scientific investigation was *a knowledge of the variety of mutations*

found in American alligator populations. There is no mention of studies of mutations provided in the information given in the passage.

(2) It is *not* true that the information that was most essential in preparing to carry out this scientific investigation was *the arrangement of the DNA bases found in the TRPV4 protein.* Proteins such as TRPV4 are constructed of chains of amino acid molecules, not DNA bases.

(4) It is *not* true that the information that was most essential in preparing to carry out this scientific investigation was *a knowledge of previous cloning experiments conducted on alligators and other reptiles.* There is no mention of cloning experiments provided in the information given in the passage.

38. **2** The movement of the calcium ions into certain cells is most likely due to *the action of TRPV4 proteins on the cells involved with sex determination.* This fact is clearly stated in the second sentence of the second paragraph of the passage.

WRONG CHOICES EXPLAINED:

(1), (3), (4) The movement of the calcium ions into certain cells is *not* most likely due to *the destruction of the TRPV4 when it contacts the cell membrane, the sex of the alligator embryo present in that particular egg,* or *the action of receptor proteins attached to the mitochondria in alligator sex cells.* None of these phenomena are described in the passage and are unlikely to be the cause of calcium ion movement into certain alligator cells.

39. **2** Environmental changes such as global warming could affect species such as the American alligator because even slight increases in environmental temperature could *lead to an overabundance of males and few, if any, females.* The passage states in the second sentence of the first paragraph that increasing environmental temperatures as little as 3 degrees Celsius (from 30° C to 33° C) tips the probabilities of sex determination in alligators in favor of the production of all males.

WRONG CHOICES EXPLAINED:

(1) It is *not* true that environmental changes such as global warming could affect species such as the American alligator because even slight increases in environmental temperature could *lead to an overabundance of females and few, if any, males.* According to the passage, just the opposite effect could be expected if environmental temperatures rise.

(3) It is *not* true that environmental changes such as global warming could affect species such as the American alligator because even slight increases in environmental temperature could *cause the breakdown of the TRPV4 protein in*

female alligators. No information is provided in the passage that would lead to this inference. In general, proteins are sensitive to environmental temperatures and many begin to denature (deform and become inactive) at temperatures above 40° C.

(4) It is *not* true that environmental changes such as global warming could affect species such as the American alligator because even slight increases in environmental temperature could *increase the rate at which calcium ions exit the sex cells*. No information is provided in the passage that would lead to this inference.

40. **2** *Building dams to control the flow of water in rivers in order to produce electricity* is the human activity that can have a *negative* impact on the stability of a mature ecosystem. The process of hydroelectric dam construction typically involves the removal of vegetation from, and the flooding of, mature forested areas and freshwater ecosystems. This activity destroys natural habitats and disrupts animal breeding and migration cycles, often permanently. These are generally considered to be negative consequences of such projects and must be weighed against their potential benefits.

WRONG CHOICES EXPLAINED:

(1), (3), (4) *Replanting trees in areas where forests have been cut down for lumber*; *preserving natural wetlands, such as swamps, to reduce flooding after heavy rainfalls*; and *passing laws that limit the dumping of pollutants in forests* are *not* the human activities that can have a *negative* impact on the stability of a mature ecosystem. Each of these responses represents a human activity that can have a positive effect on the ecosystem.

41. **4** Energy in this ecosystem passes directly from the Sun to *autotrophs*. Autotrophs (self-feeders) found in this pond ecosystem (such as grass, sedge, cattail, pickerelweed, water lily, duckweed, and algae) are all capable of absorbing solar energy and using it to convert carbon dioxide and water into glucose and oxygen. Most autotrophs employ the process of photosynthesis to absorb solar energy and incorporate it into the chemical bonds of glucose.

WRONG CHOICES EXPLAINED:

(1), (2), (3) Energy in this ecosystem does *not* pass directly from the Sun to *herbivores*, *consumers*, or *heterotrophs*. Each of these categories includes animal species that consume preformed nutrients, either plant or animal tissue, for their energy. The term *heterotroph* means "other feeder" and is the category that includes herbivores and consumers.

42. **4** The risk to the horses during the roundups compared to the entire loss of plants and soils is considered *a trade-off*. The death of 1% of the mustangs, though regrettable, can be considered a small price to pay for the preservation of the prairie lands that could potentially become overgrazed and irreparably damaged by these animals. Weighing such alternatives in decision making is known as a trade-off between two alternatives and their outcomes.

WRONG CHOICES EXPLAINED:

(1) The risk to the horses during the roundups compared to the entire loss of plants and soils is *not* considered *selective breeding*. Selective breeding is a technique used by animal and plant breeders in which organisms with desirable traits are cross-bred with the hope of producing offspring that also display those traits. This is not the process described in the question.

(2) The risk to the horses during the roundups compared to the entire loss of plants and soils is *not* considered *a technological fix*. A technological fix is a general term that usually refers to a method of solving a problem using a computer program or other modern technology. This is not the process described in the question.

(3) The risk to the horses during the roundups compared to the entire loss of plants and soils is *not* considered *direct harvesting*. Direct harvesting is a human practice in which plants or animals are removed (harvested) from a natural environment for their economic value. This is not the process described in the question.

43. **3** The number of organisms that an area of land can sustain over a long period of time is known as *its carrying capacity*. The carrying capacity of an environment is the maximum number of individuals of a species that can be supported in an area given limited food and other resources. In this example, the amount of food (grass and other vegetation) available on the prairie environment would provide a check on the mustang population, limiting the number of horses that could be naturally maintained in the area over time.

WRONG CHOICES EXPLAINED:

(1) The number of organisms that an area of land can sustain over a long period of time is *not* known as *its finite resources*. A finite resource is a general term that refers to any biotic or abiotic factor that is considered nonrenewable. This description does not match the biological definition of carrying capacity.

(2) The number of organisms that an area of land can sustain over a long period of time is *not* known as *ecological succession*. Ecological succession refers to the series of changes that occur to the dominant plant community of an

area that has been significantly affected by an environmental disruption. This description does not match the biological definition of carrying capacity.

(4) The number of organisms that an area of land can sustain over a long period of time is *not* known as *evolutionary change*. Evolutionary change is a term that refers to the process of natural selection that results in new or altered species characteristics leading to species change. This description does not match the biological definition of carrying capacity.

PART B–2

44. One credit is allowed for correctly marking an appropriate scale, without any breaks in the data, on the axis labeled "Number of Bats." [1]

45. One credit is allowed for correctly plotting the data for big brown bats on the grid, connecting the points, and surrounding each point with a small circle. [1]

46. One credit is allowed for correctly plotting the data for little brown bats on the grid, connecting the points, and surrounding each point with a small triangle. [1]

Summer Brown Bat Count

Key
⊙ =Big brown bats
△ =Little brown bats

Number of Bats

7000
6500
6000
5500
5000
4500
4000
3500
3000
2500
2000
1500
1000
500

2009   2010   2011   2012   2013

Years

47. **1** *A few of the bats possessed an immunity to the WNS disease and produced offspring that were immune* is the statement that best explains this increased survival rate. This statement is consistent with the theory of natural selection, which is thought to be the major biological mechanism that drives the evolution of Earth's species. It hypothesizes that some bats carried genetic mutations that

gave them a natural immunity to WNS and therefore provided them with an adaptive advantage over other bats that lacked this adaptation. Surviving bats passed on this favorable trait to future generations. If this hypothesis is true, after several generations, the majority of the bats will be immune to WNS.

[NOTE: The experiment as described in the passage included no genetic testing to determine whether or not a mutation providing immunity against WNS was present in the healthy bats but not in the diseased bats. Without this information, it is impossible to say with certainty that this statement is accurate with respect to the surviving bats' resistance to WNS.]

WRONG CHOICES EXPLAINED:

(2) *The bats needed to reproduce in greater numbers, otherwise they would have died out completely* is *not* the statement that best explains this increased survival rate. This statement implies that the bats possess reasoning ability common to humans but not to the vast majority of Earth's species. It is more likely that the reproductive process in bats is an instinctive behavior that cannot be altered by the bats' free will.

(3) *The people that performed the recent counts did not identify the bats correctly and were counting bats of a different species* is *not* the statement that best explains this increased survival rate. This statement implies that the scientists conducting this count were not competent to recognize the similarities and differences between these two bat species and to record scientific data properly. Such an implication is highly unlikely to be the case.

(4) *The original decline of the bat population due to WNS was a natural occurrence and is part of a natural cycle* is *not* the statement that best explains this increased survival rate. This statement makes a case for the concept that fungal infections and other diseases come and go within a species depending on varying environmental conditions, the immune response of the bats, and the life cycle of the infectious agent.

[NOTE: Insufficient information is given in the passage to eliminate this concept as a possible contributing factor in the WNS infection in bats. For this reason, this response may be considered a potential correct answer to the question as written, although it is not the best answer.]

48. One credit is allowed for correctly explaining why this may have a positive effect on the control of WNS in bats. Acceptable responses include but are not limited to: [1]

- *The bat houses are more easily monitored than the natural bat habitats, and the sick bats could be removed or treated.*
- *The factors of temperature and humidity could be more easily controlled in the constructed bat houses.*
- *The bat houses can be sterilized, and the disease will be less likely to be transmitted,*
- *The amount of contact among members of the bat population would be less, slowing the spread of the disease.*
- *Bat houses would keep bats away from infected caves/mines/bats.*

49. **2** *The starfish is important in maintaining the biodiversity of this ecosystem* is the conclusion that can be made regarding the role of the starfish in this ecosystem. Biodiversity is a measure of the number of different, compatible species that inhabit an area. In 1963, prior to the start of the experiment, scientists counted approximately 17 species in the experimental tract. However, only about two species were present ten years after the removal of starfish. In the control tract that included starfish, the number of species present after ten years rose from 17 to 20 species. Assuming all other factors were identical in these two tracts, it is likely that the loss of the starfish had a negative effect on the biodiversity in the experimental tract. It can therefore be concluded that starfish have an important role in maintaining biodiversity in this ecosystem.

WRONG CHOICES EXPLAINED:

(1), (3) *The biodiversity of this ecosystem increased within ten years as organisms adjusted to the loss of the starfish* and *when the starfish were removed, the ecosystem decreased in stability and increased in biodiversity* are *not* the conclusions that can be made regarding the role of the starfish in this ecosystem. The data in the graph show that the biodiversity of the ecosystem decreased, not increased, over the ten-year study period. A decrease in biodiversity is known to result in a decrease of stability in an environment.

(4) *Biodiversity in this ecosystem is not dependent on the presence of starfish* is *not* the conclusion that can be made regarding the role of the starfish in this ecosystem. On the contrary, the data in the graph demonstrate that the loss of the starfish had a significant impact on the biodiversity of this ecosystem.

50. **3** The ecologist would most likely be asked about *what effect the production of the fuel will have on the environment.* An ecologist is a scientist who specializes in the science of ecology (study of the environment). Ecologists understand

and can speak knowledgeably about the impacts of an industry's actions on the living things that inhabit the area where the fuel production facility will be built and operate.

WRONG CHOICES EXPLAINED:

(1) The ecologist would *not* most likely be asked about *the cost of producing the fuel compared with the profit when the fuel is sold*. This question would most probably be asked of a business accounting expert, not an ecologist.

(2) The ecologist would *not* most likely be asked about *whether the fuel will be widely accepted by customers*. This question would most probably be asked of a business marketing expert, not an ecologist.

(4) The ecologist would *not* most likely be asked about *the time it will take to produce large amounts of the fuel*. This question would most probably be asked of a chemical engineering expert, not an ecologist.

51. One credit is allowed for correctly explaining why biomass is considered a renewable energy source. Acceptable responses include but are not limited to: [1]

- *Biomass is continually being produced by plants, animals, and humans.*
- *More plants or trees can be grown to replace those used for fuel.*
- *Humans will always be generating food wastes and garbage.*
- *Biomass is an energy source that is quickly replaced by natural processes.*
- *When available supplies of biomass fuels are exhausted, more can be produced relatively quickly.*

52. One credit is allowed for correctly stating *one* specific advantage and *one* specific disadvantage of the use of biofuels as an energy source. Acceptable responses include but are not limited to: [1]

Advantage:

- *Fossil fuel use will decrease.*
- *Less waste/garbage will end up in landfills.*
- *Less gasoline will be used.*
- *Biofuels are renewable.*
- *Biofuels can be used to power cars and generate electricity.*

Disadvantage:

- *There will be fewer farm crops available as food.*
- *Burning biofuels will still produce some pollution.*
- *Burning wood for heat produces carbon dioxide and particulate matter.*
- *Collecting and transporting biofuels can be costly.*
- *Harvesting native plants/trees as biofuel will alter natural ecosystems.*

53. One credit is allowed for correctly identifying *two* raw materials necessary for photosynthesis. Acceptable responses include: [1]

- *Carbon dioxide, CO_2, $O=C=O$*
- *Water, H_2O, $H-O-H$*

54. One credit is allowed for correctly stating *one* reason why photosynthesis is necessary for animals to survive. Acceptable responses include but are not limited to: [1]

- *Photosynthesis provides the raw materials/glucose and oxygen for cell respiration.*
- *A product of photosynthesis is glucose, which can be converted to other energy-containing compounds that animals use for food.*
- *Animals are heterotrophic, so they are dependent on plants to manufacture food.*
- *Photosynthesis produces oxygen gas needed by animals for respiration.*
- *Only green plants can use photosynthesis to convert solar energy into food energy.*

55. One credit is allowed for correctly identifying *one* abiotic factor in the pond ecosystem and explaining how this abiotic factor would affect the frogs in the pond. Acceptable responses include but are not limited to: [1]

- *Sunlight—The Sun provides energy to the plants so they can perform photosynthesis/produce food/release oxygen that the frogs need to survive.*
- *Oxygen—Frogs use oxygen for aerobic respiration.*
- *Temperature—If the average environmental temperature gets too low or too high, it could cause the frogs to die out.*
- *Soil—Soil anchors the plants that the frogs use to hide from predators.*
- *Water—Frogs depend on abundant water in their environment for their survival/breeding/habitat.*
- *Acidity/pH—The frog's enzymes function best in a neutral acidity (pH = 7) environment.*

PART C

56. One credit is allowed for correctly stating how the TR4 fungus threatens homeostasis within the banana plant. Acceptable responses include but are not limited to: [1]

- *The TR4 fungus interferes with the transport of water and other materials within the banana plant.*
- *The fungus that attacks the banana plant interferes with the plant's normal functions, and the plant basically dies of thirst.*
- *The fungus prevents water from reaching the leaves, preventing photosynthesis.*
- *The plant cannot get nourishment from water and nutrients.*

57. One credit is allowed for correctly explaining why the entire Cavendish banana crop worldwide is particularly vulnerable to the TR4 fungus. Acceptable responses include but are not limited to: [1]

- *All of the Cavendish banana plants are genetically identical, so none of them have adaptations to fight off the disease.*
- *There is no genetic diversity among the Cavendish bananas to provide immunity against TR4.*
- *Without genetic variation that provides defense against TR4, the banana plants are more likely to be killed by the fungus.*
- *Once the fungus is transferred to an uninfected banana grove, it spreads rapidly through all the trees.*
- *The crop is grown as a monoculture, so all the banana plants are susceptible to TR4.*
- *Soil infected by TR4 can easily be transferred to uninfected groves on the boots of workers.*

58. One credit is allowed for correctly describing *one* other possible way that the growers may be able to combat the disease if the fungus cannot be stopped by chemical treatment of the soil. Acceptable responses include but are not limited to: [1]

- *Look for a biological control that would attack the fungus.*
- *Genetically engineer the bananas so that they are not affected by the fungus.*
- *Do not allow people to bring contaminated boots to uninfected banana fields.*
- *Develop a "clean room" protocol to ensure that workers and others are completely decontaminated before entering and after leaving any Cavendish banana grove.*

59. One credit is allowed for correctly explaining why having children eat foods high in calcium and iron is a scientifically valid recommendation. Acceptable responses include but are not limited to: [1]

- *If there is a high amount of calcium in the diet, it is less likely that lead will be used in the formation of enzymes.*
- *Having a lot of calcium and iron available in the cells will make it more likely that these will be incorporated into enzymes when they are synthesized.*
- *A greater concentration of iron and calcium will make it more likely that they will diffuse from the blood vessels, through cell membranes, and into the cells.*

60. One credit is allowed for correctly describing how the presence of lead in the body could interfere with the ability of enzymes to function. Acceptable responses include but are not limited to: [1]

- *Lead could give the enzyme a different shape/molecular structure, which could change the enzyme's function.*
- *If lead replaces calcium or iron in the enzyme molecule, the enzyme will not have the right shape to do its job.*
- *The enzyme changes shape, which interferes with its function in the cell.*
- *If the enzyme is changed and can no longer catalyze key cell reactions, the health of the child/person may be adversely affected.*

61. One credit is allowed for correctly identifying *one* type of cell that would be expected to have numerous calcium channels and supporting the answer. Acceptable responses include but are not limited to: [1]

- *Nerve cells—Nerve/brain cells are known to be damaged by lead in the body. Therefore, nerve cells likely contain numerous calcium channels that admit lead ions instead of calcium ions.*
- *Brain cells—Lead poisoning in adults/children often leads to learning/ memory/concentration/hearing/speech problems.*
- *Bone/connective tissue—Calcium is needed for healthy bone/joint development. However, lead interferes with that development/slows bone growth/ causes joint pain in adults/children.*
- *Muscle—Lead poisoning can result in poor muscular coordination/muscular spasms.*

62. One credit is allowed for correctly stating *one* reason why the mutation in Manx cat embryos causes them to have such short tails. Acceptable responses include but are not limited to: [1]

- *Manx genes cause abnormalities in the number and/or shape/size of bones in the spine.*
- *The mutation causes the spine to be shorter in Manx cats, so there could be too few or smaller tail bones formed.*
- *The mutation interferes with the development of the spine.*
- *The Manx mutation in the embryo is replicated so that all cells, including those in the spine, have a copy of the mutation.*

63. One credit is allowed for correctly explaining how the genes that the kittens inherit determine whether they will have a normal tail or a short tail. Acceptable responses include but are not limited to: [1]

- *If a kitten gets one mutated gene from one of the parents, it will have a short tail or no tail.*
- *If a kitten receives a normal gene from each parent, it will develop a normal tail.*
- *Each parent has a Manx gene and a normal gene, so kittens will be born with either a Manx tail or a normal tail depending on whether they inherit one or two normal genes.*
- *Heterozygous kittens will develop Manx tails, and homozygous normal kittens will develop normal tails. Homozygous Manx embryos will die before birth because that allelic combination is lethal.*

64–66. Three credits are allowed for correctly discussing how sweet sensitivity in hummingbirds has developed. In your answer, be sure to:

- Identify the initial event responsible for the new sweet-sensing gene.   [1]
- Explain how the presence of the sweet-sensing gene increased in the hummingbird population over time.   [1]
- Describe how the fossil record of hummingbird ancestors might be used to learn more about the evolution of food preferences in hummingbirds.   [1]

Acceptable responses include but are not limited to: [3]

The initial appearance of the hummingbird sweet-sensing gene was most likely the result of a random mutation of its savory-sensing gene. [1] It is likely that hummingbirds that were able to detect and feed on sweeter nectar were more successful and produced more offspring than hummingbirds that did not have this ability. [1] A study of the shapes of fossil hummingbird beaks might reveal a series of physical changes in the beaks that made sweet-sensing hummingbirds increasingly efficient or successful at gathering food. [1]

The most likely reason for the appearance of this trait was an alteration of the base sequence in one or more DNA strands in the hummingbird genome. [1] Hummingbirds that were able to sense and ingest sweeter nectars gained more energy than those that could not, increasing the sweet-sensing birds' ability to survive and reproduce. [1] By studying fossil evidence of ancestral hummingbirds and other species, scientists might be able to learn more about the environment that hummingbirds lived in and any changes in food types that may have occurred and selected for the sweet-sensing gene. [1]

67. One credit is allowed for correctly explaining why taking folic acid early in pregnancy is important to the prevention of neural tube defects. Acceptable responses include but are not limited to: [1]

- *Early in pregnancy, folic acid promotes the normal development of the brain and spinal cord.*
- *Folic acid is essential for normal development of the central nervous system.*
- *The brain and spinal cord form early in pregnancy. Folic acid helps them to develop normally.*
- *Folic acid reduces the risk of an embryo developing neural tube defects.*

68. One credit is allowed for correctly describing how a fetus receives folic acid and other essential materials directly from its mother for its development. Acceptable responses include but are not limited to: [1]

- *Materials diffuse from the mother's bloodstream into the bloodstream of the fetus.*
- *Materials are transported from mother to fetus across the tissues of the placenta.*
- *Essential materials, such as folic acid, are exchanged between the mother and the fetus within the structure of the placenta.*
- *The placenta allows the exchange of materials, including folic acid, between the mother's blood and her developing fetus.*

69. One credit is allowed for correctly identifying *one* factor, other than the lack of folic acid, that may interfere with the proper development of essential organs during pregnancy. Acceptable responses include but are not limited to: [1]

- *Genetic mutations in the fetal tissues*
- *Mother's use of alcohol/drugs/tobacco during pregnancy*

- *Mother's exposure to environmental toxins*
- *Bacterial/viral infections that occur in the mother during pregnancy*
- *Poor diet/nutritional deficiency while the fetus is developing*

70. One credit is allowed for correctly explaining why adding folic acid to foods is an advantage to people other than pregnant women. Acceptable responses include but are not limited to: [1]

- *Folic acid is essential for normal growth and development of cells in children and adults.*
- *The body is continually producing new cells for the purpose of tissue repair/growth, and folic acid is essential to these processes.*
- *A deficiency of folic acid may interfere with a person's ability to repair damaged tissues/organs.*
- *Children use folic acid to add new cells that are needed to help their bodies grow normally.*

71. One credit is allowed for correctly circling *Yes* and supporting the answer. Acceptable responses include but are not limited to: [1]

- A *and* B *share a more recent common ancestor (H) than do* A *and* D.
- A *and* B *evolved from* H, *while* D *evolved from* G.
- *The most recent common ancestor of* A *and* B *is* H; *the most recent common ancestor of* A *and* D *is* K.
- D *is on an evolutionary line that split from that of* A *in the distant past (at* K). *The long period of time since that split occurred makes it unlikely that* A *and* D *share as much DNA in common as do* A *and* B.

72. One credit is allowed for correctly stating *one* possible cause for the extinction of species *E*. Acceptable responses include but are not limited to: [1]

- *An environmental change occurred that did not favor the survival of species* E.
- *Species* E *could not successfully compete against other similar species in a changed environment.*
- *The environment changed, and species* E *was not adapted to this change.*
- *An extinction level event such as an asteroid strike changed the environment drastically, and members of species* E *could not survive the change.*
- *Plant species that species* E *used for food became extinct, so all the members of species* E *died off from starvation because they could not adapt to new food sources.*

PART D

73. **3** *No. Additional tests should be done to test for other chemical similarities* is the conclusion that is correct, based on the results of this comparison alone, to determine if there is enough information to conclude which of the other three species is most closely related to *Botana curus*. Although the banding patterns of *B. curus* and species Z appear to be nearly identical, enough differences can be seen in the electrophoresis results (middle band) to warrant additional tests to determine the degree of relationship between these species.

WRONG CHOICES EXPLAINED:

(1) *Yes. Only species X has the same bands as* Botana curus is *not* the conclusion that is correct, based on the results of this comparison alone, to determine if there is enough information to conclude which of the other three species is most closely related to *Botana curus*. The banding pattern of species *X* is actually quite different from that of *B. curus*, so they are not likely to be closely related.

(2) *Yes. Species Z has only two of the bands that* Botana curus *has* is *not* the conclusion that is correct, based on the results of this comparison alone, to determine if there is enough information to conclude which of the other three species is most closely related to *Botana curus*. The banding pattern of species Z shows three, not two, bands that are similar though not identical to those of *B. curus*. So, it is inconclusive as to whether or not they are closely related.

(4) *No. Other rainforest plant species should be tested* is *not* the conclusion that is correct, based on the results of this comparison alone, to determine if there is enough information to conclude which of the other three species is most closely related to *Botana curus*. The conclusion called for in this experiment involves only these four species and not additional, unidentified species.

74. **1** Among the birds living there, the finches most likely to experience a drastic *decrease* in population size would be the *warbler finches*. The information provided in the diagram shows that warbler finches are 100% dependent on animal (insect) food for their nutritional requirements. If all the insects on this island are killed by the infection, warbler finches will have no food source remaining and will have to migrate to another island that still has insects or die of starvation.

WRONG CHOICES EXPLAINED:

(2), (3), (4) Among the birds living there, the finches most likely to experience a drastic *decrease* in population size would *not* be the *cactus finches*, the *large ground finches*, or the *medium ground finches*. The information provided in the diagram shows that these finch species are mainly dependent on plant food for their nutritional requirements. If all the insects on this island are killed by the infection, these finch species will still have adequate plant food sources remaining and so will survive in the short term.

75. **3** A variety of species of Galapagos finches evolved from one original species long ago through the process of *natural selection*. Natural selection, sometimes referred to as survival of the fittest, postulates that certain members of a species are better adapted to their environment than others and that these organisms are more likely to survive and pass on their favorable adaptations to future generations. It has long been thought that an original, ancestral finch species arrived on the Galapagos from the South American mainland and that genetic variation in that ancestral finch promoted the development of new finch species via natural selection as time and environmental selection pressures allowed.

WRONG CHOICES EXPLAINED:

(1) It is *not* true that a variety of species of Galapagos finches evolved from one original species long ago through the process of *asexual reproduction*. Asexual reproduction is a type of reproductive activity in which a single parent organism produces genetically identical offspring by mitotic cell division. Asexual reproduction minimizes genetic change, which limits evolutionary change. So, asexual reproduction is unlikely to be the force behind the speciation of finches in the Galapagos.

(2) It is *not* true that a variety of species of Galapagos finches evolved from one original species long ago through the process of *ecological succession*. Ecological succession refers to the series of changes that occur to the dominant plant community of an area that has been significantly affected by an environmental disruption. This description does not explain the biological evolution of finch species in the Galapagos.

(4) It is *not* true that a variety of species of Galapagos finches evolved from one original species long ago through the process of *selective breeding*. Selective breeding is a technique used by animal and plant breeders in which organisms with desirable traits are cross-bred with the hope of producing offspring that also display those traits. This description does not explain the biological evolution of finch species in the Galapagos.

76. **2** If scientists want to determine the similarities in the DNA fragments in several plant species, they should *analyze electrophoresis results*. Electrophoresis is a laboratory technique in which DNA fragments from an organism are placed in and drawn through an electrified gel to produce a unique banding pattern of the DNA for that organism. Based on the reference to DNA fragments in the question, electrophoresis is the only reasonable methodology available to perform this analysis.

WRONG CHOICES EXPLAINED:

(1), (3), (4) It is *not* true that if scientists want to determine the similarities in the DNA fragments in several plant species, they should *add salt water to cells from each plant, compare seed structures of the plants,* or *examine their chromosomes with a microscope.* These techniques are inadequate to analyze accurately the similarities and differences among DNA fragments in several plant species.

77. One credit is allowed for correctly identifying the dependent variable in this experiment. Acceptable responses include but are not limited to: [1]

- *The time it takes each team to row a specific distance on a specific course under identical conditions*
- *The speed at which the teams complete the same course under the same conditions*
- *Which team finishes in the fastest time in a head-to-head competition*

78. One credit is allowed for correctly stating *one* biological reason for the decrease in the number of squeezes during the second trial. Acceptable responses include but are not limited to: [1]

- *The student squeezed the clothespin less the second time because the muscles in his or her hand began to fatigue.*
- *The student squeezed the clothespin fewer times because her or his hand muscles had less oxygen.*
- *The student squeezed the clothespin less the second time because waste products were building up in his or her cells.*
- *As more squeezes occurred, cellular respiration switched over from aerobic respiration to less efficient anaerobic respiration, so less ATP energy was available to the muscles.*
- *The buildup of lactic acid in the student's hand muscles caused a burning sensation and made continued squeezing more difficult and painful.*

79. One credit is allowed for correctly stating *one* reason why the heart rate increased during exercise. Acceptable responses include but are not limited to: [1]

- *The faster heart rate helped to lower the level of carbon dioxide in the bloodstream.*
- *Increased blood flow helped to maintain homeostasis throughout the body.*
- *Muscles required more nutrients and oxygen, which were delivered by the circulatory system.*
- *Carbon dioxide entered the bloodstream and was detected by the central nervous system, which stimulated the heart to beat faster.*

- *The CNS sent a command to the heart muscle to increase its rate of contraction based on elevated CO_2 concentration in the blood.*

80. One credit is allowed for correctly explaining why a paramecium would require contractile vacuoles while a similar protozoan living in salt water would *not*. Acceptable responses include but are not limited to: [1]

- *Excess water would diffuse into the freshwater paramecium, but the saltwater organism would lose water.*
- *The saltwater organism would lose water to its environment/dehydrate instead.*
- *In freshwater organisms, the higher water concentration outside causes water to enter cells. This is the opposite of what happens in saltwater organisms.*
- *For a freshwater paramecium, the net diffusion of water molecules is from the environment into the cell. For a saltwater protozoan, the net diffusion of water molecules is out of the cell to the environment.*

81. **1** Row *1* in the chart correctly describes what would be expected to occur in the potato cells with regard to both the starch and water molecules. Diffusion of water (osmosis) is the movement of water molecules across a membrane with the concentration gradient from a region of high relative concentration of water to a region of low relative concentration of water. Distilled water is the most highly concentrated form of water. So, more water will enter the potato cell through its cell membrane than will leave the cell. Starch molecules are too large to pass through a cell membrane and will therefore be retained inside the cell.

WRONG CHOICES EXPLAINED:
   (2), (3), (4) Rows 2, 3, and 4 in the chart do *not* correctly describe what would be expected to occur in the potato cells with regard to both the starch and water molecules. Each of these rows contains one or more descriptions that are inconsistent with the correct answer above.

82. **2** *The indicator would remain amber in color if starch molecules were not present in the water in the beaker* is the statement that correctly describes a possible result if starch indicator is added to the water in the beaker one hour after the potato cube was added. Starch molecules are too large to pass through a cell membrane from inside the cell to the beaker water. So, no starch will be present in the beaker water with which the indicator can react. For this reason, the indicator in the beaker water will retain its amber color.

WRONG CHOICES EXPLAINED:

(1) *The indicator solution would turn to an amber color in the water if starch molecules were present in the water in the beaker* is *not* the statement that correctly describes a possible result if starch indicator is added to the water in the beaker one hour after the potato cube was added. Starch indicator is typically amber in color. If starch molecules were present in the beaker water, the indicator would change to a blue-black color, not remain an amber color.

(3) *The indicator would change to a black color if starch molecules were not present in the water in the beaker* is *not* the statement that correctly describes a possible result if starch indicator is added to the water in the beaker one hour after the potato cube was added. Starch indicator is typically amber in color. If starch molecules were *not* present in the beaker water, the indicator would remain an amber color, not change to a blue-black color.

(4) *The indicator solution would remain black in color if starch molecules were present in the water in the beaker* is *not* the statement that correctly describes a possible result if starch indicator is added to the water in the beaker one hour after the potato cube was added. Starch indicator is typically amber in color. If starch molecules were present in the beaker water, the indicator would change from an amber color to a blue-black color.

83. One credit is allowed for correctly describing how this information could specifically be used to determine if water moved during the investigation. Acceptable responses include but are not limited to: [1]

- *If the cube's mass increased, it would indicate that water had moved into the potato cells.*
- *A positive change in mass would indicate a change in the water content of the potato cells.*
- *If the mass increased, water had probably moved into the cells.*
- *The cube soaked up water and became heavier.*

84. One credit is allowed for correctly selecting *one* row in the chart and explaining how the systems in that row work together during exercise. Acceptable responses include but are not limited to: [1]

Row 1:

- *These systems work together to take in and move oxygenated blood to the muscles for their use.*
- *The respiratory system takes in oxygen, which is passed into the circulatory system, which then takes the oxygen to the muscles.*
- *The respiratory and circulatory systems work together to remove carbon dioxide from the muscles.*

Row 2:

- *These systems work together to move wastes away from muscle tissues to be released into the environment.*
- *Muscle cells produce wastes and circulatory organs transport these wastes to excretory organs to be excreted.*

Row 3:

- *These systems work together to take in and digest food and move soluble nutrients to the muscles for their use.*
- *The digestive system breaks down food into nutrients, and the circulatory system transports nutrients to muscle cells for energy.*

85. One credit is allowed for correctly stating *one* way the bag or its contents will have changed by the next day and supporting the answer. Acceptable responses include but are not limited to: [1]

- *The bag will become smaller because the water will diffuse from inside the bag to outside the bag.*
- *Water will move from higher concentration inside the bag to lower concentration outside the bag.*
- *The membrane bag will decrease in size due to osmosis of the water out of the bag.*
- *The size of the bag will become smaller due to water loss.*
- *The salt concentration inside the bag will have increased as water moved out of the bag.*

Standards/Key Ideas	August 2019 Question Numbers	Number of Correct Responses
Standard 1		
Key Idea 1: The central purpose of scientific inquiry is to develop explanations of natural phenomena in a continuing and creative process.	33, 58, 59	
Key Idea 2: Beyond the use of reasoning and consensus, scientific inquiry involves the testing of proposed explanations involving the use of conventional techniques and procedures and usually requiring considerable ingenuity.	36	
Key Idea 3: The observations made while testing proposed explanations, when analyzed using conventional and invented methods, provide new insights into natural phenomena.		
Laboratory Checklist	31, 32, 44, 45, 46	
Standard 4		
Key Idea 1: Living things are both similar to and different from each other and from nonliving things.	4, 7, 8, 17, 20, 22, 23, 35, 38, 61, 70	
Key Idea 2: Organisms inherit genetic information in a variety of ways that result in continuity of structure and function between parents and offspring.	11, 12, 14, 19, 37, 39, 62, 63	
Key Idea 3: Individual organisms and species change over time.	47, 57, 64, 65, 66, 71, 72	
Key Idea 4: The continuity of life is sustained through reproduction and development.	16, 26, 28, 34, 67, 68, 69	
Key Idea 5: Organisms maintain a dynamic equilibrium that sustains life.	5, 10, 13, 15, 25, 27, 30, 53, 54, 56, 60	
Key Idea 6: Plants and animals depend on each other and their physical environment.	1, 2, 3, 9, 18, 41, 43, 49, 55	
Key Idea 7: Human decisions and activities have a profound impact on the physical and living environment.	6, 21, 24, 29, 40, 42, 48, 50, 51, 52	
Required Laboratories		
Lab 1: "Relationships and Biodiversity"	73, 76	
Lab 2: "Making Connections"	77, 78, 79, 84	
Lab 3: "The Beaks of Finches"	74, 75	
Lab 5: "Diffusion Through a Membrane"	80, 81, 82, 83, 85	